PL/I
PROGRAMMING FOR ENGINEERING AND SCIENCE

PRENTICE-HALL INTERNATIONAL, INC., *London*
PRENTICE-HALL OF AUSTRALIA, PTY., LTD., *Sydney*
PRENTICE-HALL OF CANADA, LTD., *Toronto*
PRENTICE-HALL OF INDIA PRIVATE LIMITED, *New Dehli*
PRENTICE-HALL OF JAPAN, INC., *Tokyo*

PL/I
PROGRAMMING FOR ENGINEERING AND SCIENCE

DAVID R. STOUTEMYER

PRENTICE-HALL, INC.
Englewood Cliffs, N.J.

© 1971 by
PRENTICE-HALL, INC.
Englewood Cliffs, New Jersey

All rights reserved. No part of this book except the programs may be reproduced in any way, or by any means, without permission in writing from the publisher. The programs may be reproduced for use, but not publication, without permission.

Current printing (last digit):
10 9 8 7 6 5 4 3 2 1

ISBN 0-13-676528-9

Library of Congress Catalog Card Number: 74-146683
Printed in the United States of America

To Karen

PREFACE

My own first exposure to programming was so discouraging that I did not brave a second exposure until three years later. This first exposure consisted of a noncredit course directed towards a general audience. The content of this course consisted of a rigorous, categorical treatment of the language syntax. No applications were presented. I dropped out with the impression that computers were not nearly as useful as they were touted to be.

My second experience was almost as frustrating. Upon a recommendation, I read a book which dealt with nothing but number systems, organization of a computer, and computer circuitry. This was all very interesting, but I still did not know of what use it all was. From these experiences, I have rather strong convictions about how programming should be taught.

1. For motivational purposes, extensive treatment of how a computer works should be postponed until after the first exposure to programming.

2. To teach programming without also teaching meaningful and useful applications is not only a lost opportunity, it is demotivating. Moreover, these applications should be relevant to the student's discipline. This requires separate texts and courses for engineering and science, business, and humanities majors.

Accordingly, there is only a brief introduction on how a computer works, and this text is directed specifically towards engineers and scientists. Consequently, the student is presumed to have a mathematical aptitude. However, he need not have taken courses beyond trigonometry and analytic geometry. The sections and exercises requiring calculus or higher mathematics are optional and marked as such.

This text uses a case study approach with the emphasis on applications rather than on a thorough and rigorous definition of the language. PL/I is the most comprehensive language yet developed, but I have avoided the temptation to display its full range and depth. Instead, I have concentrated on the language features which are particularly useful for engineering and scientific programmers.

One of the unique features of this text is the early introduction and extensive use of subordinate procedures. Wherever possible, the case studies and exercises make use of previously developed subordinate procedures so that the student will gain a thorough understanding of the techniques and motivations for using them. Most texts treat subordinate procedures only briefly at the end; so this subject is often hurried or omitted in a course due to lack of time. This leaves the student unable to use subroutine libraries, which is one of the most important things for him to learn.

Another strong feature of this text is the many unique exercises that are relevant to a great variety of engineering and scientific disciplines. These exercises are chosen to suggest fruitful computer applications and to provide further motivation for engineering and science students.

The text is suitable for any PL/I compiler. Specific details are given in footnotes for the IBM 360 F, D, Model 20 and Student PL compilers together with the IBM 1130 SL1 compiler.

In writing this text, I was fortunate to have an excellent technical editor right down the hall. Consequently, he contributed more than is usually the case. Without him, the text would have been intelligibile only to me. Any remaining deficiencies are due to my stubborness or fatigue. Thank you, Frank Kuo.

I am particularly grateful to Karen Eng for typing the manuscript, to Max Starr and James Thompson for punching and testing the programs, and to the University of Hawaii Computer Center for providing the testing facilities. I am also indebted to Don Avery, John Shupe, and Howard Harrenstein for their support and encouragement, and to Gretchen Schule, Linda Drake, Otis Powell, Cleve Moler, and Ned Weldon for their helpful suggestions.

David R. Stoutemyer

CONTENTS

INTRODUCTION: THE ANATOMY OF A COMPUTER   1

    I.1   Analog vs. Digital, 1
    I.2   Card Programmed Calculators, 2
    I.3   Binary vs. Decimal, 5
    I.4   Stored Program, 6
    I.5   Programming Languages, 7
    I.6   Compile vs. Execute, 8
    I.7   Hardware, 8
    I.8   Batch vs. Time-shared Processing, 10
Test Your Understanding, 12
    I.9   An Important Note, 12

CASE STUDY 1: HEAT TRANSFER   14
    The PROCEDURE, Assignment, PUT DATA, END and DECLARE Statements

    1.1   A Sample Program, 14
    1.2   The PROCEDURE Statement, 15
    1.3   The Assignment Statement, 15
    1.4   The PUT DATA Statement, 16
    1.5   The END Statement, 17
Exercises, 17
    1.6   Identifiers, 18
    1.7   Heat Transfer Example, 18
    1.8   Floating-point Numbers, 19
    1.9   Fixed-point Numbers, 20
    1.10 The First Letter Convention, 20
    1.11 The Declare Statement, 20
    1.12 Default Lengths, 22

1.13 Precision and Mixed Expressions, 22
1.14 Priority of Arithmetic Operations, 24
1.15 Output, 25
Summary, 26
Exercises, 27

CASE STUDY 2: TRIANGLE SOLUTION 30
Function Procedures, Mathematical Built-in Functions, and the PAGE Option

2.1 Function Procedures, 31
2.2 The Base Attribute, 34
2.3 Argument-parameter Agreement for SL1, 34
2.4 Argument-parameter Agreement for Student PL, 34
2.5 Argument-Parameter Agreement for the Model 20, 35
2.6 Argument-parameter Agreement for the F and D Compilers, 35
2.7 The RETURNS Attribute, 37
Exercises, 38
2.8 Built-in Functions, 40
2.9 The Page Option, 42
Summary, 42
Exercises, 43

CASE STUDY 3: SURVEYING TRAVERSE 48
The GET LIST, PUT LIST, and GO TO Statements, the SKIP Option, Comments, and Character String Constants

3.1 The GET LIST Statement, 48
3.2 The GO TO Statement, 50
3.3 The ENDFILE Condition, 51
3.4 The Null Field, 52
Exercises, 52
3.5 Comments and the SKIP Option, 53
3.6 Absolute and Relative Errors, 55
3.7 The PUT LIST Statement and Character String Constants, 56
Summary, 58
Exercises, 58

CASE STUDY 4: BEARING SELECTION 61
The IF Statement and Arithmetic Built-in Functions

4.1 The IF Statement, 61
4.2 Flow Charts, 63

4.3 Some Additional Built-in Functions, 64
Exercises, 67
4.4 The *And* and *Or* Operators, 70
4.5 Noninteger Fixed-point Variables, 72
4.6 Nested if Statements, 74
Summary, 74
Exercises, 75

CASE STUDY 5: SOLUTION OF QUADRATIC EQUATIONS SUBROUTINES — 79

5.1 Quadratic Equations Example, 81
5.2 Subroutines in the IBM PL/1 Scientific Subroutine Package, 83
5.3 Optional Language Feature: Internal Procedures, 88
5.4 Optional Language Feature: The External Attribute, 90
Summary, 91
Exercises, 91

CASE STUDY 6: DATA REDUCTION — 95
The ON ENDFILE Statement, Double Precision, and the ADD, MULTIPLY, DIVIDE and PRECISION Functions

6.1 The ON ENDFILE Technique, 97
6.2 Least Squares Example, 98
Exercises, 98
6.3 Chopoff Error, 100
6.4 Double Precision, 104
6.5 The ADD, MULTIPLY, DIVIDE, and PRECISION Built-in Functions, 105
6.6 Optional Mathematically Advanced Application: Numerical Integration, 106
6.7 Optional Mathematically Advanced Application: Differential Equations, 107
Summary, 108
Exercises, 109

CASE STUDY 7: SOLUTION OF ALGEBRAIC EQUATIONS — 113
The DO Group, the ELSE Clause, and the STOP Statement

7.1 The ELSE Clause, 114
7.2 The Null ON Statement, 116
7.3 Functions As Parameters, 117
7.4 The DO Group, 118

7.5 The STOP Statement, 120
7.6 Dummy Arguments, 123
7.7 DO Groups and Nested IF Statements, 125
7.8 Optional Mathematically Advanced Application: Infinite Series, 127
Summary, 128
Exercises, 129

CASE STUDY 8: TABLE PRODUCTION   **139**
The DO Loop

8.1 The DO Loop, 139
8.2 Flow Chart for DO Loops, 141
8.3 Optics Example, 142
Exercises, 143
8.4 Multiple Specifications, 143
8.5 Nested Loops, 144
8.6 Optional Language Feature: The WHILE Clause, 147
Summary, 148
Exercises, 149

CASE STUDY 9: INTERPOLATION   **153**
Singly Subscripted Variables, the INITIAL and STATIC Attributes

9.1 Statistics Example, 154
9.2 Storage Allocation, 154
9.3 Subscripted Parameters and Arguments, 155
9.4 Repetitive Input and Output Specifications, 156
Exercises, 157
9.5 Bounds and Extent of a Subscripted Variable, 158
9.6 The Initial and Static Attributes, 160
9.7 A General Purpose Interpolation Procedure, 161
Summary, 162
Exercises, 163

CASE STUDY 10: SORTING AND MERGING   **166**
The BEGIN Block and Array Expressions

10.1 Maximization Example, 166
10.2 Sorting Example, 167
Exercises, 168
10.3 Merging Example, 169
10.4 Adjustable Bounds, 170
10.5 Input and Output of Entire Arrays, 171
10.6 The Begin Block, 171
10.7 Array Expressions, 172

10.8 Array Manipulation Functions, 173
10.9 Subscripted Statement Labels, 174
10.10 Optional Difficult Application: Indirect Sorting, 175
Summary, 176
Exercises, 177

CASE STUDY 11: MATRIX MULTIPLICATION   181
Multiply Subscripted Variables and Array Cross-sections

11.1 Declaration of Arrays, 182
11.2 Array Expressions, 182
11.3 Array Manipulation Functions, 183
11.4 Array Cross Sections, 183
11.5 Input and Output of Arrays, 184
Exercises, 186
11.6 Matrix-vector Multiplication, 187
11.7 Matrix-matrix Multiplication, 191
11.8 The GET DATA Statement, 193
Summary, 194
Exercises, 194

CASE STUDY 12: LINEAR EQUATIONS   199
Further Practice with Subscripted Variables

12.1 Gaussian Elimination, 200
12.2 Ill-conditioned and Singular Matrices, 206
Exercises, 209
12.3 Iterative Solution of Linear Equations, 210
12.4 Symmetric Matrices, 214
Exercises, 215

CASE STUDY 13: BOOLEAN ALGEBRA   219
Bit-Strings

13.1 The Not Operator, 219
13.2 Bit-String Constants, 220
13.3 Bit-String Variables, 220
13.4 Concatenation, 222
13.5 Mixtures of Bit-strings and numbers, 222
13.6 Water Distribution Example, 222
Exercises, 224
13.7 Subscripted Bit-string variables, 225
13.8 The Aligned Attribute, 227
13.9 The ALL Function and the ANY Function, 228
13.10 STRING and SUBSTR Built-in Functions and Pseudovariables, 228

13.11 The VARYING Attribute and the LENGTH Function, 229
13.12 The BOOL Function, 230
13.13 Binary Arithmetic Constants, 231
Summary, 233
Exercises, 234

CASE STUDY 14: PATTERN RECOGNITION AND GRAPHICAL OUTPUT — 237
Character Strings

14.1 Character-String Variables, 237
14.2 The INDEX Function, 238
14.3 Concatenation and the SUBSTR Function, 239
14.4 The TRANSLATE Function, 239
14.5 Pattern Recognition Example, 240
14.6 The SUBTR Psuedovariable, 241
14.7 A Weather Map Example, 245
14.8 Comparison of Character Strings, 247
Summary, 248
Exercises, 249

CASE STUDY 15: BEARING SELECTION REVISITED — 252
Structures

15.1 Declaration of Structures, 253
15.2 Qualified Names, 254
15.3 Structure Expressions and Assignment, 255
15.4 The BY NAME Option, 256
15.5 The LIKE Attribute, 257
15.6 Arrays of Structures, 258
Summary, 258
Exercises, 259

CASE STUDY 16: LIST PROCESSING — 262
POINTER Variables, CONTROLLED and BASED Storage

16.1 CONTROLLED Storage, 262
16.2 The ALLOCATION function, 263
16.3 Pointers and the ADDR function, 264
16.4 BASED Storage, 264
16.5 The NULL Function, 266
16.6 Queue Example, 268

16.7 Subordinate Procedures for List Processing, 268
16.8 Multiple Links, 269
16.9 Arrays of Pointers, 273
Summary, 276
Exercises, 277

CASE STUDY 17: COMPLEX ARITHMETIC
Complex Variables

282

17.1 The COMPLEX attribute, 282
17.2 Polynomial Roots Example, 284
17.3 Transfer Function Example, 285
17.4 Comformal Mapping Example, 285
Summary, 289
Exercises, 289

CASE STUDY 18: LINEAR EQUATIONS REVISITED
A Package of Subroutines

292

18.1 The LU Decomposition, 293
18.2 Doolittle's Method, 295
18.3 Crout's Method, 296
18.4 Scaled Partial Pivoting, 299
18.5 Iterative Refinement, 304
18.6 Computing the Determinant, 306
18.7 Computing the Inverse, 307
18.8 Symmetric, Positive Definate Matrices, 308
Exercises, 310

APPENDIX: ANSWERS TO SELECTED EXERCISES   313

INDEX   353

PL/I
PROGRAMMING FOR ENGINEERING AND SCIENCE

INTRODUCTION:
THE ANATOMY OF A COMPUTER

In the early days of automotive travel, an understanding of automechanics was a virtual prerequisite to driving. Besides the dangerous crank starter, there was the necessity of grinding the valves and carrying two spare tires for even a modest Sunday drive. Today, developments such as the electric starter and the automatic transmission have enabled the mechanically disinclined to drive. Nevertheless, at least a vague notion of automechanics is helpful if one wishes to achieve maximum economy or performance.

A similar situation exists for computer programming. An understanding of the electronic and organizational design was essential in the early days of programming. Only a full-time specialist could program. Today, programming languages such as PL/I make it possible to program without understanding how a computer works. Nevertheless, as with the automobile, at least a vague understanding is helpful if one wishes to achieve maximum economy or performance. Consequently, we will begin with an explanation of how computers work. Although brief, this explanation will be enough for our purposes.

I.1 ANALOG VS. DIGITAL

Digital computers are most easily understood by tracing their development from the desk calculator. The ordinary mechanical desk calculator consists of mechanisms for adding, subtracting, multiplying, and dividing. A mechanical counter such as an automotive mileage indicator or *odometer* typifies the sort of mechanism involved.

The one-tooth input gear of an odometer drives a ten-tooth gear on the first digit wheel. Consequently, the first digit wheel *jumps* one digit at the end of each revolution of the input gear. Similarly, a one-tooth gear on the first digit wheel drives a ten-tooth gear on the second digit wheel. As a result, the second digit wheel jumps one digit at the end of each revolution of the first digit wheel. In other words, a digit is carried to the second wheel whenever the first wheel jumps forward from 9 to 0. Similarly, the second digit wheel drives the third at a ratio of ten to one and so on.

This explanation is somewhat simplified, but the important feature is that the counter is *digital* because it changes by discrete jumps. Perhaps you have had the experience of watching an odometer jump from 99999.9 to 00000.0 miles. This phenomenon is called *overflow*. The counter does not have the capacity to store seven digits. Disreputable used car dealers often take advantage of this fact.

In contrast to the odometer, a speedometer needle rotates smoothly which makes it an *analog* indicator. This may seem an advantage because a smoothly rotating needle or wheel can represent a continuous set of numbers rather than only ten. However, the position of a smoothly rotating wheel can be read easily to only about 0.1 to 1 per cent; so by including enough digits, a digital representation of a number can always be made more accurate than an analog representation. This is the reason desk calculators are generally more accurate than slide rules.

Even when the gears are driven by an electric motor, as they are in most desk calculators, the moving parts limit speeds to about one second for addition or subtraction and five seconds for multiplication or division. Recent desk calculators perform these operations thousands of times faster by digital electronic circuits. However, the overall operation speed is at most only twice as fast because each operation generally requires the punching in of at least one number, the punching in of at least one instruction ($+$, $-$, $\div$, or $\times$), and perhaps also the writing of an answer or intermediate result on a worksheet. Consequently, there is a limit of about five seconds per operation regardless of the machine speed. Actually, 60 seconds per operation is a typical day-long average for a skilled operator. The human operator is the bottleneck. He is also the source of many errors.

I.2 CARD PROGRAMMED CALCULATORS

Card programmed calculators were developed to reduce this time consumption and error by eliminating all human participation during the calculation process. Punched cards were first used in the 1700's by Jacquard to control automatic weaving looms. Herman Hollerith adapted the idea for numerical processing of the 1890 census data.

Sec. I.2 *Card Programmed Calculators*

The standard computer card bearing the injunction "do not fold, spindle, or mutilate" is a fixture of modern life. The pattern of holes in each column denotes a certain number, letter, or symbol. For example, the upper card in Fig. I-1 displays the Binary Coded Decimal (BCD) PL/I 48-character set, and the card below it displays the Extended Binary Coded Decimal Interchange Code (EBCDIC) PL/I 60-character set. The patterns of holes are *interpreted* at the top for the sake of humans. Machines read the hole patterns by passing the cards under a row of metallic brushes or photocells to generate corresponding patterns of electrical pulses.

Card programmed calculators were quite useful for situations where the same fixed set of instructions had to be performed over and over again on different

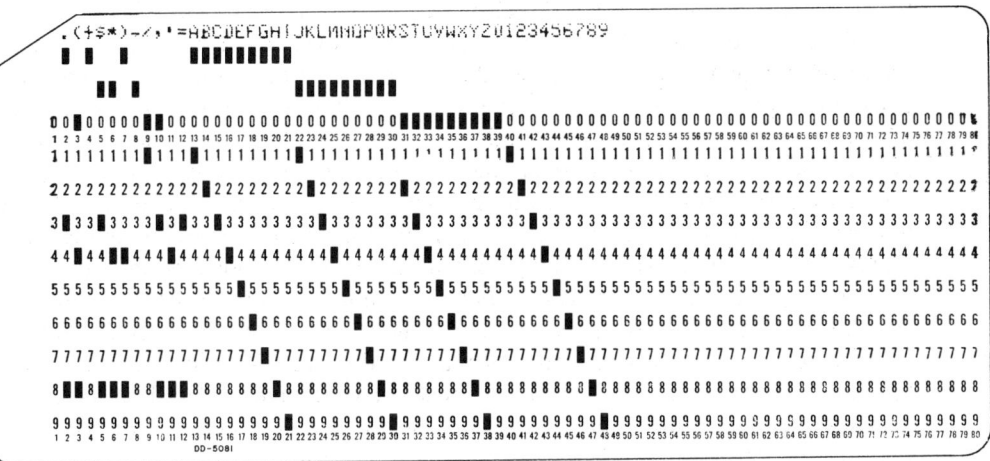

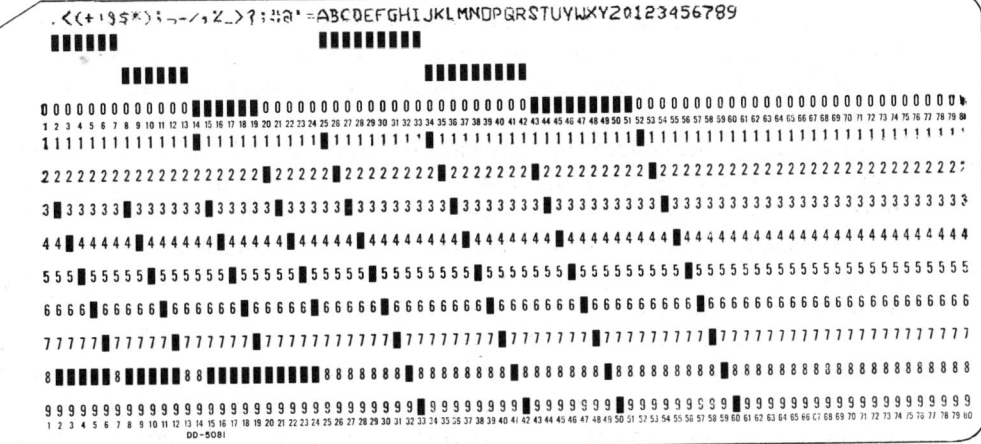

Fig. I-1 PL/I BCD 48-character code and EBCDIC 60-character code.

numbers. The instructions were encoded on punched cards which were repetitively cycled through the machine automatically. This deck of cards was called the *program* deck. The numbers to be sequentially operated upon were encoded on a separate deck of cards called the *data* deck. Also, it would not do to have a slow and careless human copy the numbers from a row of digit wheels; so the answers were automatically printed on a roll of paper. In accounting applications the paper often consisted of blank checks or bills.

Humans still had to prepare the program and the data, but the program could be used many times, and one machine could process the data prepared by many keypunch operators. It is a division of labor. The machine does not wait while the keypunch operator prepares data, nor does the keypunch operator wait while the machine calculates. Also, the keypunch operator does not constantly alternate between data and instructions; therefore he can work more quickly than the operator of a desk calculator. It is often possible, moreover, to arrange for the data to *originate* in punched form to save additional time. For example, a payroll time clock could punch as well as print the time. A ballot, an examination, or a census questionnaire could have perforated dots which could be punched out with a pencil. The observations of an experiment could be directly punched at a keypunch in a laboratory, thus eliminating intermediate writing. Also, some applications do not require any data. As an example, a program which prepares mathematical tables can generate its own equally-spaced values of the independent variable.

Besides the equipment for card input and printed output, the card programmed calculator must have another component which the desk calculator does not have —a memory. To understand the reason for this, consider how the expression "$36.528 \times 9.543 - 537.96 \times 5.7375$" must be evaluated on a desk calculator. The intermediate result of the first product must be noted on scratch paper before clearing the "accumulator" or "register" to form the second product. After the second product has been formed in the accumulator, the first product may be reentered from the scratch paper and added to the number in the accumulator. Consequently, the card programmed calculator needs a memory to serve as its scratch paper for intermediate results. In fact, it may need to store several intermediate results at once.

Remarkably, the concept of these card programmed calculators was developed over 100 years ago by an Englishman named Babbage. His entirely mechanical invention was programmed by punched cards, it printed its answers, and it stored its intermediate results in numerous rows of gears. Unfortunately, machining methods were not accurate enough at that time, and he died before successfully completing the machine.

The first successful card programmed calculator was electromechanical because it used banks of relays as the memory device. Each combination of activated relays represented a different number. In fact, relay circuits were used for the arithmetic too, because they were somewhat faster than the gear mechanisms. However, a relay has only the two states "off" and "on" rather than ten. These states may be associated with the 0 and 1 of binary arithmetic. For example, a bank of three relays with the states on, on, and off (1, 1, and 0) may represent the number 6 (one 4 + one 2 + no 1). Alternatively, four relays may be made to directly represent the digits 0 through 9 by assuming various combinations of on and off. Machines may be made both ways; therefore it is important to know the advantages and disadvantages of each.

1.3 BINARY VS. DECIMAL

A binary machine must generally perform a preliminary conversion of the data from decimal to binary and a terminal conversion of the results from binary to decimal. This conversion consumes time, and it may result in an anomaly in the last digit of the number. This anomaly arises when decimal fractions in the data do not have finite binary fraction representations. For example, 0.6 decimal corresponds to the nonterminating binary fraction 0.1001100110011 $\cdots$ (1 half + 0 fourths + 0 eighths + 1 sixteenth + $\cdots$). Having only a finite number of digits, the machine will *truncate* or chop off this fraction to a finite number of digits. Subsequent conversion of the number back to decimal will result in the number 0.599999999 $\cdots$ truncated to a finite number of digits. This fraction anomaly can be very annoying in accounting applications which deal with dollars and hundredths of dollars. However, the data for engineering and scientific applications are usually either whole numbers or arbitrary fractions which need not be in even hundredths. "0.599999999" is, after all, in error by only about one part in a billion. It is unreasonable to insist that a numerical result be exactly 0.6 when a physical measurement is unlikely to be exactly 0.6. Also, binary arithmetic has lower *roundoff error* than decimal. The roundoff error for a binary operation is ± 1 after the last digit whereas the roundoff error for a decimal operation is ± 5 after the last digit.

Decimal machines use their resources less efficiently because four relays actually have 16 possible code combinations rather than ten ($2 \times 2 \times 2 \times 2 = 16$). Also, the arithmetic is generally slower because the circuitry is more complicated. The extra speed of binary arithmetic more than compensates for the extra time taken for the preliminary and terminal conversion when there is a great deal of calculation for each item of data, as is typical of many engineering and scientific applications. On the other hand, a decimal machine is faster when there is a great deal of data and relatively little calculation for each item, as is typical of accounting applications.

Regardless of whether a card programmed calculator is binary or decimal, its basic limitation is that the instructions can be followed in only one predetermined order. Many calculations require the ability to jump backward or forward in a list of instructions. Often the jump must be conditional, depending upon the outcome of some intermediate calculation. Plugboard calculators were developed to overcome this restriction. The program was wired on a plugboard much like that used by a telephone operator. Relays could be mounted on the plugboard to activate different portions of the program depending upon the successive states of the relays. As might have been expected, these plugboard calculators took more time and skill to program than did the card programmed calculator. Many especially versatile calculators could be programmed by both plugboards and cards.

The next important development was a vast improvement in hardware. The relays were replaced by vacuum tubes. Having no moving parts, the vacuum tubes could be switched on and off far more rapidly than relays. Also, a number of devices such as acoustic delay lines and electrostatic storage were developed to permit relatively fast, low cost, high capacity memories. Calculations such as the solution of many simultaneous linear algebraic equations require a great deal of memory space, and the cost of large vacuum tube memories is prohibitive.

I.4 STORED PROGRAM

The most important development of all was in concept rather than hardware. John Von Neumann suggested that the instructions be numerically encoded and stored in the memory along with the intermediate numerical results. This way, the instructions may be loaded by cards, eliminating the difficult and time-consuming wiring operation, yet retaining its flexibility. Also, this scheme represents a more flexible allocation of resources than a fixed division between circuitry devoted to instructions and circuitry devoted to intermediate results. For the same amount of circuitry, one may have a more elaborate program or more storage of intermediate results (but not both). However, the most important advantage of all is that being stored in the same way as numbers, the instructions can be made to arithmetically alter themselves during the course of a program! This possibility opened up a whole new world of opportunities. More than any other feature, this *stored program* capability is what distinguishes computers from calculators. It is such a powerful feature that some manufacturers have introduced limited program storage capability into their desk calculators, making them "desk-top computers."

The instructions may be stored sequentially in a contiguous portion of the memory with the intermediate results stored either before or after the instructions. The instructions are then successively brought to a special *instruction register* where they are separated into parts—one part indicating what operation is to be performed ($+$, $-$, $\div$, or $\times$) and the other part indicating what memory locations to get the operands from and what memory location to put the result in. For example, 03-54-32-78 might mean "divide the number in location 54 by the number in location 32 and put the result in location 78." The number 54 in the second pair of digits triggers the circuitry which copies the number at location 54 into the arithmetic unit. The number 32 in the third pair of digits triggers the circuitry which copies the number at location 32 into the arithmetic unit. Next, the number 03 in the first pair of digits triggers the division circuitry. Then finally, the number

78 in the fourth pair of digits triggers the circuitry which copies the quotient from the arithmetic unit into location 78, erasing whatever may have been there previously. The location numbers are often called *addresses.*

There must also be an instruction for an *unconditional jump.* For example, 05-12-xx-xx might mean "transfer to the instruction at address 12" (the last two pairs of digits being irrelevant). There must be another instruction for the *conditional jump.* For example, 06-19-47-62 might mean "transfer to the instruction at address 19 if the number at address 47 is greater than the number at address 62." There must also be input and output instructions. For example, 07-58-xx-xx might mean "copy the next number in the data deck into location 58," and 08-89-xx-xx might mean "copy the number at address 89 on the printer."

As we can see, this "machine language" programming is quite tedious and fraught with opportunity for making errors. The problem must be broken down into mundanely elementary steps, and the programmer must remember the numeric instruction codes together with the numeric addresses of his various instructions and intermediate results. To make matters worse, the numbers are much longer than our artificial example, and they are often in binary or some closely related system such as *octal* (base 8) or *hexadecimal* (base 16). Machine language programming was responsible for the discouragement of a good number of brave souls who tried to program on a part-time or occasional basis as a supplement to their professional skills. Many of these disheartened pioneers may be found in the ranks of the hard-core computer haters. The halls of industry and academe still ring with their curses.

I.5 PROGRAMMING LANGUAGES

Assembly languages, then *procedural languages* were developed to make programming faster and easier. With assembly languages, mnemonic alphabetic names are used for the instructions and storage locations. For example, "CLA X" might mean "CLear the accumulator and Add the contents of location X." A machine language program called the *assembler* translates the assembly language program into machine language, replacing the mnemonics with numeric instruction codes and addresses. Although mnemonics are much easier to remember than numbers, assembly language is still closely related to the particular organizational design of a computer; so assembly languages vary considerably among machines. Also the problem must still be broken down into elementary steps.

Procedural languages such as PL/I are more universal and easier to use.[1] Lengthy formulas need not be broken down into elementary steps, and the instructions use standard English words rather than acronyms. Typical PL/I instructions

[1] PL/I is an acronym for "Programming Language One," and it is also sometimes written as PL/1, PLI, or PL1.

are "Y = 56.7 − X/(X + 9.8574);" and "IF X > Y THEN GO TO INSTRUCTION 3;." The case studies will explain these statements in detail, but as is clearly evident, they resemble ordinary algebra and English.

I.6 COMPILE VS. EXECUTE

A machine language program called the *compiler* translates procedural language programs into machine language, breaking the instructions down into elementary steps and assigning numeric addresses and instruction codes. A program must be completely compiled before it can be run or executed. This compilation often takes considerably longer than the execution; so a machine language version of the program is often saved on punched cards or some other medium. The compilation can then be skipped for all subsequent uses of the program. However, machine language versions of student exercise programs are rarely saved because the objective is to gain wide experience by using many programs once rather than a few programs many times. Nevertheless, students are usually asked to write their programs as if they were to be used many times, thus simulating a realistic industrial programming experience. A procedural language program is often called a *source program*, and a machine language program is often called an *object* or a *binary program*. Languages and programs are called *software*.

I.7 HARDWARE

While software was improving, hardware was too. For example, the vacuum tubes in the control and arithmetic circuitry were replaced with faster, less expensive transistors; then with microcircuits. The acoustic delay and electrostatic memories were replaced with faster, less expensive magnetic memories of various designs. The earliest of these was the *magnetic drum*. It consists of an iron oxide coated cylinder which rotates rapidly under a row of electromagnetic heads. As with an ordinary tape recorder, the heads may write and read magnetic signals on the drum. Every number passes under a head once per revolution, but other devices such as *core* storage are even faster.

Core storage consists of stacks of wire grids strung with tiny ferrite "doughnuts" which can be magnetized clockwise or counterclockwise by passing currents in the same direction through pairs of crosswires. This operation corresponds to *writing* a binary zero or one. The direction of magnetization can be sensed by a third wire passing through the doughnuts, and this operation corresponds to *reading* a binary zero or one.

Thin film storage is even faster. It consists of a sheet of thin magnetic material adjacent to a grid of crossed wires. Similar to core storage, the area adjacent to an intersection of two wires can be magnetized in either of two directions by

Sec. I.7   Hardware   9

passing currents in the proper directions through the two wires. Integrated circuits are becoming inexpensive enough so that even faster solid state "registers" may be used for at least a portion of the memory.

Thin film and core storage are more expensive than drums per unit of storage.[2] Consequently, drums are still used as a low cost, high capacity *backup*, *secondary*, or *auxiliary* storage. *Disks* and *tape* are also frequently used in this manner. Disks consist of a stack of iron-oxide coated phonograph-like discs that rotate rapidly between electromagnetic heads which can move radially in and out. These devices are more capacious but slower than drums because the heads may need to be moved to locate a number.

Magnetic tape storage is the least expensive of all, but it is quite slow if one

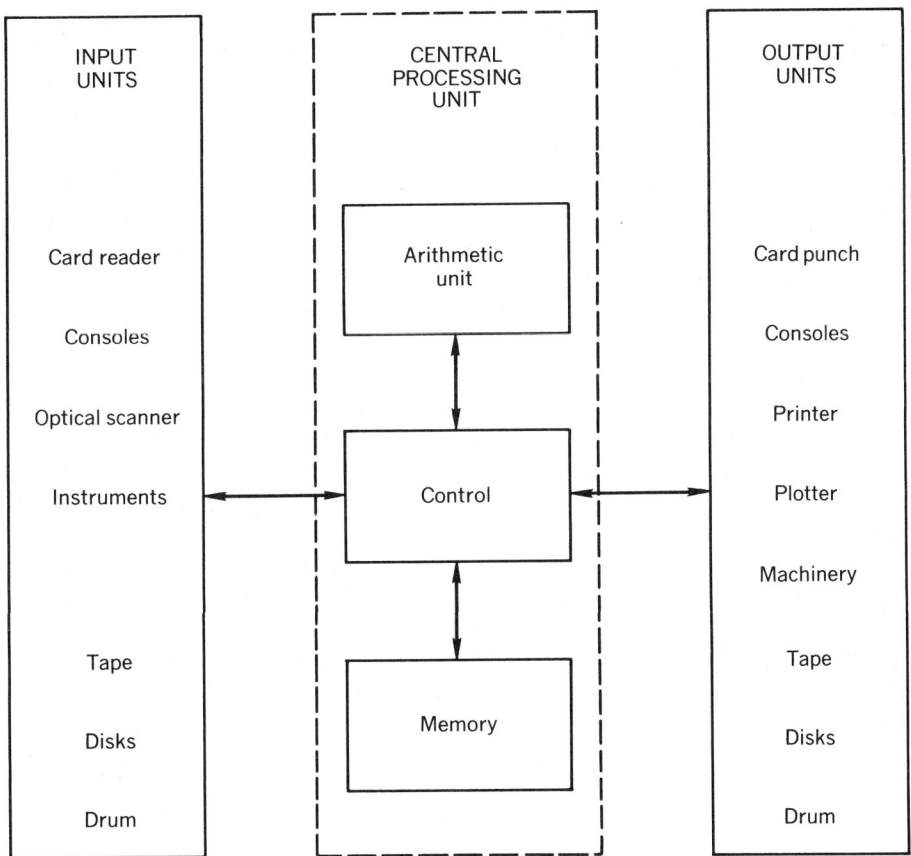

Fig. I-2   The components of a computer.

[2] Storage capacity is often measured in *bits*: A bit is the amount of storage required to hold 1 binary digit. Other units are the *byte*, which is 8 bits, and the *word*, which varies from computer to computer but is generally in the range of 16 to 60 bits.

wishes to read its contents in a random order rather than sequentially. Magnetic tape can also be considered an input–output, or I/O, medium since reels of tape may be generated by various instruments, shipped about, and used by various machines. Drums and disks are also usually considered I/O media since I/O-like instructions are required to transfer data between them and the main memory. Computers may also be connected directly to instruments, servomotors, machine tools, and drafting machines. The relationship between the parts of a computer is summarized in Fig. I-2.

I.8 BATCH VS. TIME-SHARED PROCESSING

A computer may process programs on an individual basis, a *batch* basis, or a *time-shared* basis. The individual basis was the original way, but it is quite rare now. With individual processing the programmer gives his program, data deck, and instructions to the operator. The operator puts the program deck in the card hopper. He then types in instructions on an electric typewriter, telling the computer what kind of program it is and what to do with it; for example, "compile and punch a machine language version of this PL/I program" or "assemble and execute this assembly language program." Next he puts in the data deck, if any, and types in a corresponding instruction for it. After the computer is through, he collects any printed or punched output and gives this to the programmer along with the original decks. Programmers may do their own operating with simple computers which are used infrequently.

As we see, the computer is idle between customers and also while the operator loads cards, types in instructions, and deliberates. This idle time would represent an extravagant waste of money for all but the very smallest of computers. An installation should have enough work to keep it operating continuously—at least during normal working hours.

Batch processing was developed to enable computers to run continuously. With batch processing, each deck of cards is accompanied with supplementary *control cards* which tell the computer what kind of a deck it is and what to do with it. A whole waiting line or *queue* of jobs[3] is then maintained in the card hopper. The control cards often give the computer accounting information such as what financial account to charge the computer time towards. It is important to realize that these control cards are not PL/I. The program that interprets these cards is called a *monitor* or *operating system*.

The trouble with batch processing is that the waiting time might be quite long, regardless of how quickly the program runs. There are often a number of jobs waiting besides those in the card hopper. An overnight wait or *turn-around* time is common for low-priority jobs. Even for high-priority jobs, the turn-around is

---

[3] A *job* is a program plus its control cards and data (if any).

generally too long to wait for the results at the computer; so batch processing has the additional disadvantage of requiring two trips to the computer—even more if the job is not completed when expected.

Time-shared processing was developed to overcome these disadvantages. The programmers type in their jobs at remote consoles. With large computers, over 100 conveniently located, geographically scattered consoles may be used simultaneously. The computer services each one for a few seconds, rotating between them. Any jobs not completed in that time are transferred to auxiliary storage in their states of semi-completion, to be recalled and further worked upon when their turns come up again. Batch processing is often run simultaneously to take advantage of any slack time. Some time-shared systems compile and possibly also execute each instruction as it is typed in. Others are like batch processing in that the entire program is typed in before compilation is begun; then the compilation is completely finished before execution is begun.

A programmer's deliberations and typing are so slow compared with processing speeds that each user of a time-shared system feels as though he is getting nearly full-time use of the computer. The turn-around time is essentially zero because a programmer has his answer in little more than the time it takes to keypunch a job for batch processing.

Although time-sharing is usually more efficient for the user, it is less efficient for the computer because there is extra computer time spent shuffling programs around. Also, the input–output equipment is generally more expensive with time-sharing. What is often done is to develop a program on a time-shared console, then to use batch processing for the routine production runs—especially when the programs take a long time to execute. The routine monthly use of a proven payroll program, for example, is more suitable for batch processing. Because of the extra expense, many computer installations do not have time-sharing facilities.

As with batch processing, there are generally supplementary non-PL/I instructions which tell the computer what kind of program or data will be typed in and what to do with it. Since the details vary considerably from installation to installation, they are not treated in this text. These details may be learned from an instructor or a programmer at a specific installation.

Regardless of which type of processing is used, it is difficult to write an errorless program on the first attempt. The art of finding and correcting these errors is called *debugging*. The compiler and the operating system offer quite a bit of help in this process by giving the user *diagnostic error messages* when the program does not make sense grammatically or when conditions such as an attempted division by zero arise. The details of these diagnostics and other helpful debugging aids vary considerably from compiler to compiler. The details for a specific compiler may be found in the appropriate programmer's guide manual, but are more easily learned from an instructor or programmer.

TEST YOUR UNDERSTANDING

1. What is the difference between analog and digital?
2. What components distinguish a card programmed calculator from a desk calculator?
3. What are the comparative advantages and disadvantages of binary and decimal computers? Find out whether your computer is binary, decimal, or both.
4. Why is a memory necessary in a computer or card programmed calculator?
5. What are the advantages of a stored program?
6. What is an address?
7. What is an assembler?
8. What is a compiler? Find out what compilers are used with your computer.
9. What is the difference between software and hardware?
10. What is auxiliary storage? Find out the types of storage used with your computer.
11. What is a bit?
12. Find out the word size for your computer. Find out the size of the main and auxiliary memories for your computer.
13. What is an operating system?
14. What is the difference between individual and batch processing?
15. What is the difference between time-shared and batch processing?
16. What is debugging?

I.9 AN IMPORTANT NOTE

The material in this book has been arranged to span the interests and mathematical abilities of practicing engineers and scientists as well as freshman through graduate engineering and science students. In order to do this, the sections and exercises have been grouped into several classifications. Except for some optional sections and exercises, this text requires no more than a previous exposure to the concepts of coordinates, graphs, and trigonometry. However, there are additional sections and exercises labeled "Mathematically Advanced" which are included to provide additional interest for those who are taking or have taken calculus or higher mathematics. There are also sections and exercises labeled "Additional Language Feature" which introduce unessential PL/I language features closely related to the essential ones introduced in the corresponding case study. These sections and exercises are included for the sake of those who have had previous programming experience. Beginning programmers are well advised to avoid confusion by sticking to the essential language features.

Finally, as a challenge to those who find the basic exercises too easy, there are some especially difficult ones which do not necessarily involve additional language features or advanced mathematics. These exercises are labeled "Difficult" to forewarn the easily disheartened.

The answers to the exercises marked with the symbol ▼ are listed at the back of the text. These exercises frequently precede related exercises, and so it is advisable to do these exercises first and check the answers before attempting the following ones.

For easy reference, each case study has its new language features summarized at its end.

Machine independence is one of the objectives of PL/I. This feature enables programs written for one machine to be used on another without modification. To achieve this independence, PL/I does not define features which are likely to depend upon the individual characteristics of a computer. Such features are said to be *implementation defined* or *implementation restricted*. An example is the maximum number of digits in a number. These restrictions should be known before programs are written for a particular computer.

Manufacturers are introducing and improving their PL/I compilers at such a great rate that any attempt at a complete summary would be futile and distracting. However, as an example of typical implementation restrictions, those for the IBM 360 D, F, and Model 20 compilers are given in footnotes. The Model 20 compiler is for the 360 Model 20; the D compiler is for the medium size 360s with a disk or tape operating system, and the F compiler is for 360s with main memories large enough to contain the operating system. The footnotes also describe the IBM distributed Student Language One (SL1) compiler for the 1130 and Student PL compiler for the 360. These are for fast processing of student-type programs which are typically small, simple, and run only once. Space is provided at the end of each case study for notes on the implementation restrictions for other compilers or on the inevitable changes in the model 20, D, F, SL1, or Student PL compilers.

There is no specific case study at which the actual running of programs should be started. Programs may be run as early as Case Study 1, but the first programs really worthy of a computer occur in Case Study 3.

CASE STUDY 1

HEAT TRANSFER[1]

The PROCEDURE, Assignment, PUT DATA, END and DECLARE Statements

In general, a program consists of instructions or *statements* grouped into a main procedure and possibly also several subordinate procedures. The major purpose of this grouping is to enable programmers to incorporate standard pre-programmed subordinate procedures into their programs. For example, the calculation of standard mathematical functions, the standard statistical reduction of data, and the solution of simultaneous equations occur quite frequently as components of larger programs. Consequently, it is worthwhile to have these components programmed and compiled once and for all in a form suitable for use in any program.

The utility of this approach will become more apparent in Case Study 2 where we will begin to study subordinate procedures. The first Case Study will be limited to programs which have only a main procedure.

1.1 A SAMPLE PROGRAM

As a preliminary example of a PL/I program, the following main procedure calculates p from the formula $p = a(b + c)/q$ for the specific case when $a = -1.40$,

---

[1] The material in these case studies depends heavily upon the Introduction; so it is suggested that you read it before beginning here.

$b = 27{,}200$, $c = 0.573$, and $q = 673$:

```
FRMULA:   PROCEDURE OPTIONS (MAIN);
   P = -1.4*(27200 + .573)/673;
   PUT DATA (P);
END;
```

This program consists of four statements which are much like English sentences. They each tell the computer certain properties of the program or tell it to perform certain operations. However, statements end in semicolons rather than periods to avoid ambiguity with the decimal point. Also, lower-case letters are not used. This is a concession to economy because the expense increases with the number of characters which must be encoded. As a matter of fact, some implementations may not include the colon and semicolon in their hardware. If so, they are punched or typed as "..". and ",." respectively, without the quotes.[2]

1.2 THE PROCEDURE STATEMENT

The first statement is called a **PROCEDURE** statement. Notice that we use capital letters in referring to words which have special significance in the PL/I language. These words are called *keywords*. This procedure is designated to have the **MAIN** *option*. More generally, main **PROCEDURE** statements have the form:

procedure name: PROCEDURE   OPTIONS   (MAIN);

Lower-case letters indicate that the programmer substitutes a name of his choice. It is the usual practice to choose a name which is suggestive of the application, like **FRMULA, ORBIT,** or **PAYROL**.[3]

1.3 THE ASSIGNMENT STATEMENT

The second statement is called an *assignment statement*. It illustrates several rules:

1. *Multiplication is indicated by an asterisk.* An "X" might cause confusion with

---

[2] Implementations which do not include the colon and semicolon are said to be restricted to the *48-character set* as opposed to the *60-character set*.
[3] Procedure names must not exceed seven characters for the F compiler, or six characters for the Model 20 and D compilers, or five characters for the SL1 compiler. The D and F compilers permit the main procedure name to be omitted. "OPTIONS (MAIN)" is omitted with SL1. The PROCEDURE and END statements are omitted from a Student PL main procedure.

a variable name; a dot might cause confusion with a decimal point; and juxtaposition such as "$-1.4(27200 + 82.6)$" must be reserved for other purposes.

2. *Division must be indicated by a diagonal slash.* Most input media are limited to one line at a time; so a horizontal line would be beyond their capabilities. The symbol "$\div$" is not used for the historical reason that it is not on the standard typewriter keyboard.

3. *Commas may not be used in numbers.* Commas must be reserved for other purposes.

4. Leading zeros may be used or not, according to taste. For example ".573" and "0.573" are both valid, but the latter is less likely to be misread by a human.

5. Trailing zeros may be used or not after a decimal point. For example, "-1.4," "-1.40," and "-1.4000" are all valid. Leading or trailing zeros may use extra computer time and space; so they are generally omitted in this text.

6. *The left side of an assignment statement must consist of a simple variable.* It may not be a more general mathematical expression. In order to be used as an assignment statement, equations in the form "mathematical expression = mathematical expression" must be manipulated into the form "variable = mathematical expression." The reason for this is that an assignment statement is not an equation! An assignment statement means "assign the value of the expression on the right to the storage location associated with the name on the left."[4]

1.4 THE PUT DATA STATEMENT

The third statement in our example is a PUT DATA statement.[5] The computer does not automatically print a record of each calculation. It puts out only what it is explicitly directed to. Note that statements are executed in the order in which they are written; so it is essential to have the assignment statement before the PUT DATA statement in our example. Otherwise, the proper value of P would not have been established by the time the value of P was put out. Consequently, as output we would get whatever value was left in P's storage location by the previous program. Moreover, there would be no indication that this value was incorrect!

---

[4] Although algebraic manipulation is not built into PL/I, there is a general-purpose IBM PL/I version of a program for this called FORMAC.

[5] We may use "PUT LIST (P);" for the D and SL1 compilers which do not support the PUT DATA statement. We may use "OUTPUT = P;" for Student PL which does not support the PUT DATA statement. For the Model 20 compiler, which supports neither PUT DATA nor PUT LIST, we may use "PUT FILE (OUT) EDIT (P) (E(15,6));."

1.5 THE END STATEMENT

The END statement marks the end of the procedure just as the PROCEDURE statement marks the beginning. The beginning and the end are obvious in this example, but beginnings and ends are scattered throughout programs which consist of more than one procedure.

For legibility, we list one statement per line with outline-like indentation, but the statement boundaries are arbitrary. We often list several short, closely related statements on one line; and we often need more than one line for lengthy statements. The spacing within statements is arbitrary too. However, there may be no spaces within words or numbers, and there must be at least one mathematical symbol, punctuation symbol, or space between the words and numbers. As an illustration, our example procedure could be written as follows:

```
                    FRMULA:PROCEDURE              OPTIONS(MAIN    )
;P=-1.4*(   27200+.573)    /673   ;   PUT   DATA  (P );     END   ;
```

This procedure is perfectly valid. It is also an excellent example of how to antagonize whoever is trying to read your program.

When a program is transferred from scratch paper to an actual input medium, there may be further physical limitations and implementation restrictions which must be observed. For example, with the standard 80 column cards, the lines of PL/I obviously cannot exceed 80 characters. Actually, most implementations restrict PL/I statements to columns 2 through 72—column 1 being reserved for control cards and columns 73 through 80 being reserved for *sequence numbers*. Sequence numbers optionally may be added by machine after a program is completely debugged. They are used as a precaution with long programs, because if a deck is accidentally scrambled, a mechanical sorting machine can sort them by sequence number.

EXERCISES

(The answers to the problems marked with a "▼" here and throughout the text are given in the Appendix.)

Write complete programs which evaluate the following formulas and put out the answers:

▼1. $E = CQ^2/2$, the energy stored in a linear capacitor as a function of charge for $C = 0.0001$ and $Q = 0.0025$.

2. $V = RT/P$, the specific volume of a gas as a function of its gas constant, temperature, and pressure for $R = 55.2$, $T = 1260$, and $P = 876.5$.

3. $V = abc$, the volume of a rectangular parallelepiped for the specific case when the lengths of its edges are $a = 4.62$, $b = 8.791$, and $c = 0.353$.

▼4. The following program supposedly calculates b from the formula $\dfrac{b + 0.62}{b - 57.3} = 18.6$, then prints out the result. Correct any mistakes.

```
SOLVE:    PROCEDURE   OPTIONS  (MAIN)
          PUT  DATA  (b)
          (b + .62) ÷ (b - 57.3) = 18.6
```

5. The following program supposedly calculates Y from the formula 45,679 = YX for the specific case when X = 3.78, then prints out the result. Correct any mistakes.

```
HYPERB:   PROCEDURE  OPTIONS  (MAIN).
          PUT  DATA  (X).
          45,679 = YX.           THE  END
```

1.6 IDENTIFIERS

Actually, a variable name may have more than one letter. The maximum number of characters is implementation defined.[6] Variable names, procedure names, and keywords are all examples of identifiers. Identifiers may be comprised of *alphabetic characters* (A through Z together with $, #, and @), *numeric characters* (the digits 0 through 9), and the special *break character* (_). However, the first character must be alphabetic. "$," "#," and "@" would be used primarily for business applications. Imbedded blanks are not allowed, but the break character may be used to improve legibility.[7]

In standard mathematical usage, juxtaposition implies multiplication; so variable names are essentially limited to one character. Consequently, subscripts, superscripts, and supplementary foreign alphabets must be used to achieve sufficient variety. Moreover, the names are rarely suggestive of the physical quantities involved; thus they must be memorized or else a glossary must be used. The multiple-character capability of PL/I is a useful feature and it is easy to see why its inventors do not want juxtaposition to imply multiplication. It is also advisable to use names suggestive of the application—abbreviated where necessary to avoid unwieldy expressions or an undue amount of typing.

1.7 HEAT TRANSFER EXAMPLE

As an example, the following program uses the formulas $Q_1 = H \cdot (T_{body} - T_{sur})$ and $Q_2 = F \cdot (T_{body}^4 - T_{sur}^4)$ to calculate the heat flow rate from a body to its surroundings by convection and radiation respectively, for the specific case when

---

[6] 31 for the F, D, and Model 20 compilers; six for the SL1 compiler.
[7] "#", "@", and "_" are not included in the 48-character set.

Sec. 1.8   *Floating-Point Numbers*   19

$T_{body} = 815.3°$ Kelvin, $T_{sur} = 530°$ Kelvin, $H = 3{,}720{,}000$ calories/hour-° Kelvin, and $F = 0.00000097$ calories/hour-° Kelvin^4:[8]

```
HEAT:   PROCEDURE OPTIONS(MAIN);
        T_BODY = 815.3;   T_SUR = 530;
        Q1 = 3.72E6 * (T_BODY - T_SUR);
        Q2 = 9.7E-7 * (T_BODY**4 - T_SUR**4);
        PUT DATA (T_BODY, T_SUR, Q1, Q2);
END;
```

1.8 FLOATING-POINT NUMBERS

Engineers and scientists often use scientific notation for numbers which have very large or very small magnitudes compared with 1. This compact notation is more legible and convenient for such numbers. As we can see from the example above, $3720000 = 3.72E6$ and $0.00000097 = 9.7E-7$. In programming terminology, as opposed to logarithm terminology, the number to the left of the E is called the *mantissa* and the integer to the right is called the *exponent*.

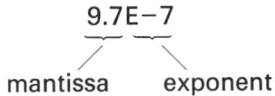

mantissa   exponent

The number of digits in the mantissa is called the *length*. The exponent indicates how many places the decimal point in the mantissa must be shifted to the right or left to arrive at the ordinary nonscientific notation. Actually, the numbers 3.72E6 and 9.7E-7 would be represented as 3.72×10^6 and 9.7×10^{-7} in true scientific notation, but it is uneconomical to design raised character capability into the input and output equipment. The E takes the place of "$\times 10$." Numbers written with an E are called floating-point because the decimal point may "float" relative to the significant digits. For example, 372E4 and 0.9E-6 would be acceptable substitutes in our program. Note that no blanks may appear anywhere within the number. Also, E3 and 5E are invalid because a mantissa or exponent may not be implied. Within the machine, floating-point numbers are stored in normalized form. The exponent is adjusted if necessary so that the mantissa has a magnitude between 0.1 and 1 (unless the number is zero). However, the number may be written in any mathematically equivalent form.

There are implementation restrictions on the magnitude of a floating-point constant and on its maximum length.[9] The magnitude restrictions are usually so

---

[8] You do not need to be familiar with a formula in order to program it. Many professional programmers have no understanding of the physical principles underlying their programs.

[9] For the D, F, and Student PL compilers, the magnitude must be either zero or between about 10^{-78} and 10^{+75}, and the length must not exceed 16. For the SLI compiler, the magnitude must be either zero or between 10^{-998} to 10^{+998}, and the length must not exceed 15. For the Model 20 compiler, the magnitude must be either zero or between 10^{-50} and 10^{+49}, and the length must not exceed 15.

generous that they are of no concern, unless we use absurd physical units such as Angstroms for interplanetary distances or light-years for intermolecular distances. The length is usually more than enough to accurately represent any conceivable physical measurement.

1.9 FIXED-POINT NUMBERS

Constants written in nonscientific notation are called *fixed-point* because the decimal point is fixed relative to the significant digits. Thus, 815.3 and 550 are fixed-point constants. 550 is a special case called a *fixed-point integer* because it has no fractional digits. Although it is a whole number, 550.0 is not a fixed-point integer constant because it has a fractional digit. The total number of digits is called the *length*, and the number of fractional digits is called the *scale factor*. The length together with the scale factor is called the *precision*. There is an implementation restriction on the maximum length of a fixed-point constant.[10]

1.10 THE FIRST-LETTER CONVENTION

The values of variables are also stored in either fixed or floating-point form. We may make a variable be a fixed-point integer by making its name start with the letters I, J, K, L, M, or N. We may make a variable floating-point by making it start with one of the other alphabetic characters. This *first-letter convention* is not as arbitrary as it may appear because i, j, k, l, m, and n are the letters most frequently used for integers in mathematics—for example, "n equations."

If an appropriate name for a variable begins with the wrong kind of letter, the name may be "fudged" to comply with the first-letter convention. For example, we could use "THE_INTENSITY" rather than "INTENSITY" if we wished it to be floating-point. Similarly, we could use "NFACTORIAL" rather than "FACTORIAL" if we wished it to be fixed-point.

1.11 THE DECLARE STATEMENT

The first-letter convention is an example of a *default* assumption. PL/I has many other default assumptions too, such as the lengths of variables. The first-

---

[10] 15 digits for the D, F, Model 20, and SL1 compilers. Student PL considers any constants written with a decimal point to be floating-point and permits fixed-point integers in the range −2147483648 to 2147483647.

letter convention takes effect by default unless overridden by a DECLARE statement. We could include in our procedure the statement "DECLARE INTENSITY FLOAT;" or "DECLARE FACTORIAL FIXED;". The keywords FIXED and FLOAT are called *attributes* because they describe attributes or characteristics of the variables. More specifically, they are the two alternatives for the *scale* attribute. If we used "INTENSITY" and "FACTORIAL" in the same program, we could use the statement "DECLARE INTENSITY FLOAT, FACTORIAL FIXED;". If we also wanted "LENGTH" to be a floating-point variable, we could use the statement "DECLARE (LENGTH, INTENSITY) FLOAT, FACTORIAL FIXED;". This technique is called *attribute factoring*.[11] FLOAT applies to all variables within the parentheses before it. Note that commas are used to separate variables, but are not used to separate variables from their attributes. More generally, then, the DECLARE statement may have the form:

```
DECLARE  (list of variables separated by commas) FLOAT,
         (list of variables separated by commas) FIXED;
```

DECLARE statements may be placed anywhere between the PROCEDURE and END statements, but they take effect immediately upon entry into a procedure. Consequently, it is advisable to place DECLARE statements immediately after the PROCEDURE statement to emphasize this fact.

Some programmers prefer the technique of fudging the names, and others prefer the technique of overriding the convention. Overriding the convention enables one to use exactly the name desired, but it may involve more typing. Also, there is some value to associating automatically the letters I through N with integers. When reviewing a procedure, it enables immediate identification without reference to a DECLARE statement.

Some programmers like to declare all variables, including those which comply with the first-letter convention anyway. At the expense of extra programming, this makes a neat summary of all the variables, and it lessens the chance of inadvertently establishing an inappropriate attribute. Whichever technique one eventually gravitates toward is a matter of individual taste. This text will illustrate a variety of techniques.[12]

It is all right to use a keyword as a variable or procedure name. Keywords are recognized as such only when used in the proper context. Consequently, we do not have to learn all the keywords in order to avoid improper use.[13] In fact, we may use a keyword as both a keyword and a variable name in the same procedure, provided the variable is explicitly declared. However, this makes the procedure difficult to read; so a good rule is to *never use an identifier two ways in one procedure*.

[11] SL1 does not support attribute factoring.
[12] The Student PL and Model 20 compilers have no first-letter convention, so all variables must be declared.
[13] There are a few reserved words for the F, D, and Model 20 compilers using the 48-character set. Keywords are reserved in SL1 and Student PL.

1.12 DEFAULT-LENGTHS

There are implementation-defined *default-lengths* for fixed and floating-point variables.[14] In order to conserve memory space and computation time, these lengths are generally less than the maximum allowed for constants. Default fixed-point arithmetic should be used when applicable because it is faster than default-length floating-point arithmetic. Default fixed-point variables are usually applicable when quantities assume only whole values, such as when counting. However, the limited magnitude and the restriction to integers are unsuitable for applications such as the formula evaluations of the procedures **FRMULA** and **HEAT**. Floating-point variables are appropriate for such applications. The default-length fixed and floating-point variables are adequate for most engineering and scientific applications, but later we will learn how to establish lengths up to the maximum allowed for constants. We will also learn how to establish fixed-point noninteger variables.

1.13 PRECISION AND MIXED EXPRESSIONS

In order to avoid unnecessary loss of precision, whenever two floating-point values are combined, the result will have a length not less than the greater of the two. For example, if **X** is a default floating-point variable, the result of **X + 3E6** will also be default length. The **OVERFLOW** *condition* is raised if the magnitude of a result exceeds the maximum allowed by the implementation. The numerical result is unpredictable and the computer will put out an appropriate message and terminate execution. The **UNDERFLOW** condition is raised if the magnitude is less than the minimum allowed by the implementation but not zero. The numerical result is set equal to zero, and the computer will put out an appropriate message and continue execution.

Whenever fixed and floating-point values are combined in an arithmetic expression, the fixed-point value is converted to floating-point without loss of precision before the result is evaluated.

Whenever two fixed-point values are combined in an expression, the precision

[14]

compiler	Floating-point variables		fixed-point variables	
	magnitude must be 0 or between	decimal length	range if undeclared	range if declared FIXED
F & D	10^{-78} to 10^{75}	6	-32767 to 32768	-99999 to 99999
Student PL	10^{-78} to 10^{75}	6	must be declared	-2147483648 to 2147483647
Model 20	10^{-50} to 10^{49}	6	must be declared	-99999 to 99999
SL1	10^{-999} to 10^{998}	7	-9999999 to 9999999	-9999999 to 9999999

of the result is not less than whatever is required in order to avoid unnecessary loss of precision. For example, the result of 56.783 + 92.2 will be 148.983. This rule has two consequences which must be guarded against:

1. An arithmetic combination of two fixed-point values which have a great number of whole or fractional digits could require more digits than the maximum allowed by the implementation. This condition is called **FIXEDOVERFLOW**. The numerical result is unpredictable, and the computer will put out an appropriate message and terminate execution of the program. Most of the variables in engineering and scientific programs will be floating-point, and most constants will be combined with variables rather than other constants; but *if two values which both require many digits to represent in fixed-point notation are combined, floating-point should be used for at least one of them.*

2. In order to prevent unnecessary loss of precision, the length of a fixed-point quotient is the maximum allowed by the implementation. For example, if this maximum is 15 digits, .2/3 will have the value .066,666,666,666,666. It is highly probable that any subsequent combination with another fixed-point value will raise the **FIXEDOVERFLOW** condition! For example, 1 + .2/3 would raise the **FIXEDOVERFLOW** condition. Consequently, we should *avoid writing expressions in a way which causes the result of a fixed-point quotient to be combined with another fixed-point value.* Use a floating-point value for one of the constituents—for example, 1E0 + .2/3, 1 + .2E0/3, or 1 + .2/3E0.[15]

A fixed-point value is converted to floating-point if it is assigned to a floating-point variable. Similarly, a floating-point value is converted to fixed-point if it is assigned to a fixed-point variable. If the value has more fractional digits than the variable can accommodate, the excess will be chopped off upon assignment. For example, the statement "N = 583.79;" will give N the value 583 if N is a default fixed-point variable. This technique is quite useful for isolating the whole part of a number. This technique may also be used to round a positive value to a whole number if 0.5 is added to the right side: "N = X + 0.5;". If the whole part of a number is too large for the fixed-point variable to which it is being assigned, the excess leftmost digits will be chopped off without notice. Later, we will learn how to detect this condition and how to do something about it. For the time being, we must simply be careful not to assign large values to fixed-point variables.

Some programmers like to save the computer a conversion by using floating-point for all of their constants which are combined with floating-point values or assigned to floating-point variables. This may be done by simply adding an "E0" to any fixed-point constant. However, such conversions are performed only once at compile time. Most computers perform each such conversion for less than 0.0001¢ whereas an average keypunch operator paid typical wages costs more than 0.1¢ for each "E0" punched. The computer is cheaper and less error prone.

[15] The Student PL compiler chops off the fractional digits from the division of two fixed-point integers.

Returning to the heat transfer program, another feature to note is that variables may be used in statements—provided their values have been established by previous statements. Thus, we would get erroneous results without warning if the first two assignment statements were interchanged with the last two. (However, the first two may be interchanged with each other and the last two may be interchanged with each other.) Note, by the way, that we do not include the physical units in a program. Computers deal only in pure numbers. It is up to the programmer to make a consistent physical interpretation.

This procedure also indicates that a double asterisk is used for *exponentiation* or raising a number to a power. Again we can see that the restriction to one line has necessitated a deviation from standard mathematical notation. The power may be fractional, and the base and power may be either a constant or a variable. Note that there may be no spaces between the two asterisks. Also, be careful not to confuse exponentiation and scientific notation. For example,

$$\text{2E4 is } 2 \times 10^4 \text{ or } 20{,}000 \qquad \text{whereas 2**4 is } 2^4 \text{ or } 16$$

Note too, that variables may not be used for the mantissa and exponent in scientific notation and that the exponent must be an integer.

$$\text{AEN} \neq A \times 10^N$$
$$1.5\text{E}3.5 \neq 1.5 \times 10^{3.5}$$

On the other hand, 1.5*10**6 may be used for 1.5×10^6, but it is unwieldy and time-consuming compared with 1.5E6.

1.14 PRIORITY OF ARITHMETIC OPERATIONS

Also implicit in the heat transfer program is the fact that certain operations have higher priority than others. The operations must be performed sequentially; so it is important to know the natural order.

The following chart summarizes the hierarchy for arithmetic operations:

first priority	**, prefix —, and prefix +
second priority	* and /
third priority	infix — and infix +

Equal-priority operations are performed from right to left for the first priority and from left to right for the other priorities.

These rules may seem complicated, but actually they are the same rules that we always use. For example, how should we interpret -5^{-3}? Most people would interpret this as $-(5^{-3})$. In other words, the *prefix* — before the 3 is performed before the exponentiation, which is performed before the prefix before the 5. Thus "—5**—3" gives the expected result. It is natural, then, that among exponentiation, prefix —, and prefix +, the order is from right to left. (Actually, there is no

Sec. 1.15   *Output*   25

need to ever use a prefix +.) If we want $(-5)^{-3}$ we may use "(−5)∗∗−3." As another example, it is natural to interpret $8q^6$ as $8(q)^6$ and $5/-d^p$ as $5/-(d)^p$; thus it is natural that multiplication and division have a lower priority than exponentiation and prefixes:

$$8*Q**6 \quad \text{means} \quad 8*(Q**6)$$
$$5/-D**P \quad \text{means} \quad 5/-(D**P)$$

Mathematically, the expression a/bc is ambiguous because division and multiplication have equal priority. However, PL/I had to make a rule to cover such situations, and since we ordinarily read from left to right, the rule is that the operations are performed from left to right. Similarly, although the order of additions in the expression "$3 \times 10^6 + .2/3 + 2$" is mathematically ambiguous and irrelevant, the order is from left to right and highly relevant in the PL/I expression "3E6 + .2/3 + 2." As it stands there would be no **FIXEDOVERFLOW**, but there would be if the order were from right to left. Finally, note that it is natural to consider multiplication and division as having higher priority than addition and subtraction. No one would interpret the above expression as (3E6 + .2)/(3 + 2). Parentheses are used to alter the natural order in PL/I the same way as they are in mathematics.

There are no brackets or braces ([ ] or { }), but PL/I permits multiple parentheses. For example:

$$A*(B + C*(D + E/(F + G))) \quad \text{means} \quad a\left[b + c\left(d + \frac{e}{f+g}\right)\right]$$

Because of the restriction to one line, we must be more certain of the hierarchy for PL/I than for ordinary mathematical notation. However, the rules are essentially the same, so they are not hard to learn. Some programmers follow the old adage "when in doubt, parenthesize," but this clutters the program, decreasing the legibility. In the long run, it leads to more mistakes than it prevents. You may as well learn once and for all to exploit the priority system.

1.15 OUTPUT

The final new feature illustrated by the heat transfer program is the more general form for the PUT DATA statement:

PUT DATA (list of variables separated by commas);[16]

---

[16] For the D and SL1 compilers use PUT LIST (list of variables separated by commas);. The output is similar but without variable names. For Student PL use a separate output statement of the form "OUTPUT = variable;" for each variable. For the Model 20 compiler use "PUT FILE (OUT) EDIT (list of variables separated by commas) (8E(15,5));" together with a declaration such as "DECLARE OUT FILE PRINT ENVIRONMENT (MEDIUM (SYSLST), 2203) F(120));". The PUT EDIT statement is not discussed in this text; so consult the appropriate reference manual if you wish to learn more about it.

Note that we have put out the given information as well as the results. This is good practice if the results are to be seen by someone other than the programmer himself, or if they are ever to be seen other than immediately after the program is run. Otherwise the subsequent need for referral to the program itself is a nuisance to the programmer and a definite annoyance to anyone not interested in the details of the program. Note that we assigned the given values to variables because we cannot list constants in a **PUT DATA** statement.

The output from a printer or on a remote console would look something like the following:

```
T_BODY= 8.15299E+02    T_SUR= 5.30000E+02
Q1= 1.06131E+09        Q2= 3.52051E+05;
```

The exact spacing and number of items per line depend upon the implementation, the length of the variable names, and whether they are fixed or floating-point. The values of fixed-point variables will appear in fixed-point form (e. g., N = 785), and the values of default-length floating-point variables will appear in default-length floating-point notation. Note, however, that the mantissa is renormalized to have a magnitude between one and ten unless it is zero, and for uniformity the exponent is printed with a sign and two digits even if the sign is positive and the exponent is zero. Plus signs for the mantissa are suppressed.

SUMMARY

(The blanks are implementation restrictions to be filled in by you.)

Typical Program

```
procedure name:   PROCEDURE   OPTIONS   (MAIN);
    DECLARE (list of variables separated by commas) FLOAT,
            (list of variables separated by commas) FIXED;
    variable = expression;
       . . .
    PUT DATA (list of variables separated by commas);
END;
```

Identifiers are formed from a maximum of ____ adjacent characters, the first of which must be alphabetic (A-Z, $, #, and @) and the remainder of which must be alphabetic, numeric (0 – 9), or the special break character (_).

Keywords are identifiers which have special meaning in PL/I.

Procedure names are identifiers which have the additional restriction of ____ characters maximum.

Variable names are identifiers which by default represent a fixed-point integer value in the range _____ to _____ if the names start with the letters I, J, K, L, M, or N. Otherwise by default they represent floating-point values with a length of ____ decimal digits and a magnitude of zero or within the range _____ to _____.

Arithmetic constants may have no commas or imbedded blanks.

Fixed-point constants may have a maximum length of _____ digits. The scale factor is the number of fractional digits.

Floating-point constants consist of a mantissa followed by an exponent. The mantissa may have a maximum length of _____ digits and the magnitude of the number must be either zero or within the range _____ to _____.

Arithmetic expressions may be formed with constants, variables, and any number of parenthetical levels. Precision is never lost unnecessarily.

first priority	**, prefix −, prefix +	(right to left ⟵)
second priority	*, /	(left to right ⟶)
third priority	+, −	(left to right ⟶)

Conditions

UNDERFLOW is raised when a floating-point result is less than the minimum magnitude but not zero. The numerical result is set to zero, and the computer will put out a message, then continue execution.

OVERFLOW is raised when a floating-point result is greater than the maximum magnitude. The numerical result is unpredictable, and the computer will put out a message, then terminate execution.

FIXEDOVERFLOW is raised when a fixed-point result is greater than the maximum length. The numerical result is unpredictable, and the computer will put out a message and terminate execution. To avoid this condition:

1. when combining two values which both require many digits in fixed-point notation, use floating-point for at least one of them.
2. avoid writing your expressions in a way which causes the result of a fixed-point quotient to be combined with another fixed-point value.

EXERCISES

6. For your implementation, which of the following are valid variable names, and which are valid procedure names?

nucleus	▼REACT	▼H2004
PL/I	NERVE	ISOTOPES
RH+	$LOSS	TEST#
T BODY	@COST	Y'
N!	BLOOD__TYPE	BROWN&SHARPE

7. For your implementation, which of the following floating-point constants is valid, and what numbers do they represent in nonscientific notation?

38E9	▼1.68E−4	▼76E3.5
−3E4	E3	5.6 E3
−4E−$	1E	3E80
53E 7	8EN	0.00068E−77
5.783856473897E4	0E0	1E0

8. For your compiler, which of the following fixed-point constants are valid, and what are their lengths and scale factors?

869	−00.3860	.50
59386.	∞	−0.00000000045
456,378	0.00	1234567890123456

▼9. Write an appropriate DECLARE statement to make MAGNITUDE, TIME, INTERVAL, and LAG be floating-point variables and to make PERCENTILE, SCORE, and ITEM be fixed-point variables.

▼10. Same as exercise 9 except fudge the names instead of declaring them.

11. Write an appropriate DECLARE statement to make NORM, KILOGRAMS, FORCE, MAGNITUDE, and INTERVAL be floating-point variables and SIDES, FACES, ANGLES, and NODES be fixed-point.

12. Same as exercise 11 except fudge the names instead of declaring them.

13. What conditions, if any, will be raised by your implementation for the following statements where all variables are established by the first-letter convention?

(a) X = 8E40/4E−40;
(b) Y = 5.678E−77−5.677E−77;
(c) Z = 3E3−3E3;
(d) P = 1234567890*1000000;
(e) U = 1234567890 + 0.1234567;
(f) V = (8.1/0.9)*1234.5678;
(g) W = 5.6 + 4/2E0;
(h) KIND = 2/3;
(i) L = 456E6 + 328E6;
(j) A = 56E3*(8/9)*56000;

14. None of the following PL/I expressions correctly represents its mathematical counterpart. Rewrite the PL/I expressions correctly:

Mathematical Expression	Incorrect PL/I Counterpart
▼(a) $a(b + c)$	A(B + C)
(b) $x \div y$	X ÷ Y
(c) $\dfrac{p \cdot q}{h}$	$\dfrac{P \cdot Q}{H}$
(d) $p(h + \bar{A})$	p*(h + Ā)
(e) $\dfrac{a_6 + b' \times 10^6}{10^{3.5}}$	(A6 + BPRIME + 10**6)/1E3.5
(f) $5{,}400{,}000A + 0.000023Q$	5.4E**6 *A + 2.3 E−5 * Q
(g) $\dfrac{a^6 - 10^{3.5}}{10^n}$	(AE6 − 1EN)
(h) $10^{-5} + 5 \times 10^0$	E−5 + 5E
(i) $(-q)^{-6}$	−Q**−6
(j) $\dfrac{(8q)^b}{p}$	8*Q**B/P
(k) $x^{n+5} y$	X**N+5 *Y

(l) $\dfrac{a+b}{c-d}$ A + B/C - D

(m) $\dfrac{q \cdot h}{p \cdot d}$ Q*H/P*D

(n) $\left(\dfrac{b+c}{x+y}\right)^{5m}$ (B + C)/(X + Y) ** 5*M

(o) $a/\{c[d + e/(f + g)]\}$ A/{C*(D + E/(F + G)]}

(p) $\dfrac{x}{\dfrac{y}{z+w} + 0.00005}$ X + Y/Z + W + .5E-4

(q) $\dfrac{x}{a + \dfrac{x}{b + \dfrac{x}{c+x}}}$ X/(A + X/b + X/(C + X))

15. Write the following PL/I expressions in unambiguous standard mathematical notation or indicate that the PL/I expression is meaningless:

▼(a) (A + 2E14)/B**-N   (f) A/B/C/D*E + F
▼(b) A + *X   (g) -X**-Y**-Z
(c) D + N/E*P   (h) A*(X + B*(X + C*(X + D))
(d) A**B**C   (i) B**1/3
(e) (A + B)/(C - 6.3E-9/   (j) A + B** /(H + 6)
    (E + F))

16. Write PL/I equivalents for the following mathematical expressions, using no unnecessary parentheses:

▼(a) $36{,}000{,}000 + \dfrac{x}{y - \dfrac{z^{-5}}{6+q}}$ (e) $\left(\dfrac{6q}{p \cdot d}\right)^{1/3}$

(b) $0.0000007b^5$ (f) $(-w)^{(z^2)}$
(c) $(0.000007b)^5$ (g) $(w^{-z})^2$

(d) $(0.000007b)^{n+5}$ (h) $\dfrac{\dfrac{a}{b} - \dfrac{c}{d}}{\dfrac{e}{f} - \dfrac{g}{h}}$

17. The total mechanical energy of a projectile is given by the formula $E = M(gh + V^2/2)$ where M is the mass, g is the acceleration of gravity, h is the height, and V is the velocity. Write a procedure which will calculate and put out E for the specific case when $M = 15.6$, $g = 32.17$, $h = 528$, and $V = 95.7$.

18. Same as exercise 17 except have the procedure also put out the values of M, g, h, and V.

19. The ideal compressor outlet temperature and efficiency for a gas turbine are $T_2 = T_1(P_2/P_1)^{(\gamma-1)/\gamma}$ and $e = 1 - (P_1/P_2)^{(\gamma-1)/\gamma}$, respectively; where T_1 is the inlet temperature, P_1 is the inlet pressure, P_2 is the outlet pressure, and γ is the ratio of specific heats. Write a procedure to calculate T_2 and e, putting them out together with T_1, P_1, P_2, and γ for $T_1 = 500$, $P_1 = 12$, $P_2 = 68$, and $\gamma = 1.4$.

CASE STUDY 2

TRIANGLE SOLUTION

Function Procedures, Mathematical Built-in Functions, and the PAGE Option

The only arithmetic operations that can be performed exactly with a finite number of steps are addition, subtraction, multiplication, division, the evaluation of integer powers, and the evaluation of integer roots such as the square and cube roots. Even these operations usually cannot be performed exactly if the numbers are repeating fractions such as $0.333 \cdots = 1/3$ or irrational such as $1.414 \cdots = \sqrt{2}$. Other operations such as finding the logarithm or sine of a number must be approximated by using a finite number of the above-mentioned fundamental arithmetic operations. For this reason, they are called *transcendental* functions. We can deduce the exact value for a few special cases such as sine (30°) or $\log_{10}$ (100), but in general, we must resort to a formula which only approximates the function. This is inescapable, and even the entries in standard tables are produced in this manner. There are standard techniques for devising such approximations, and there are references with a great number of useful approximations already developed. For example, the following so-called *Chebyshev* polynomial is adapted from a formula by Hastings:[1]

$$\text{sine}(A) \cong 0.999892A - 0.16596A^3 + 0.007603A^5$$

for $-\pi/2 \leq A \leq \pi/2$ where A is in radians. (One radian equals 57.2958 degrees.) Any angle can be reduced to this range for the sake of finding its sine. The maximum relative error in this approximation is about one part in ten thousand. This is rather impressive for only three terms, and Hastings also gives a five-term approximation with a maximum error of only about five parts per billion.

[1] Cecil Hastings, Jr., *Approximations for Digital Computers* (Princeton, N.J.: Princeton University Press, 1955), p. 138.

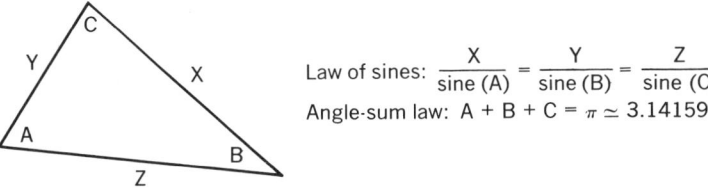

Fig. 2-1 Some trigonometric relationships.

Approximations such as these may be substituted wherever transcendental functions are needed in a program. For example, given angle A = 1.235 radians, angle B = 0.571 radians, and side Y = 157.3 feet for the triangle shown in Fig. 2-1, the following procedure uses the angle–sum law and the law of sines to calculate the remaining angle and sides:

```
TRINGL:   PROCEDURE OPTIONS (MAIN);
   A = 1.235;   B = .571;   Y = 157.3;
   COEF = Y/(.999892*B - .16596*B**3 + .007603*B**5);
   X = COEF*(.999892*A - .16596*A**3 + .007603*A**5);
   C = 3.14159 - A - B;
   Z = COEF*(.999892*C - .16596*C**3 + .007603*C**5);
   PUT DATA (A, B, C, X, Y, Z);
   END;
```

2.1 FUNCTION PROCEDURES

Note that the compiler does not recognize Greek letters such as π and that the computer does not have the value of π permanently stored. COEF is an intermediate variable which is established to avoid writing (and having the computer evaluate) the formula for the coefficient Y/SINE(B) twice. Nevertheless, it is tiresome to rewrite the approximation every time the sine of a different angle is needed. To avoid this essentially repetitive programming, the approximation may be written once as a separate *subordinate* procedure which is used three times by the main procedure—each time with a different angle:

```
SINE:   PROCEDURE (A);
   RETURN (.999892*A - .16596*A**3 + .007603*A**5);
   END;
```

The main procedure could then be written:

```
TRINGL:   PROCEDURE OPTIONS (MAIN);
   A = 1.235;   B = .571;   Y = 1.573;
   COEF = Y/SINE(B);
   X = COEF*SINE(A);
   C = 3.14159 - A - B;
   Z = COEF*SINE(C);
   PUT DATA (A, B, C, X, Y, Z);
   END;
```

The procedure named SINE is subordinate because it must always be used by another procedure rather than alone. Moreover, it is called a *function procedure* because it is used in TRINGL like a mathematical function—for example, SINE(B). In fact, the parentheses around B and the lack of an arithmetic operation joining SINE and (B) are what alert the compiler to assume that there is a procedure named SINE to use at this point. The appearance of the name SINE followed by parentheses in the main procedure is said to *invoke* the function procedure. SINE and TRINGL are submitted together as a program. It does not matter which comes first. Execution always begins at the first main procedure of a program, skipping over any subordinate procedures. Subordinate procedures are entered only when invoked. Whenever a subordinate procedure is invoked, control diverts to that subordinate procedure. Control reverts to the point of invocation when a RETURN statement is reached in the subordinate procedure. The value of the expression in parentheses after the RETURN statement is returned to be used at the point of invocation.

The name A used in the *definition* of SINE is called a *parameter*. The name B used in the first *invocation* is called an *argument*. Parameters do not have an address or a value of their own. They assume the identity of whatever argument is being used. During the second invocation the argument also happens to have the name A, but this is coincidental, and it will not confuse the computer. The names in subordinate procedures of this type are completely independent of names in any other procedure.

The names of a corresponding argument and parameter do not have to be the same.

This independence enables us to use SINE in any procedure which needs this trigonometric function. We never again have to concern ourselves with the details of SINE or the particular name of its parameter. We may use the argument names which are suitable to our particular application, and we may coincidentally use the parameter name in any context whatsoever without worrying about interference.

To clarify these concepts, let us consider the sequence of operations in TRINGL as depicted in Fig. 2-2. When execution reaches SINE(B) in the main procedure, control and the address of the argument are passed to the procedure named SINE. The subordinate procedure then associates that address with its parameter A, uses the value stored at that address to evaluate the expression, and then returns the result to the main procedure where it is used in place of SINE(B). Execution then proceeds in the main procedure until it reaches SINE(A). At this point, control and the address of the argument A are passed to the subordinate procedure. The subordinate procedure then associates the address of the argument A with its parameter A, evaluates the formula, and returns the resulting value to the main procedure where it is used in place of SINE(A). Execution then proceeds in the main procedure until it reaches SINE(C). At this point, control and the address of the argument C are passed to the subordinate procedure. The subordinate procedure then associates the address of that argument with parameter A, evaluates the formula, and returns the result to the main procedure where it is used in place of SINE(C).

Function procedures may have any number of statements within them, and

Sec. 2.1 Function Procedures 33

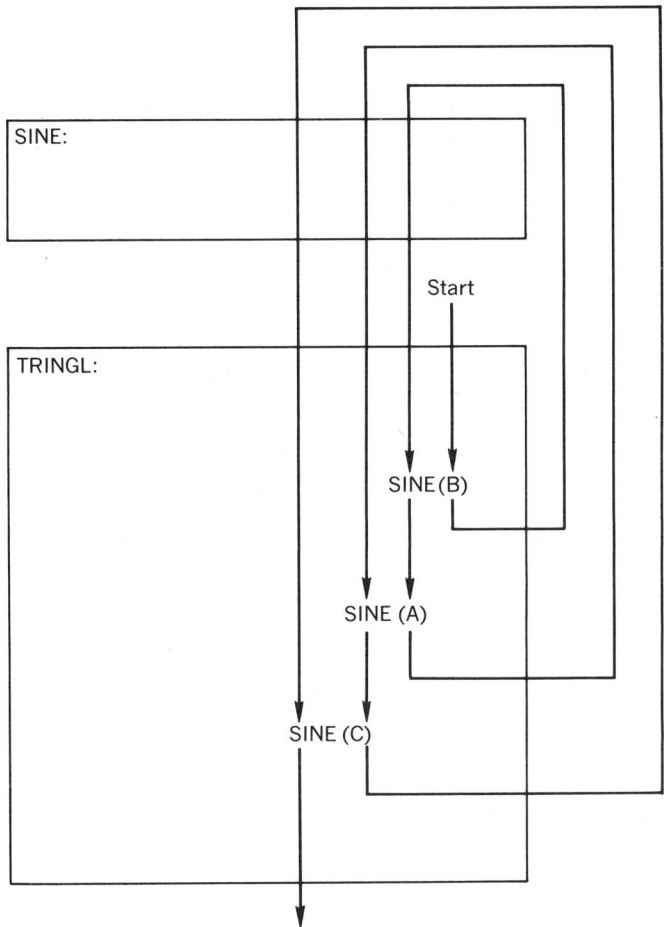

Fig. 2-2 Alternation of execution between the procedures TRINGL and SINE.

they may depend upon more than one variable. For example, the following procedure evaluates the function $y(a, b, n) = (a + b/2)^n + 1/(a + b/2)^n$:

```
Y:   PROCEDURE (A, B, N);
     TERM = (A + B/2)**N;
     RETURN (TERM + 1/TERM);
     END;
```

Here, an intermediate variable TERM was established to avoid having to write, and having the computer evaluate, the expression (A + B/2)**N twice.[2] We must be sure when invoking functions of more than one variable to list the

---

[2] Some compilers are sophisticated enough to recognize common subexpressions within a statement. However, few are sophisticated enough to recognize them among two or more statements; so it is good practice to learn to recognize and precalculate common subexpressions.

same number of arguments as parameters—and in the same order. For example, the invocation Y(C, D, M) would give an incorrect result if we wanted to evaluate $(d + c/2)^m + 1/(d + c/2)^m$. Also note that the invoking procedure does not pass information as to how arguments are encoded—for example, FIXED vs. FLOAT. This fact is an unavoidable consequence of the ability to compile procedures separately. Only the machine language addresses of the arguments are passed, and the subordinate procedure must assume that each argument is the same type as the corresponding parameter. Consequently, *the implied or declared attributes of the parameter and argument must agree*. Otherwise, the result may be incorrect without notice. We would get an error, for example, using SINE(K) in an expression unless K was declared FLOAT because the subordinate procedure would misinterpret two of K's digits as being an exponent. If we wanted to find the sine of an integer variable K, it could be done by using a statement such as "FLT__K = K;"—then using "SINE(FLT__K)." An expression may also be used as an argument provided the resulting attributes of the expression agree with those of the parameter. For example, we may use SINE(A + 3), provided A is a default floating-point variable. However, the use of expression arguments requires considerations discussed in the following sections.

2.2 THE BASE ATTRIBUTE

Numbers may have either a BINARY or DECIMAL *base*. We could, for example, have the statement "DECLARE A FLOAT DECIMAL, B FLOAT BINARY, C FIXED DECIMAL, D FIXED BINARY;". As is indicated in the following sections, not all compilers support all four types of arithmetic data.

2.3 ARGUMENT-PARAMETER AGREEMENT FOR SL1

SL1 supports only DECIMAL. Arguments and parameters do not need to have the same precision, but they must have the same scale (FIXED or FLOAT). The result of an expression is fixed-point if all of the values in it are fixed-point. Otherwise, the result is floating-point.

2.4 ARGUMENT-PARAMETER AGREEMENT FOR STUDENT PL

The BINARY or DECIMAL attributes may not be declared in Student PL. All fixed-point numbers are stored as 31-bit binary integers, and all floating-point numbers are stored with mantissas of six hexadecimal digits. Consequently, arguments and parameters need to have only the same scale (FIXED or FLOAT.) The result of an expression is fixed-point if all of the values in it are fixed-point.

Otherwise, the result is floating-point. Remember that any constant with a decimal point is considered floating-point in Student PL.

2.5 ARGUMENT-PARAMETER AGREEMENT FOR THE MODEL 20

The Model 20 supports only DECIMAL. All floating-point numbers are stored with mantissas of either six or 15 decimal digits. Fixed-point numbers require one byte (eight bits) for the sign and first digit plus one byte for every additional two digits or fraction thereof. Default-length for fixed-point variables is five digits, and default-length for floating-point variables is six digits. Arguments and parameters must have the same scale, and use the same amount of storage. Consequently, for a default-length fixed-point parameter, we may use a default-length fixed-point variable, a fixed-point integer constant with exactly four or five digits (which may require leading zeros such as 00738), or a fixed-point integer expression which requires exactly three bytes of storage. Fixed-point expressions are always allocated enough storage space to prevent unnecessary loss of precision, but the exact rules are not presented here because they are somewhat complex and they are of little interest to scientific programmers. It is safest and simplest to simply *avoid using constants or expressions as arguments for fixed-point parameters.*

For default-length floating-point parameters we may use any expression which has at least one floating-point value and does not result in a length greater than six digits. The length will not exceed six digits if no value exceeds six digits and no two fixed-point values are combined. For example, 5E19 and 3.65928E−7 + .0058*3E−2 would be acceptable whereas 5E2 + .0000579 and 5E2 + 2/3 would not. (Remember that to minimize loss of precision, a fixed-point quotient is fifteen digits long.)

If this rule is not perfectly clear, it is safest to simply avoid using constants or expressions as arguments for floating-point parameters.

The Model 20 requires that the *invoking* procedure declare the ENTRY attribute for any programmer-defined invoked procedure. For example, TRINGL on page 31 must include the declaration

DECLARE SINE ENTRY;

Procedure names are called *entries* because they designate the point of entry into procedures.

2.6 ARGUMENT-PARAMETER AGREEMENT FOR THE F AND D COMPILERS

The F and D compilers support both the BINARY and DECIMAL attributes, but floating-point numbers are stored with mantissas of six or fourteen hexadecimal digits whether declared DECIMAL or BINARY. The default is DECIMAL for the sake of discussing and specifying lengths even though there is no *true* decimal. For example, it is easier to appreciate a length of about 6 decimal digits than it is to appreciate the actual length of six hexadecimal digits, which is 21 to 24 binary digits. For uniformity with default DECIMAL FLOAT variables, it is better to avoid FLOAT BINARY declarations even though floating-point base disagreement causes no actual error.

The rule for floating-point argument-parameter agreement is the same as for the Model 20.

FIXED DECIMAL numbers are stored the same as for the Model 20; so the rule for argument-parameter agreement is the same.

FIXED BINARY numbers are stored with a length of 31 bits for the D compiler and with a length of 15 or 31 bits for the F compiler. Default length for the F compiler is 15 bits. Undeclared variables starting with the letters I through N default to FIXED BINARY whereas variables declared merely FIXED default to DECIMAL. Binary is more efficient for most engineering and scientific applications; so it is usually best to avoid declaring merely FIXED. Instead, we should use the first-letter convention or declare FIXED BINARY.

A fixed-point constant will be converted during *compilation* to the type it is combined with or assigned to. If not combined with or assigned to a floating-point or a fixed binary value, the constant will be stored as written—in fixed decimal. As examples using the first-letter convention:

1. N = 56; (56 will be converted to fixed binary and assigned to N during compilation.)

2. N = K + 56; (56 will be converted to a fixed binary during compilation.)

3. N = K + 56*32; (56*32 will be multiplied in fixed decimal and the result converted to fixed binary during compilation.)

4. X = 56; (56 will be converted to floating-point and assigned to X during compilation.)

5. X = Y + 56*32; (56*32 will be multiplied in fixed decimal and the result converted to floating-point during compilation.)

As we can see, all constants in assignment statements will be stored either as fixed binary or floating-point if all variables are either fixed binary or floating-point. However, an *argument* consisting of a fixed-point constant or a combination of only such constants will remain decimal. Thus, for the procedure named Y on p. 33, the invocation Y(A, B, 5) will result in an error. The constant, 5, will be stored as fixed decimal using one byte, whereas the parameter N is fixed binary by default. For the D compiler, we could use an expression such as N + 85000 as an argument for a fixed binary parameter since the 85000 would be converted to binary during compilation. However, the result requires more than fifteen bits (85000 exceeds 32767;) so for the F compiler there would be a precision disagreement with a default-length fixed binary parameter. It is safest simply to avoid using a constant or expression as an argument for any kind of fixed-point parameter.

Argument-parameter agreement for the F and D compilers is somewhat complicated because of the variety of data types. If one does not wish to learn the rules, he may simply avoid using constants or expressions. As an alternative, he may declare the *parameter* attributes after the ENTRY attribute in the *invoking* procedure. Then a copy with the correct attributes is automatically made for any arguments with the wrong attributes. For example, the procedure TRINGL of p. 31 may include the declaration

        DECLARE SINE ENTRY (FLOAT);

If we then had an invocation such as

        SINE(N)

where N was fixed binary, a floating-point copy of N's value would automatically be used. Similarly, in any procedure which invokes the procedure Y of p. 33, we may include the declaration

        DECLARE Y ENTRY (FLOAT, FLOAT, FIXED BINARY);

Some programmers recommend always declaring the entry parameter attributes in the invoking procedure as a precaution.[3] However, this is unnecessary if one is familiar with the agreement rules or avoids constants and expressions as arguments.

2.7 THE RETURNS ATTRIBUTE

Unless declared otherwise, a function returns a value of the type implied by its first letter for the F and D compilers; and the invoking procedure treats the value as the type implied by its first letter. Thus, L__SINE would be an inappropriate choice for the name of our sine function. To override this first-letter convention we may list the RETURNS attribute in the PROCEDURE statement of the function and in the DECLARE statement of the invoking procedure:

        L__SINE: PROCEDURE (A) RETURNS (FLOAT);
            .
            .
        END;

[3] SL1 and Student PL do not support the ENTRY attribute, and the Model 20 does not permit parameter attributes to be listed after the ENTRY attribute.

```
          TRINGL: PROCEDURE OPTIONS (MAIN);
             DECLARE L_SINE RETURNS (FLOAT);
                    .
                    .
                    .
          END;
```

The RETURNS attribute implies the ENTRY attribute, but not the parameter attributes.[4]

EXERCISES

▼1. The torque in ft-lbf versus speed in RPM of a certain induction motor is given by the equation

$$T = \frac{61.4}{\frac{1800 - S}{133} + \frac{133}{1800 - S}}$$

Write a function procedure for this equation and use it in a main procedure to calculate and put out the torque for $S = 0$, 450, 1710, and 1800 RPM.

2. The scattering cross section of an electron in square micro-angstroms is given by the formula

$$A = \frac{0.0666 r^4}{(r^2 - 1)^2}$$

where r is the ratio of the radiation frequency to the bound natural frequency of the electron. Write a function procedure for this equation, and use it in a main procedure to calculate and put out the cross section and r, for $r = 1$, 10, 100, and 1000.

▼3. The wavelengths, in angstroms, of the hydrogen spectrum are given by the formula

$$L = \frac{911.8}{\frac{1}{n^2} - \frac{1}{m^2}}$$

where n and m are two of the integer quantum numbers. Write a corresponding function procedure; and use it in a main procedure to calculate and put out the wavelengths for $n = 2$ together with $m = 3$, 4, and 5.

---

[4] The Model 20 does not support the first-letter convention; so one *must* declare the RETURNS attribute. The same is true for Student PL except the invoking procedure knows the returned attributes and the word RETURNS with parentheses is omitted from the function PROCEDURE statement, e.g., "SINE: PROCEDURE (A) FLOAT;". SL1 does not support the RETURNS attribute. The attributes of the value returned are those of the expression in the RETURN statement rather than those implied by the first letter. These attributes are known to the invoking procedure.

4. According to Van der Waal's equation of state, the reduced pressure of a gas is given by the formula

$$P = \frac{8T}{3V - 1} - \frac{3}{V^2}$$

where T is the reduced temperature and V is the reduced volume. Write a corresponding function procedure and a main program which uses it to calculate and put out P, T, and V for $T = 3$ together with $V = 1, 2$, and 3.

5. (Mathematically Advanced) The zeroth order Bessel function may be approximated by the rational expression

$$J_0(x) = \frac{-6.09x^2 + 345.5}{x^4 + 2.56x^2 + 345.5}$$

Write a corresponding function procedure, and a statement using it to evaluate the expression $Q = J_0(x + y) + 3\ J_0(x/y)/J_0(y)$. Note that a rational expression is the most general arithmetic approximation possible involving only addition, subtraction, multiplication, division, and integer powers because any expression involving these may be *rationalized* to a ratio of two polynomials.

6. The accompanying experimental data suggest an approximate quadratic relationship of the form $y = ax^2 + bx + c$. Find the relationship and write a corresponding function procedure.

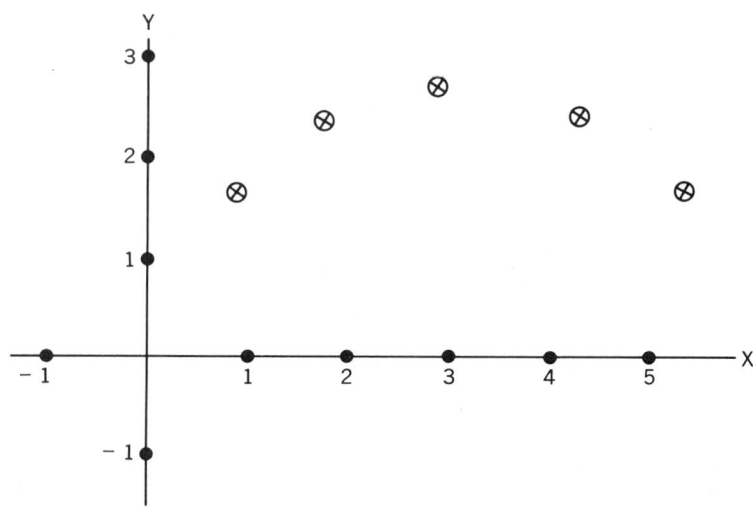

7. Write a function procedure for a common formula relevant to your discipline.

2.8 BUILT-IN FUNCTIONS

The important point about subordinate procedures is that once written, they may be used many times by a procedure and/or by many procedures. Consequently, it saves programming time to write recurring tasks as subordinate procedures. In fact, most computer facilities maintain subordinate procedure libraries, but the elementary transcendental functions are needed so frequently that, better yet, they have been built right into the PL/I language! We need only know the names of those we wish to use. We do not have to submit subordinate procedure decks, and we do not need to be familiar with the approximation formulas or the details of the built-in procedures which use them.

The built-in trigonometric functions are SIN(a), SIND(a), COS(a), COSD(a), TAN(a), TAND(a), ATAN(r), ATAND(r), ATAN(y, x) and ATAND(y, x). These names stand for the sine, cosine, tangent, and arctangent functions with the "D" suffix meaning that the argument or result is to be in degrees rather than radians. The arctangent function requires further explanation. Referring to Fig. 2-3, the tangent of angle A is defined as the ratio r = y/x. Consequently, we may say that A equals the angle whose tangent is r: A = ATAN(r) or A = ATAND(r), depending upon whether we want the answer in radians or degrees. Unfortunately, the arctangent function is not unique. Given the ratio 0.5, for example, how can we determine whether it corresponds to a negative y divided by a negative x (third quadrant) or a positive y divided by a positive x (first quadrant)? Similarly, given the ratio -0.5, how can we determine whether it corresponds to a negative y divided by a positive x (fourth quadrant) or a positive y divided by a negative x (second quadrant)? However, the angle can be placed in the proper quadrant if we restrain ourselves from dividing y by x, letting the arctangent procedure do the division after inspecting the signs of y and x. This is the reason for the two-argument version of the arctangent function. The one-argument version gives only the so-called *principal value*—the answer corresponding to the first or fourth quadrant. However, the one-argument version is faster; so we should use the form A = ATAN(y/x) or A = ATAND(y/x) if we know ahead of time that $-\pi/2 < A < \pi/2$, or $-90° < A < 90°$. Note that the single-argument version does not work for angles of $\pm 90°$ because then the value of x will cause an attempted division by zero which raises the ZERODIVIDE condition, causing termination of the program. The two-argument version will work in every case except the ambiguous one when y = 0 and x = 0. Sometimes the arctangent function arises in nontrigonometric contexts. In such cases the one-argument radian version is usually called for. For example, the cumulative Cauchy probability distribution function is given by $P(u) = 0.5 + \mathrm{atan}(u)/\pi$.

Some other built-in mathematical functions are LOG10(x), LOG(x), and LOG2(x). Although you may be most familiar with the logarithm to the base 10, it is useful primarily as an aid to *manual* multiplication, and therefore is of little use

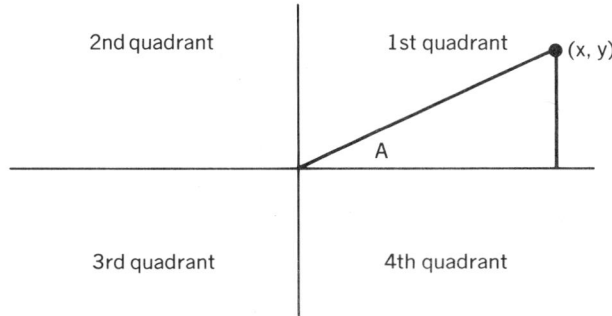

Fig. 2-3 Angle and quadrants in the x-y plane.

with digital computers. LOG(x) stands for logarithm to the base 2.718 ···, which is called the *natural* logarithm. This natural base is so important that it is often denoted in mathematics by the symbol *e*. Do not confuse this with the use of "E" in PL/I for scientific notation. The computer does not automatically associate the value 2.718 ··· with "E." The logarithm to the base 2 is used primarily in information theory. Note that the three log functions automatically return a floating-point value in spite of their first letter. Moreover, all of the built-in mathematical functions will accept arguments greater than default length, and they will automatically convert fixed-point arguments to floating-point. Accordingly, we do not need to declare the ENTRY or RETURNS attributes.[5]

We may raise any number to any power by *exponentiation* using the double asterisk. However, e^x occurs so frequently that there is a special faster built-in *exponential* function—EXP(x). Actually, the double asterisk evaluates fractional or negative powers by converting to this natural base. For example, A**X is evaluated as EXP(X*LOG(A)). Consequently, we should use EXP(X) rather than 2.71828**X to evaluate e^x because 2.71828**X needlessly causes the computer to evaluate $\log_e(e)$ which is 1.

The computer raises a number to a positive fixed-point integer power by repetitive multiplication. For example, to evaluate X**5, the computer will multiply X by itself, then multiply the result by itself, then multiply by X. This is generally faster than using logarithms and exponentials; so we should avoid using fractional places or scientific notation if, in fact, the power is a whole number. For example, B**4 and N**10 are better than B**4.0 and N**1E1.

PL/I does not recognize the square root symbol, $\sqrt{}$, but the square root may be taken by raising to the 0.5 power: for example, X**0.5. However, the square root occurs so frequently that a special faster function, SQRT(x), has been devel-

---

[5] For the Model 20, we *must* not declare the ENTRY or RETURNS attribute, and the arguments must be floating-point.

oped for it. There are other built-in functions too, which we will introduce as the need arises.[6]

2.9 THE PAGE OPTION[7]

Computers generally put out a great deal of reference information besides whatever is called for in PUT DATA statements. Many implementations do not have the output from the PUT DATA statements begin on a separate page after the reference information. This reference information is invaluable for debugging a program, but it would only clutter and distract from a report or paper. Consequently, it would be nice to insure that the answers are put out on a fresh page so that they can easily be separated from the reference information. A fresh page also makes the answers easier to find initially. To put the answers on a fresh page, we may insert the PAGE *option* between the PUT and the DATA of the PUT DATA statement.[8] To illustrate this, the following program evaluates and puts out $y = \log_e (\sqrt{x^4 + a^2}) + e^a$ for the specific case when $x = 41.567$ and $a = 1.2 \times 10^{-3}$:

```
PRCTCE:    PROCEDURE OPTIONS (MAIN);
    X = 41.567;   A = 1.2E-3;
    Y = LOG(SQRT(X**4 + A*A)) + EXP(A);
    PUT PAGE DATA (X, A, Y);
END;
```

This example points out the fact that arguments may themselves involve functions. We may, for example, have COS(COS(LOG(G))).

SUMMARY

Typical Function Procedure

```
function name: PROCEDURE (list of variables separated by commas);
    variable = expression;
        .
        .
        .
    RETURN (expression);
END;
```

[6] Of the functions which have been discussed, Student PL supports only SIN(a), COS(a), TAN(a), ATAN(r), LOG(x), EXP(x), and SQRT(x); SL1 supports only SIN(a), COS(a), TAN(a), ATAN(r), LOG(x), LOG10(x), LOG2(x), EXP(x), and SQRT(x); and the Model 20 supports only SIN(a), COS(a), TAN(a), ATAN(r), ATAN(y, x), LOG(x), EXP(x), and SQRT(x).

[7] Unsupported by SL1 or Student PL.

[8] For the Model 20, we must use a separate statement such as "PUT FILE (OUT) PAGE;."

Exercises

Typical Function Invocations

    Y = function name (list of arguments separated by commas);
    Y = expression involving function name (list of arguments separated by commas);

Restrictions

The attributes of the returned value are implied by the first letter of the function name.

The declared or implied attributes of the corresponding arguments and parameters must agree.

Expressions may be used as arguments provided you know the rules for determining the resulting attributes of mixed expressions for your compiler.

The arguments must be listed in the same order as the parameters.

Built-in functions have no restrictions on agreement of arguments and parameters; so expressions may be used freely as arguments. Regardless of their first letter, the mathematical functions return a floating-point value. Examples are:

    SIN(a), COS(a), TAN(a), ATAN(r), ATAN(y,x), LOG(x), EXP(x), SQRT(x)
    SIND(a), COSD(a), TAND(a), ATAND(r), ATAND(y,x), LOG2(x), LOG10(x)

PUT PAGE DATA (list of variables separated by parentheses) may be used to make the output start at the top of a new page.

The ZERODIVIDE condition is raised by an attempt to divide by zero, and it causes termination of the program.

EXERCISES

8. Using floating-point variables, write assignment statements for the following formulas:

    (a) $X = Y \cdot \dfrac{\sin(A)}{\sin(B)}$ (angles in degrees)

    (b) $Z = \sqrt{X^2 + Y^2 - 2XY\cos C}$ (angle in radians)

    (c) included angle = arctan $(X_1 Y_2 - X_2 Y_1, X_1 X_2 + Y_1 Y_2)$ (answer in degrees)

    (d) flux = $\dfrac{2 \text{ fluxmax}}{\pi}$ arctan (current · inductance)

    (e) $pH = -\log_{10}[H_3O]$

    (f) $E_0 = \dfrac{RT}{F} \log K$

    (g) phase angle = arctan (imaginary part, real part)   (answer in radians)

    (h) decibels of gain = $20 \log_{10} \left(\dfrac{\text{output}}{\text{input}}\right)$

    (i) $p = \dfrac{mv}{\sqrt{1 - (v/c)^2}}$

    (j) bits of information = $-\log_2$ (probability)

(k) work = gas constant · temperature · $\log_e \left(\dfrac{\text{final volume}}{\text{initial volume}}\right)$

(l) current = leakage · $(e^{39v} - 1)$

(m) $f = \dfrac{1}{\exp\left(\dfrac{p}{1.38 \times 10^{-23}T}\right) - 1}$

▼(n) q = cotangent A (angle in degrees)

(o) p = secant A + cosecant B (angle in degrees)

▼9. (Mathematically Advanced) For the D and F compilers, the hyperbolic functions SINH(x), COSH(x), TANH(x), and ATANH(x) are also built-in. (The Model 20 supports TANH.) Why shouldn't we simply use $(e^x - e^{-x})/2$ for the hyperbolic sine and $(e^x - e^{-x})/(e^x + e^{-x})$ for the hyperbolic tangent?

10. (Mathematically Advanced) For the D and F compilers, the error function ERF(x) $= \dfrac{2}{\sqrt{\pi}} \int_0^x e^{-t^2}\,dt$ and the complementary error function ERFC(x) = 1 + ERF(x) are also built-in. ERF $\left(\dfrac{x}{\sigma\sqrt{2}}\right)$ is the probability that a zero-mean *Gaussian* or *normally* distributed random variable with standard deviation σ lies between $-x$ and $+x$. Write a main procedure which calculates this probability for $\sigma = 1$ together with $x = 1, 2,$ and 3.

▼11. Make use of the identity $\arcsin(y) = \arctan(y, \sqrt{1 - y^2})$ to write your own function procedures ASIND and ASIN for degrees and radians. Note that this will inevitably return the principal value in the first or fourth quadrants, and there is no mathematical way to resolve the ambiguity. Consequently, you should not use your arcsin function unless you are sure that $-90° \leq$ angle $\leq +90°$. It would be risky, for example, to use the arcsine function in the solution of triangles.

12. Make use of the identity $\arccos(x) = \arctan(\sqrt{1 - x^2}, x)$ to write your own function procedures ACOSD and ACOS for degrees and radians. Note that this will inevitably return the principal value in the first or *second* quadrant, and that there is no mathematical way to resolve the ambiguity. Consequently you should not use your arccos function when there is a chance that the angle will lie outside the range 0 to 180°. The arccos function is foolproof, for example, for the solution of triangles.

13. (Mathematically Advanced) Exercise 8(c) uses vector algebra to find the angle between vector$_1$ and vector$_2$ from their x and y components. The sine of the angle between them is simply the cross-product divided by their magnitudes, whereas the cosine of the angle between them is simply their dot product divided by their magnitudes. Consequently, the angle between them is given by arctan(cross-product, dot product) because the tangent is simply the sine divided by the cosine, and the magnitudes cancel. Together, the sine and cosine uniquely determine an angle although neither one alone does. The answer will be positive if vector$_2$ points within 180 degrees clockwise of vector$_1$; otherwise the answer will be negative. Write a function procedure to be used in the form ANGLE (X1, Y1, Z1, X2, Y2, Z2) which returns the angle in radians between two three-dimensional vectors.

Exercises

▼**14.** The ordinary way of expressing polynomials is computationally inefficient. For example, the polynomial "$y = 14x^5 - 8x^4 + 3x^2 + 5$" would require three additions or subtractions and nine multiplications taking into account the way integer powers are evaluated. The following *nested* rearrangement according to *Horner's rule* requires four less multiplications: $y = [(14x - 8)x^2 + 3]x^2 + 5$. One more multiplication may be eliminated if the common power x^2 is precalculated separately: $u = x \cdot x$; then $y = [(14x - 8)u + 3]u + 5$.

The general rule is:
(a) Determine the difference in power of the two highest-powered terms; then write the coefficient of the highest-powered term times the variable raised to this power difference, (e.g., $14x$).
(b) Consider the next lower-powered term, adding its coefficient and multiplying the whole quantity by the variable raised to the difference in power from the next term, e.g., $(14x - 8)x^2$.
(c) Repeat step (b) until all terms are accounted for, e.g., $[(14x - 8)x^2 + 3]x^2$ etc.
(d) Precalculate any resulting powers which are used more than once, e.g., $u = x \cdot x$.

Rewrite the SINE function at the beginning of the case study or the Bessel function in exercise 5 using Horner's rule.

▼**15.** (Difficult) For computers which divide as rapidly as they multiply, rational expressions may be evaluated more efficiently as continued fractions than as ratios of two polynomials nested according to Horner's Rule. Rational expressions may be converted to continued fraction form by the following general technique, illustrated for the example:

$$\frac{4x^4 + 4x^3 - 50x^2 + 56x + 20}{2x^4 + 2x^3 - 28x^2 + 16x + 16}$$

(a) Normalize the coefficient of the highest power in the denominator to 1.

$$\frac{2x^4 + 2x^3 - 25x^2 + 28x + 10}{x^4 + x^3 - 14x^2 + 8x + 8}$$

(b) Divide the numerator by the denominator, getting a principal part (which may be zero) and a remainder.

$$2 + \frac{3x^2 + 12x - 6}{x^4 + x^3 - 14x^2 + 8x + 8}$$

(c) Divide both the numerator and the denominator of the remainder by the normalized numerator (e.g., $x^2 + 4x - 2$), leaving the coefficient of the highest power in the numerator and a principal part together with a remainder in the denominator.

$$2 + \cfrac{3}{x^2 - 3x + \cfrac{2x + 8}{x^2 + 4x - 2}}$$

(d) Repeat step (c) until there is no remainder.

$$2 + \cfrac{3}{x^2 - 3x + \cfrac{2}{x - \cfrac{2}{x+4}}}$$

(e) Whenever possible, rewrite the principal part polynomials according to Horners' rule.

$$2 + \cfrac{3}{x(x-3) + \cfrac{2}{x - \cfrac{2}{x+4}}}$$

Continued fractions such as this are often written in the more compact notation:

$$2 + \frac{3}{x(x-3)+} \frac{2}{x-} \frac{2}{x+4}$$

This continued fraction example requires three divisions, one multiplication, and five additions or subtractions, whereas Horner's rule with the normalized rational expression requires one division, seven multiplications, and eight additions or subtractions. However, the considerable preliminary work is worth doing only if the approximation is to be used many times. Even if we never go to the trouble to derationalize an expression, this example illustrates that, contrary to our early training, we should not necessarily go to the trouble to rationalize it. For example, if a function originates in a form such as

$$\cfrac{(s-2)}{(s-8) + \cfrac{(s-3)}{(s-6)(s-5)}}$$

it is often best left in that form for computational purposes.

Rewrite the Bessel function procedure of exercise 5 using this technique.

▼16. (Difficult) Horner's rule is not the fastest method of evaluating polynomials of degree higher than four. For example, the polynomial $y = x^6 + Ax^5 + Bx^4 + Cx^3 + Dx^2 + Ex + F$ may be evaluated with only three multiplications and five additions in the two steps:

$$Q = x(x + a); \text{ then } y = [(Q + x + b)(Q + c) + d](Q + e)$$

where $a, b, c, d,$ and e may be derived from $A, B, C, D, E,$ and F by equating coefficients of like powers as given by the two formulas for y.[9] Naturally, the considerable preliminary work is worth doing only if the approximation is to be used many times. Write a function procedure using this technique for the formula $y = x^6 + 4x^5 + 8x^4 + 10x^3 + 9x^2 + 5x + 2$.

[9] For further information of this technique, see Motzkin, "Evaluation of Polynomials and Evaluation of Rational Functions," *Bull. Amer. Math. Soc.*, vol. 61, 1955, 163ff.

17. (Additional Language Feature—D and F compilers) The built-in functions FIXED(x), FLOAT(x), DECIMAL(x), and BINARY(x) provide a convenient way to convert arguments to the proper type. (FIXED(x) and FLOAT(x) are also supported by student PL.) For example, we could write SINE(FLOAT(K)) where K is fixed-point. For this exercise, make up an example illustrating the use of one or more of these functions.

18. Some implementations put out reference information on the same page *after* the output. This may be avoided by having a separate statement "PUT PAGE;" just before the END statement. Rewrite TRINGL accordingly.

CASE STUDY 3

SURVEYING TRAVERSE

The GET LIST, PUT LIST, and GO TO Statements, the SKIP Option, Comments, and Character-String Constants

Using the built-in sine function and the PAGE option, the TRINGL procedure of the previous case study becomes:

```
TRINGL:  PROCEDURE OPTIONS (MAIN);
    A = 1.235;   B = 0.571;   Y = 157.3;
    COEF = Y/SIN(B);
    X = COEF*SIN(A);
    C = 3.14159 - A - B;
    Z = COEF*SIN(C);
    PUT PAGE DATA (A, B, C, X, Y, Z);
END;
```

3.1 THE GET LIST STATEMENT

Unfortunately, for the above procedure we would have to revise the first three assignment statements, recompile, and reexecute the procedure if we ever wanted to reuse it for different values of A, B, and Y. It would be much more efficient to have a single program which would handle any values of A, B, and Y. This can be accomplished by replacing the first three assignment statements with a GET LIST

Sec. 3.1 The Get List Statement 49

statement:

```
TRINGL:  PROCEDURE OPTIONS (MAIN);
    GET LIST (A, B, Y);
    COEF = Y/SIN(B);
    X = COEF*SIN(A);
    C = 3.14159 - A - B;
    Z = COEF*SIN(C);
    PUT PAGE DATA (A, B, C, X, Y, Z);
END;
```

This procedure could then be compiled once and for all, a machine language version being saved on punched cards, magnetic tape, or another medium. The particular values of A, B, and Y would be established during *execution* rather than during *compilation*. The values could be punched on one or more data cards in the order indicated by the GET LIST statement, as is shown in Fig. 3-1[1].

The spacing of the data values is irrelevant.[2] We may put as many or as few as we wish per card, provided that they are separated by a comma and/or one or more spaces. Similar provisions apply if the data are entered at a remote console.

When execution reaches the GET LIST statement, 1.235 will be established as the value of A; 0.571 will be established as the value of B; and 157.3 will be established as the value of Y.

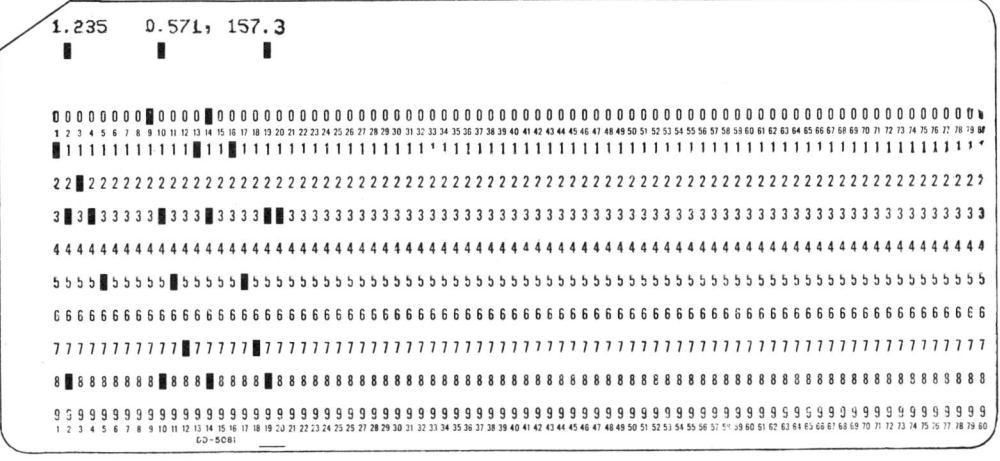

Fig. 3-1 Sample data card for the procedure TRINGL.

[1] For Student PL, which does not support the GET LIST statement, this may be indicated by the three statements A = INPUT; B = INPUT; Y = INPUT;. For the Model 20, use "GET FILE (IN) EDIT (A, B, Y) (8E(10,0));" together with a declaration such as DECLARE IN FILE INPUT ENVIRONMENT (MEDIUM (SYSIPT, 2501) F(80)).

[2] Most implementations allow all 80 columns to be used for data, in contrast to PL/I statements.

The machine language program would be submitted together with this data card, taking much less computer time than the compilation plus execution of the version without data. Only the data card need be revised each time the program is run for different values of A, B, and Y. This has the additional advantage of involving less punching than does revision of the first three assignment statements.

As we have seen, provision for establishing some values during execution generalizes a program, enabling us to use it over and over again. In the long run, this saves a great deal of computer time and keypunching time. In fact, TRINGL is not especially useful unless it *is* used over and over again. It would probably take less time to use a trig table and desk calculator than to punch and debug a program to solve a single triangle. However, it is important to realize that the values are more easily established as constants within a program which is to be run only once.

3.2 THE GO TO STATEMENT

We can improve our TRINGL procedure even more. As it stands, it would be rather clumsy to use if we wanted to solve several triangles in one day. We would have to repetitively rerun the machine language program—each time with a different set of data and each time suffering the full turn-around time. Alternatively, we could make many duplicate machine language programs and submit them together—each one with a different set of data. This would be poor practice, though, because of the prohibitive cost and effort of such extensive duplication. What is needed here is a capability for cycling or repeating portions of a program. It is the GO TO statement which provides this capability for one loading to handle any number of triangles. The GET LIST statement is given a *label* such as "IN" and the statement "GO TO IN" is inserted after the PUT DATA statement:

```
TRINGL:  PROCEDURE OPTIONS (MAIN);
    PUT PAGE;
    IN:  GET LIST (A, B, Y);
       COEF = Y/SIN(B);
       X = COEF*SIN(A);
       C = 3.14159 - A - B;
       Z = COEF*SIN(C);
       PUT DATA (A, B, C, X, Y, Z);
       GO TO IN;
 END;
```

Every time execution reaches the GO TO statement, control will loop back to the statement labeled IN. Any name may be used for this label, provided it is not used elsewhere for another purpose and provided it is formed by the same rules which apply to variable names. The data for this program will be a stream of numbers separated by commas and/or one or more blanks—for example, "0.63,

.52, 86.1, 2.16, 0.13, 2E3, 1.83. 1.01, 3E−2."[3] The first three numbers will be taken for A, B, and Y, respectively, on the first cycle; the next three numbers will be taken for A, B, and Y, respectively, on the next cycle; and so on. Each time a new set of values is gotten, they will replace the corresponding preceding values. We have used each variable name to represent *a succession of different values one at a time*. Note how different this is from standard mathematical practice where a variable stands for *either an infinite number of values simultaneously* (e.g., $y = x^2$) *or for a few values forever* (e.g., $x^2 + 8x - 3 = 0$). By using input data together with a loop we avoid having to use separate names and separate formulas for each set of values. This technique also makes one program work for *any* number of triangles—not just exactly one, exactly three, or exactly any particular number. Our program is more flexible; so our effort in writing, punching, and debugging it will be rewarded more fully.

The ability to loop is one of the fundamental features that makes computers so effective. Most computers execute an average assignment statement for the order of 0.0001 times the cost of punching it. Consequently, the benefit can be enormous for applications which use some statements many times by cycling within a program and/or by using the program many times. Problems that are not of this nature are preferably done with tables and a desk calculator or a slide rule, unless there is an especially convenient and inexpensive time-sharing system. The computer is not a glorified slide rule; instead, they complement each other. As we shall see, the computer's most fruitful applications are those which were not even dreamed of or else were accomplished only with great drudgery in the pre-computer era.

3.3 THE ENDFILE CONDITION

Returning to our example, note how the PAGE option was used separately outside the loop to avoid the waste and nuisance of having only one set of results per page.

After the last data value has been gotten, the next attempt to get a value from the *input file* will be fruitless. This raises the ENDFILE condition which causes the computer to put out a corresponding error message and then terminate execution. The computer assumes we have made an error since we told it to get a value which was not there. However, the consequent error message and termination are all right because we have accomplished our objectives by then. Note that the END statement must still be included to tell the compiler the extent of the procedure even though execution terminates before reaching the END statement.

[3] Each execution of an SL1 GET statement always starts on a new card; thus no card should have data for more than one execution. SL1 also requires that decimal points be punched on the input data for GET LIST statements. For the Model 20 GET EDIT statement in the previous footnote, we *must not* start a new card until all 8 ten-column fields are filled.

3.4 THE NULL FIELD[4]

We need not repeat any data value that is the same as for the previous case. We may indicate that the value is unchanged by using a *null field*. A null field is indicated by either the very first nonblank character in the stream being a comma, or by two commas optionally separated by blanks. For example, with TRINGL, the data "0.63 .52 86.1,,,2E3" mean that the first triangle is to have A = 0.63, B = .52, Y = 86.1 and the second triangle is to have A = 0.63, B = .52 and Y = 2E3.

EXERCISES

1. Write a procedure which uses a GO TO statement to repetitively get input values for one of the formulas in exercise 8 of Case Study 2, evaluates the formula, and puts out the input values together with the output value.

2. Assume that you have written a function procedure A(B, C) which approximates the experimentally determined dependence of A upon B and C. Write a main procedure which repetitively gets B and C, calculates A, then puts out all three quantities.

▼3. What would the output look like on your implementation for the following program and data card?

```
2 3,,,4,,5 3 2 4
```

```
NONSNS: PROCEDURE OPTIONS (MAIN);
        PUT PAGE;
   IN:  GET LIST (K, A);
        M = K**2;   A = A*K;
        PUT DATA (K, A, M);
        GO TO IN;
        END;
```

[4] Unsupported by the Model 20, SL1, and Student PL.

4. What would the output look like on your implementation for the following program and data card?

```
8E8 2E8 2 ,, -1E8 ,, 5.6 3.4
```

(punched card image)

```
EXER4:  PROCEDURE OPTIONS (MAIN);
        PUT PAGE;
 NEXT:  GET LIST (X, Y, L);
        Z = (X - Y)*L;
        PUT DATA (Z, L);
        GO TO NEXT;
        END;
```

3.5 COMMENTS AND THE SKIP OPTION

Let us consider the analysis of a surveying traverse as our next example. A traverse consists of a connected sequence of straight-line segments with measured lengths (L) and azimuths (A). The azimuth is the angle in the horizontal plane measured in degrees counterclockwise from a standard direction which we shall take to be the positive X axis. The X and Y coordinates of any vertex will simply be the coordinates of the previous vertex plus L*COSD(A) and L*SIND(A), respectively, as is shown in Fig. 3-2. Consequently, if we know the coordinates of the first point, we can find the coordinates of the other points by successively applying these formulas. The following procedure does this, introducing the comment feature and the SKIP option.

```
SURVEY:  PROCEDURE OPTIONS (MAIN);
    DECLARE L FLOAT;
    GET LIST (X, Y);   /* THE FIRST PAIR OF DATA VALUES SHOULD BE THE
        COORDINATES OF THE FIRST POINT, AND SUBSEQUENT PAIRS SHOULD BE
        THE LENGTH AND AZIMUTH OF SUCCESSIVE LINE SEGMENTS.  */
    PUT PAGE DATA (X, Y);
 GET_LINE:  GET LIST (L, A);
    X = X + L*COSD(A);
    Y = Y + L*SIND(A);
    PUT SKIP DATA (X, Y, L, A);
    GO TO GET_LINE;
END;
```

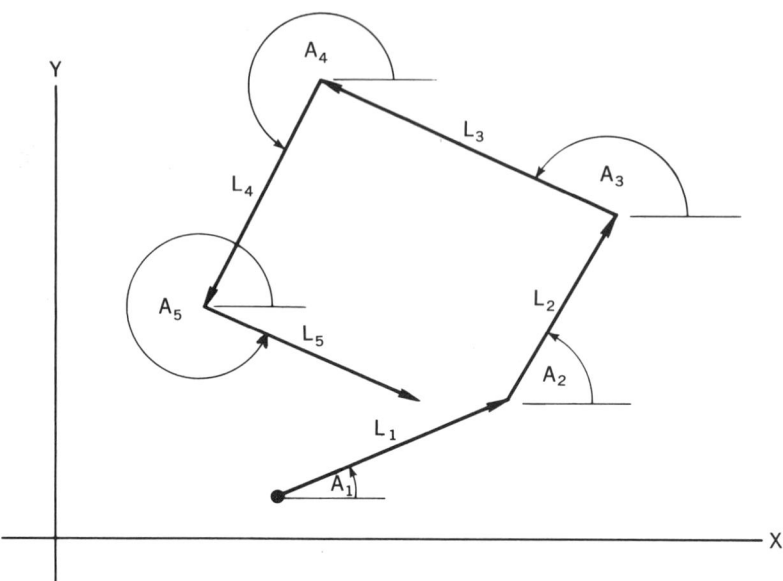

Fig. 3-2 A surveying traverse.

Comments begin with the adjacent characters /* and end with the adjacent characters */. They may appear wherever blanks are allowed, and they may contain any characters that are recognized by the hardware—for example, lower-case letters. However, comments may not include the adjacent pair */ which would end the comment prematurely. Comments are essentially parenthetical remarks which help explain a program to humans. They will be reproduced on the output listing of the program generated by the compiler, but they have no effect on the machine language program. In this program, the comment is used to indicate explicitly how to prepare the data. Although it is obvious to the original programmer, it may not be to the user—particularly if he is simply a keypunch operator. Unless the program is quite short, it often takes even the original programmer quite a while to redecipher it if it has been ignored for a few weeks. Comments are usually advisable except for short programs that will be seen only by the programmer and will not be saved for reuse. However, some programmers clutter their programs with excessive comments. This practice is distracting rather than helpful because PL/I is enough like English and algebra to be largely self-explanatory.

This procedure needs two separate GET statements because X and Y are gotten only once whereas L and A are gotten any number of times within a loop.

The assignment statements in this program emphasize their difference from mathematical equations. As an equation, "X = X + L*COSD(A)" is impossible when neither L nor COSD(A) is zero. In PL/I the equals sign means assignment instead of equality: (Take the value stored at address L times the cosine of the value stored at address A; add the value stored at address X; and store the result at address X." In this case it amounts to replacement: "Replace the value of X with its old value plus L*COSD(A)." The old value of X is lost forever. Assignment and "getting" are destructive. They erase whatever was previously stored, as does recording with an ordinary tape recorder. Use of a variable in an expression and "putting" are nondestructive. Like playing back a tape, the values are still available for subsequent use.

The SKIP option makes the output start at the beginning of a new line even if the previous line is only partially filled.[5] This option is included in the second PUT DATA statement in order to align the output into alike columns. Although implementation-dependent, the output will look something like the following:

```
X= 2.67113E+03    Y= 8.11012E+02;
X= 2.74748E+03    Y= 8.27938E+02    L= 7.82000E+01    A= 1.25000E+01;
X= 2.78249E+03    Y= 9.04070E+02    L= 8.38000E+01    A= 6.53000E+01;
X= 2.65948E+03    Y= 9.50552E+02    L= 1.31500E+02    A= 1.59300E+02;
X= 2.62449E+03    Y= 8.58425E+02    L= 9.85500E+01    A= 2.49200E+02;
X= 2.69639E+03    Y= 8.25351E+02    L= 7.91500E+01    A= 3.35300E+02;
```

Without the SKIP option, the first column would not contain all Xs, the second column would not contain all Ys, and so on. The exact appearance would depend upon the number of print positions per line and their spacing.[6] If there were four print positions per line the output would look something like the following:

```
X= 2.67113E+03    Y= 8.11012E+02;    X= 2.74748E+03    Y= 8.27938E+02
L= 7.82000E+01    A= 1.25000E+01;    X= 2.78249E+03    Y= 9.04070E+02
L= 8.38000E+01    A= 6.53000E+01;    X= 2.65948E+03    Y= 9.50552E+02
L= 1.31500E+02    A= 1.59300E+02;    X= 2.62449E+03    Y= 8.58425E+02
L= 9.85500E+01    A= 2.49200E+02;    X= 2.69639E+03    Y= 8.25351E+02
L= 7.91500E+01    A= 3.35300E+02;
```

3.6 ABSOLUTE AND RELATIVE ERRORS

With this traverse program we must be careful not to let computational errors appreciably add to the experimental errors in the data. We will speak of two types of error: The *absolute error* is defined as the true value minus the approximate value; the *relative error* is the absolute error divided by the approximate value. (The per cent error is 100 times the relative error.) Ordinarily we do not know the exact error or even its sign—if we did we could recover the exact value from the

---

[5] Unsupported by Student PL or SL1, but not necessary because each execution of an output statement starts a new line anyway. For the Model 20, we must use a separate statement such as PUT FILE (OUT) SKIP;.

[6] It is possible to vary the number of print positions and their spacing by running a special assembly language procedure with the PL/I. We have done so throughout this text in order to maximize readability.

approximate value! At best we ordinarily know only a bound or average magnitude for the absolute value of the error.

The kth decimal digit of a number is said to be *significant* if the absolute error is within $.5 \times 10^{-k}$. A normalized floating-point number is said to have k significant digits if its kth digit is significant but the next one is not. For example, .314599E1 is a three-significant-digit representation of π. Do not be fooled by the fact that the other digits are nonzero. The computer length of a number is no indication of its accuracy. Computers do not know the experimental accuracy of their numbers; so they do not follow the practice of letting the number of digits imply the accuracy. Default precision is a set length which may be greater or fewer than the number of significant digits of the data. Our example of π is a case where the length is greater than the number of significant digits, and .100000E1 standing for the exact integer 1 is a case where the length is less than the number of significant digits. Note also that the number of significant digits is a less precise concept than the error, because two numbers with the same number of significant digits can differ in relative error by a factor of almost ten.

For each application, the programmer must decide whether to be concerned with absolute error, relative error, or both. The surveying traverse is a case where we are ultimately concerned with absolute error. If a home owner in California was told that his house was 60 feet outside his property line, it would be little consolation that this is a relative error of only about 10^{-6} based on the international reference of 0° latitude and 0° longitude. For greatest accuracy, the surveyor should try to choose an origin within or relatively near the traverse. Many or all of the digits in L*COSD(A) will be lost when it is added to X if the initial magnitude of X is quite large relative to typical values of L. For example, assume the floating-point variables are stored with a length of six digits; X = 123,456 ft; and L*COSD(A) = 12.3445 ft. Four digits will be lost from L*COSD(A) when it is added to X. This loss is intolerable if the physical measurements are accurate to 0.1 ft. The loss is intolerable even if the physical measurements are accurate to only 1 ft because we want computational errors to be negligible compared with the experimental errors. These considerations have general applicability. For example, we often have the freedom to choose a reference temperature, voltage, or time. The general rule is to *choose an origin or reference within or relatively near the range of a variable if possible.*

3.7 THE PUT LIST STATEMENT AND CHARACTER-STRING CONSTANTS

As a refinement of the traverse program, we may use the PUT LIST *statement* together with *character-string constants* to avoid the repetitions "X =," "T =," "L =," and "A =" which clutter the output of the surveying program:

Sec. 3.7 The Put List Statement and Character-String Constants

```
SURVEY:  PROCEDURE OPTIONS (MAIN);
   DECLARE L FLOAT;
   GET LIST (X, Y);    /* THE FIRST PAIR OF DATA VALUES SHOULD BE THE
       COORDINATES OF THE FIRST POINT, AND SUBSEQUENT PAIRS SHOULD BE
       THE LENGTH AND AZIMUTH OF SUCCESSIVE LINE SEGMENTS.   */
   PUT PAGE LIST ('X', 'Y', 'LENGTH', 'AZIMUTH');
   PUT SKIP LIST (X, Y);
   GET_LINE:   GET LIST (L, A);
      X = X + L*COSD(A);
      Y = Y + L*SIND(A);
      PUT SKIP LIST (X, Y, L, A);
      GO TO GET_LINE;
END;
```

The output will look something like the following:

X	Y	LENGTH	AZIMUTH
2.67112E+03	8.11011E+02		
2.74748E+03	8.27938E+02	7.82000E+01	1.25000E+01
2.78249E+03	9.04070E+02	8.38000E+01	6.53000E+01
2.65948E+03	9.50552E+02	1.31500E+02	1.59300E+02
2.62449E+03	8.58425E+02	9.85500E+01	2.49200E+02
2.69639E+03	8.25351E+02	7.91500E+01	3.35300E+02

A character-string constant consists of a string of characters enclosed in single quotes. It may include any character recognized by the hardware. When listed in a PUT LIST statement, a character-string constant will be put out verbatim including any internal blanks but excluding the quotes. (We may use two adjacent single quotes if we want to represent an apostrophe within the constant. For example 'AIN''T' will be put out as AIN'T.) The maximum length of a character-string is implementation-defined, but when used as a table heading, we will generally want it to be shorter than the number of spaces between tabs.[7]

Note that the PUT LIST statement puts out the values of variables without identification. The first number on the output corresponds to the first variable in the list; the second number in the output corresponds to the second number in the list; and so on.

Since the PUT LIST statement does not require variable names to put out as identification, we may put expressions directly in the output list. This is good for "quickie" type programs where minimization of typing effort is more important than frills such as self-identifying output. For example, the first program in Case Study 1 could be written:

```
QUICKY:  PROCEDURE OPTIONS (MAIN);
   PUT LIST ( -1.4*(27200 + .573)/673);
END;
```

The value of the expression will be put out without identification. Note that we have used an abbreviation for the keyword PROCEDURE. Abbreviations save

---

[7] The maximum length is 127 for SL1, 255 for the D and Model 20 compilers and at least 1007 for the F compiler—the exact length depending upon the amount of storage.

typing time, but they reduce the legibility; therefore we will not use them in this text. To learn the valid abbreviations, consult the appropriate reference manual.[8]

SUMMARY

Typical Procedure

```
procedure name PROCEDURE OPTIONS (MAIN);
    /* comments */
    GET LIST (list of variables separated by commas);
    PUT PAGE LIST (list of character-string constants);
    PUT SKIP LIST (list of variables separated by commas);
statement label: GET LIST (list of variables separated by commas);
        .
        .
        .
    PUT SKIP LIST (list of expressions and variables separated by
        commas);
    GO TO statement label;
END;
```

Statements are normally executed in the order they are listed, but the GO TO statement enables control to skip forward or loop backward to a labeled statement.

Statement labels are identifiers; so they must be formed according to the same rules as variable names.

The SKIP *option* causes the output to begin on a new line even if the previous line is not filled.

Comments consist of a string of characters preceded by /* and followed by */. They are put out in the program listing generated during compilation, but they have no effect on the machine language version of the program.

Character-string constants consist of a string of characters enclosed in single quotes. Any character-string constants in PUT LIST statements will be put out during execution.

The PUT LIST *statement* puts out the value of variables or expressions without identification.

EXERCISES

5. Elevation, Z, may be included in the surveying traverse by also measuring the angle of elevation above horizontal, E, and modifying the formulas to:

[8] Student PL does not support abbreviations.

$$X = X + L*COSD(E)*COSD(A)$$
$$Y = Y + L*COSD(E)*SIND(A)$$
$$Z = Z + L*SIND(E)$$

Rewrite SURVEY to do this.

6. (Mathematically Advanced) For navigational purposes, SURVEY may be modified using spherical trigonometry to take into account the earth's curvature:

colatitude = arccos [cos (length/60) cos (colatitude) +
  sin (length/60) sin (colatitude) cos (azimuth)]
longitude = longitude − arcsin [sin (azimuth) sin (length/60)/sin (colatitude)]

where the length of a great circle path is in nautical miles, the azimuth is its original bearing clockwise from North, and all of the angles are in degrees. Modify SURVEY accordingly, using the ACOSD and ASIND procedures developed in exercises 11 and 12 in Case Study 2.

▼7. Using the labeling scheme of Case Study 2, if the per cent change in sides X, Y, and Z of a triangle are SX, SY, and SZ, respectively, the per cent change in angle A is given by the formula ΔA = (SX − SY) cot C + (SX − SZ) cot B. Write a well-commented procedure which repetitively gets values for SX, SY, SZ, C, and B; evaluates ΔA; and puts out all six quantities. (C and B are in degrees.)

8. If the three principal stresses are S_1, S_2, and S_3, the octahedral shearing stress is given by the formula:

$$S = \sqrt{(S_1 - S_2)^2 + (S_2 - S_3)^2 + (S_3 - S_1)^2}$$

Write a well-commented procedure which repetitively gets S_1, S_2, and S_3; evaluates S; and puts out all four values.

9. Most surveyors measure azimuth clockwise from North in degrees and minutes; and most would associate the +X direction with East. Modify SURVEY accordingly, getting in the degrees and minutes as two separate variables.

▼10. (Additional Language Feature) More precisely, the statement labels discussed in this case study are *label constants*. We may also declare *label variables*, assign them label constants, and use them in GO TO statements. As an example, we may use them to avoid the duplication of the common portion in the first of the following examples by the scheme presented in the second equivalent example:

A:_____   DECLARE X LABEL;

COMMON:_____   A:_____

_____   _____
_____   X = B;
_____   GO TO COMMON;
_____   B:_____

B:_____   _____
_____   X = A;
COMMON_DUPLICATE:_____   COMMON:_____

_____   _____
_____   _____
_____   _____

GO TO A;   GO TO X;

Indicate how this technique could be used to avoid the redundancy in the following program:

```
COMMON: _____
        _____
        _____
        _____
A:      _____
        _____
COMMON  DUPLICATE: _____
        _____
        _____
        _____
        _____
B:      _____
        _____
        GO  TO  COMMON;
```

11. (Additional Language Feature— F and SL1 compilers) The COPY option provides an easy way to get an output reproduction of the input data.

    GET LIST (list of variables) COPY;

However, the result is usually more wasteful of paper and less readable than with a separate PUT statement because the form of the data is copied directly from the input, one item per line without editing.

Rewrite SURVEY using the COPY option.

CASE STUDY 4

BEARING SELECTION

The IF Statement and Arithmetic Built-in Functions

We have seen how the ability to loop gives the computer so much power. The computer has the even more powerful capability to make decisions. We may use the IF statement to provide for such decisions in our programs.

4.1 THE IF STATEMENT

An IF statement may have the form:

IF relationship THEN statement;

The true or false relationship may have the general form:

expression comparison-operator expression

where the comparison operators are:

< = > <= >= ¬< ¬= ¬>

These operators stand for "less than," "equal to," "greater than," "less than or equal to," "greater than or equal to," "not less than," "not equal to," and "not greater than" respectively. (In mathematics, "not equal to," "less than or equal to," and "greater than or equal to" are often written $\neq$, $\leq$, and $\geq$, respectively.) The symbols in operators formed from two symbols must be adjacent and in the indicated order. For example, "$<\ =$" and "$=<$" are both incorrect.[1]

Examples of true of false relationships are:

$$A + 3 > SIN(B)$$
$$C \neg= D$$
$$Q <= P$$
$$N = 5$$

Parentheses are not needed around A + 3 because the arithmetic operators have higher priority than the comparison operators. By context, the compiler can distinguish the last case from an assignment statement.

The **THEN** clause may be any of the statements that we have discussed so far except the **PROCEDURE**, **DECLARE**, and **END** statements. If the relationship is false, execution will skip the **THEN** clause and proceed to the statement following the **THEN** clause. If the relationship is true, execution proceeds to the **THEN** clause. After the **THEN** clause is executed, execution proceeds to the statement following the clause unless the clause is a **GO TO** or **RETURN** statement, which sends the control elsewhere.

For example, suppose that we wish to put out the acceleration of gravity as a function of altitude. We may use the formula

$$G = 32.17\left(\frac{4390}{4390 + H}\right)^2$$

for positive H, where H is the height above sea level in statute miles and G is in ft/sec². However, we should use the formula

$$G = 32.17(1 + H/4390)$$

for negative H, such as in mines or under the sea. We may use the **IF** statement to decide between these two formulas as follows:

```
IF  H  >  0  THEN  PUT  LIST  (32.17 * (4390/(4390  +  H)) ** 2);
IF  H  <= 0  THEN  PUT  LIST  (32.17 * (1   +  H /4390));
```

If H is greater than zero, the first **THEN** clause will be executed, after which the second relationship will be found to be false, causing the second **THEN** clause to be skipped. If H is less than or equal to zero, the first **THEN** clause will be skipped, after which the second relationship will be found to be true, causing the second **THEN** clause to be executed.

It seems a shame to have to make two tests of H when we can deduce the result of the second test from the result of the first. At the expense of program complexity,

---

[1] For the 48-character set, the comparison operators are encoded as LT, =, GT, LE, GE, NL, NE, and NG, respectively. Operators written in alphabetic characters are reserved identifiers. SL1 does not support NL and NG.

Sec. 4.2   Flow Charts   63

the following statements suffice with one test:

```
        IF  H  >  0  THEN  GO  TO  S10;
        PUT  LIST  (32.17 * (1  +  H/4390));
        GO  TO  S20;
S10:    PUT  LIST  (32.17 * (4390/(4390  +  H)) ** 2);
S20:
```

Note how we have used the statement "GO TO S20;" to skip an unwanted portion of the program. GO TO statements are used for skipping forward as well as looping backward. PL/I provides a more general form of the IF statement that simplifies the coding of this example, but it is best to thoroughly learn the elementary IF statement first.

Suppose now that we wish to write a function procedure G(H), using the appropriate formula—depending upon H. We may do this as follows:

```
G:   PROCEDURE (H);
     IF  H > 0   THEN RETURN (32.17*(4390/(4390 + H))**2);
     RETURN (32.17*(1 + H/4390));
END;
```

We do not need two IF statements or an elaborate system of GO TO statements because the RETURN statement returns *control* as well as a *value* to the invoking procedure. If the THEN clause is executed, control is returned to the invoking procedure before the second RETURN has a chance to be executed.

4.2 FLOW CHARTS

With the IF statement, the flow of control or logic becomes complicated enough so that a graphical means of portrayal is helpful. The *flow chart* provides this means, and the flow charts in Fig. 4-1 correspond to the previous examples in this case study (see next page).

The arrows indicate the execution sequence or logic. Note that PL/I conventions such as capital letters, asterisks for multiplication, and the rigid form of the IF statement are ignored. In fact, they are usually avoided on purpose because a flow chart is supposed to be an unrestricted universal means of communication which transcends the idiosyncracies of any particular programming language. In addition to being a universal means of describing programs, flow charts provide an invaluable working tool that helps programmers get the logic of a program worked out before becoming entangled in the details.

There are no universally followed standards for flow charts, but in this text we will use ovals to indicate entries and returns, hexagons to indicate decisions (IF statements), rectangles to indicate assignments, and trapezoids to indicate inputs or outputs. GO TO statements are simply indicated by arrows. The return

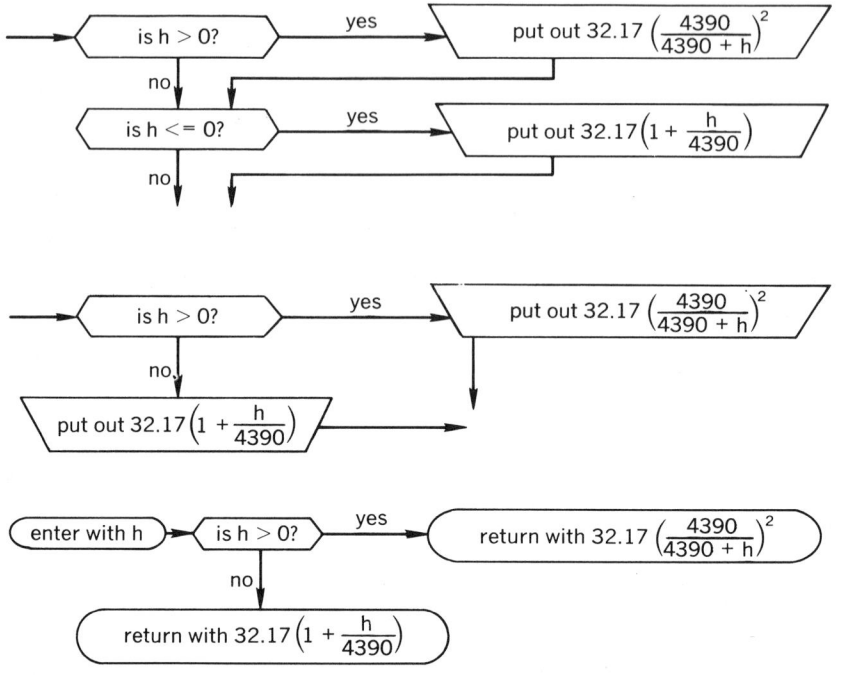

Fig. 4-1 Examples of flow charts.

from a main procedure to the operating system and the entry to a main procedure or program will not be listed when obvious.

4.3 SOME ADDITIONAL BUILT-IN FUNCTIONS

As another example of the IF statement, the following flow chart and function procedure determine the maximum of three values:

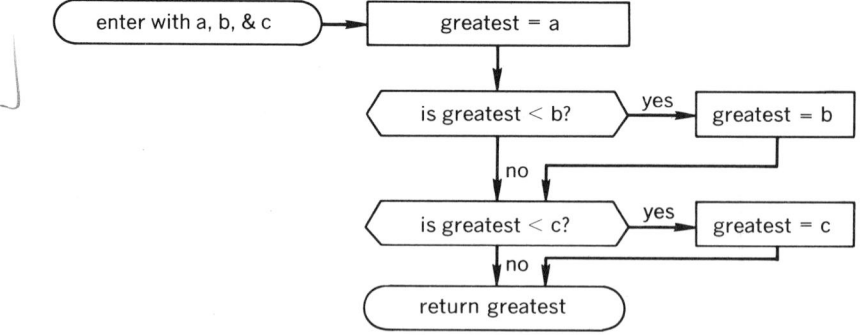

Fig. 4-2 Flow chart for determining the maximum of A, B, and C.

```
THEMAX:  PROCEDURE (A, B, C);
    GREATEST = A;
    IF  GREATEST < B  THEN GREATEST = B;
    IF  GREATEST < C  THEN GREATEST = C;
    RETURN (GREATEST);
END;
```

Note the technique of making an arbitrary assumption, and then correcting it if necessary. Actually, there is a built-in arithmetic MAX function which finds the maximum of any number of arguments—for example, MAX(A, B, C, D). Similarly, there is a built-in arithmetic MIN function which finds the minimum of any number of arguments. These two functions are the only ones which accept any number of arguments. Some other useful built-in *arithmetic* functions are the following:

ABS(x), the absolute value of x, is simply x if x is positive or −x if x is negative. This function is indicated as |x| in mathematics. For example, speed is equal to the absolute value of velocity: S = ABS(V).

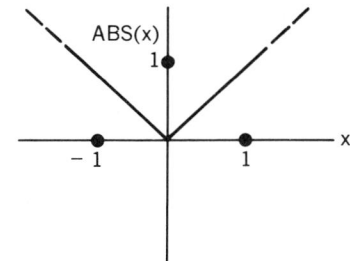

SIGN(x) is +1 if x is positive, −1 if x is negative, and 0 if x is 0. This function is pronounced *signum* to distinguish it from sine. In mathematics, this function is often abbreviated sgn. For example, the dynamic Coulomb friction force is proportional to SIGN(V) where V is the velocity.

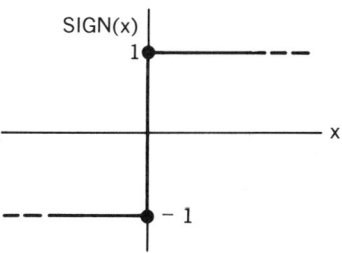

TRUNC(x) chops off the fractional digits of x. This function may be used to achieve the effect of truncation within an expression. For example, Student PL and FORTRAN compilers truncate fixed-point noninteger quantities; so we may simulate their effect on the expression M + N/2 by writing M + TRUNC(N/2).

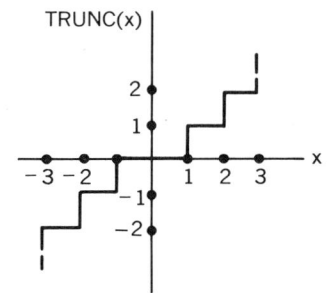

FLOOR(x) determines the largest integer which does not exceed x. Note that for negative x this gives a different result than truncation. For example, if N is a fixed-point integer variable, the statement "N = −1.5;" gives N the value of −1 whereas the statement "N = FLOOR(−1.5);" gives N the value of −2.

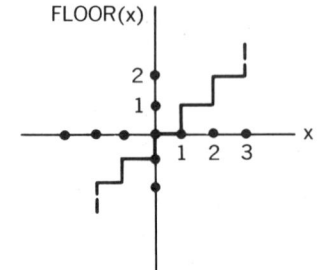

CEIL(x) determines the smallest integer which is not less than x. For example, if wall-board comes only in an integer number of feet, an architect must make his ceiling CEIL(H) high in order to accommodate a person of height H.

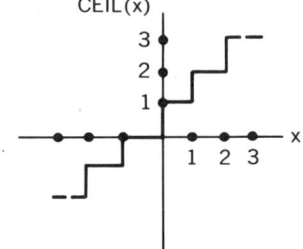

MOD(a, b) determines the positive remainder of a/b. For example MOD(23.6, 5) gives 3.6 because

$$\begin{array}{r}4 \text{ R } 3.6 \\ 5\overline{)23.6}\end{array}$$

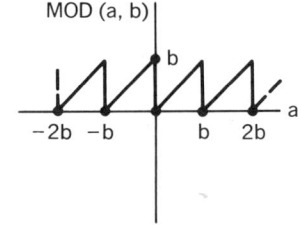

As another example, MOD(−23.6, 5) gives 1.4 because

$$\begin{array}{r}-4 \text{ R } -3.6 = -5 \text{ R } 1.4 \\ 5\overline{)-23.6}\end{array}$$

The whole part is adjusted if necessary so that the remainder is *positive*. In mathematics this function is often written *a modulo b* or *a mod b*.

For example, given the time 1617 in the European fashion, MOD(1617, 100) will isolate the number of minutes, 17.

ROUND(x, n) determines the value of x rounded to the nth fractional digit if x is a fixed-point argument. For example, ROUND (15.6364, 2) gives 15.6400. Negative n causes rounding at the −nth whole digit. For example, ROUND(1546,

−2) gives 1500. The treatment of floating-point arguments is implementation dependent.[2]

Whereas the *mathematical* functions always return a floating-point value, the *arithmetic* functions will return a fixed-point value if all of their arguments are fixed-point, and a decimal value if all of their arguments are decimal.[3]

EXERCISES

To reinforce your understanding of the IF statement and the built-in arithmetic functions, write your own flow charts and function procedures for the following, assuming default floating-point arguments and ignoring the fact that these functions are already built-in;

▼1. ABS(X)

2. THEMIN(A, B, C)

3. SIGN(X)

▼4. FLOOR(X)

5. CEIL(X)

6. TRUNC(X)

▼7. THEMOD(A, B)

8. Write a statement or statements to find the minimum absolute value of P, R, and S.

9. Write a statement or statements to find the value of P, Q, R, or S whose absolute value is minimum.

The following exercises demonstrate how the *singularity functions*, useful in dynamic system analysis, may be expressed in terms of the arithmetic functions:

▼10. The unit ramp function may be provided by the expression MAX(0, T). Write an expression for the general ramp.

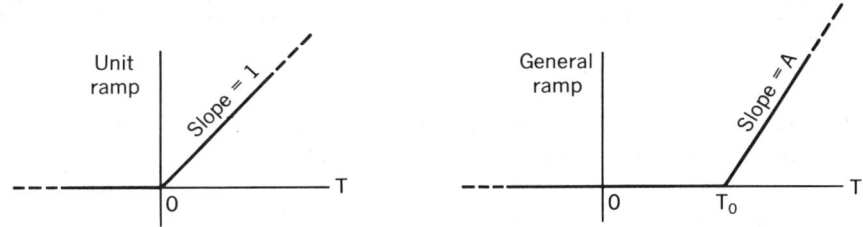

---

[2] For the D and F compiler, the last bit in the mantissa is changed to a one if it was a zero. Otherwise the number is left unchanged. The Model 20 does not permit floating-point arguments for Round.

[3] The Student PL interpreter does not support FLOOR, CEIL, or TRUNC. SL1 does not support TRUNC and ROUND; and only two arguments may be used with MAX and MIN. The Model 20 does not support SIGN or MOD.

11. The unit step function may be provided by the expression .5*(1 + SIGN(T)). Write an expression for the general step function.

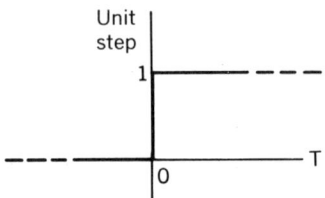
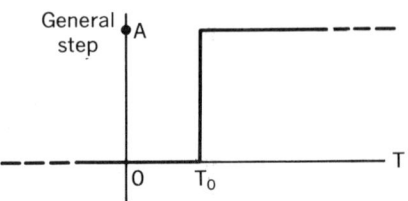

12. A rectangular pulse of unit height and width may be represented by the expression .5*(SIGN(T) − SIGN(T − 1)). Write an expression for the more general pulse.

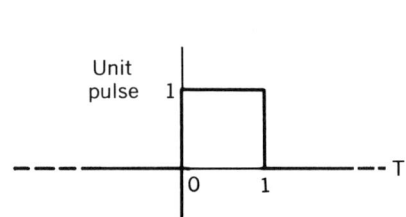
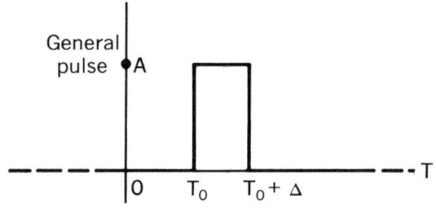

▼13. The unit sawtooth function may be provided by the expression MOD(T, 1). Write expressions for the more general sawtooth and its complement.

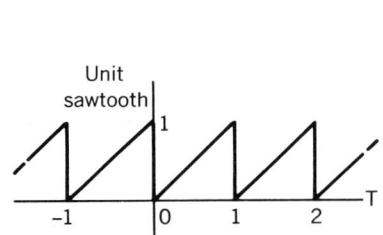
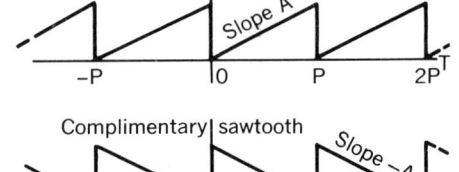

14. The unit square wave may be provided by the expression SIGN(MOD(T, 2) − 1). Write an expression for the more general square wave.

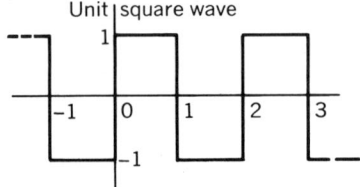
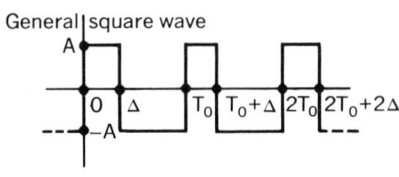

Exercises

▼15. If a function is defined by a function procedure FCN(T), the expression FCN(MOD(T, 1)) makes a periodic function from repetitions of the slice between 0 and 1. Write an expression which makes a periodic function from the slice between T_1 and T_2.

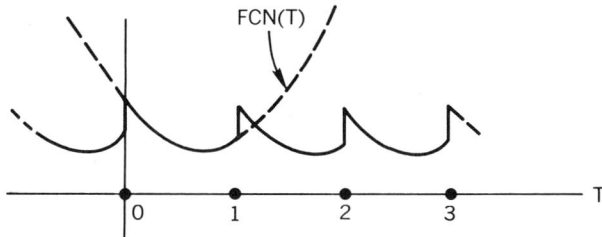

▼16. A polygonal or piece-wise linear function may be represented by a sum of general ramp functions. Do this for the example to the right.

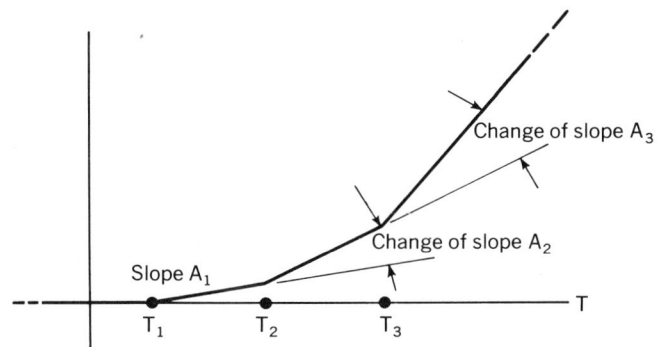

▼17. (Difficult) Most mathematical models are valid only within certain ranges of the physical variables. For example, it is customary to consider a steel compression spring to be linear: F = S · X. However, this is not true if the spring is forced outside its proper operating range. As is shown in the figure below, the spring becomes much stiffer when the coils touch, and tension cannot be supported at all due to separation from the spring seat. Rather than assume the spring is in the proper operating range, and then check the results to make sure, we may often include operation limits directly in a computer analysis:

$$F = MAX(0, \ S*X, \ Q*(X - B))$$

Show how similar limitations may be included in the equation describing the saturable inductor shown below.

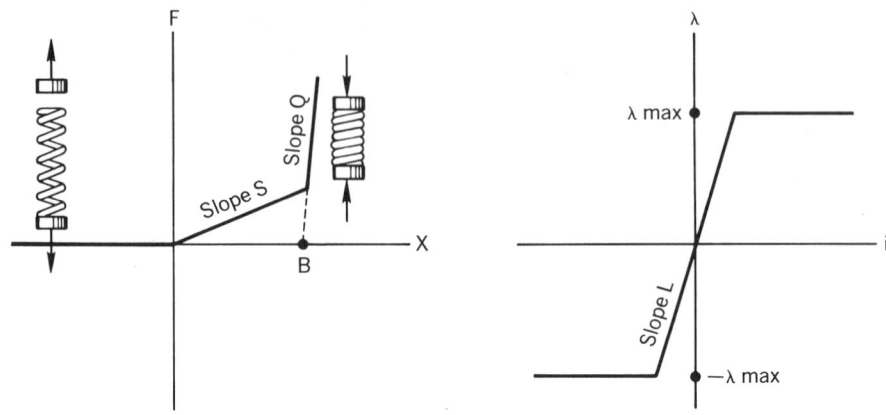

18. (Difficult) The dimensionless flow in a pipe, ϕ, as a function of the dimensionless head, η, and the dimensionless roughness, ρ, is equal to

64η if $|\eta| < 31.3$

$2000 \, \text{sgn}(\eta)\left(\dfrac{|\eta| - 30.3}{\tau - 31.3}\right)$ if $31.3 <= |\eta| < \tau$

$\text{sgn}(\eta) \sqrt{|\eta|}\left[1.14 + 4.6 \log_e\left(\rho + \dfrac{9.35}{\sqrt{|\eta|}}\right)\right]$ if $|\eta| >= \tau$

where

$$\tau = \left[\dfrac{4000}{1.14 + 4.6 \log_e(0.0133 + 0.95\,\rho)}\right]^2$$

Write a corresponding function procedure PHI(ETA, RHO).

19. (Difficult) The dimensionless head is equal to $2gD^3H/(v^2L)$, the dimensionless roughness is equal to ϵ/D, and the dimensionless flow is equal to $4Q/(\pi Dv)$, where g is the acceleration of gravity, D is the diameter of the pipe, H is the head loss, L is the length of the pipe, v is the kinematic viscosity, ϵ is the equivalent sand-grain diameter of the roughness, and Q is the flow—all in consistent units. Assuming $g = 32.17$, write a program using the PHI function of exercise 18, to repetitively get in $D, L, H, v,$ and ϵ; calculate Q; and put out all of the quantities.

This is an example of where the null field is quite useful since many of the input quantities will ordinarily remain the same for successive cases—especially v and ϵ.

4.4 THE *AND* AND *OR* OPERATORS

Designers are often faced with the problem of choosing components from a very large range of commercially available equipment. The computer can be of great help in narrowing the choice. Take, for example, the selection of ball bearings.

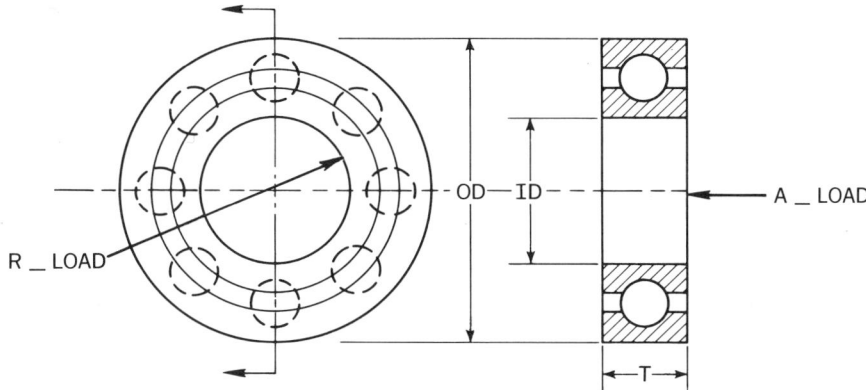

Fig. 4-3 Dimensions and loads for a ball bearing.

As indicated in Fig. 4-3, some of the relevant specifications are the identification number, radial load capacity, axial load capacity, speed limit, inside diameter, outside diameter, thickness, and cost. The input data for the program may consist of these specifications for the first bearing, followed by the corresponding specifications for the second bearing, and so on for thousands of bearings. Each time the designer wants a bearing which meets certain specifications, he may write a simple procedure which searches through this data set. Usually one program is used with many data sets, but in this instance, one data set is used with many programs. As a particular example, the following flow chart indicates how a program could select the bearings which have a speed capability of at least 5000 revolutions per minute, will fit directly on a 0.625-in. shaft, and have a thickness of no more than 0.25 in. to avoid interference with an adjacent part:

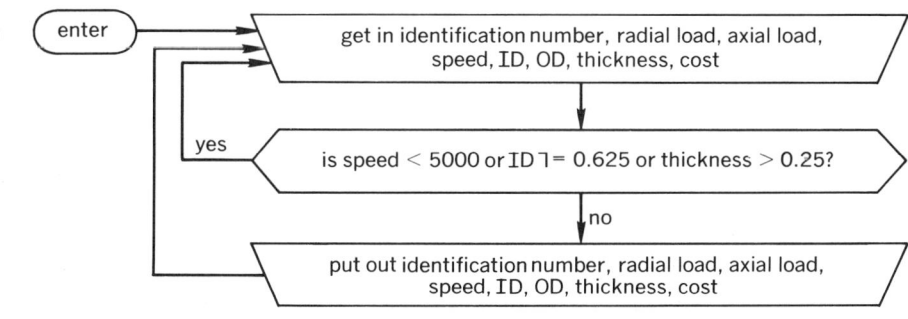

We may use three successive IF statements for the rejection criteria or we may use the *or operator* " | " to combine them in one statement:

IF (SPEED < 5000) | (ID ⌐= 0.625) | (T > 0.25) THEN GO TO NEXT;

where NEXT is the label of the GET statement.

Alternatively, we may modify the criteria into *acceptance* criteria, resulting in the following flow chart:

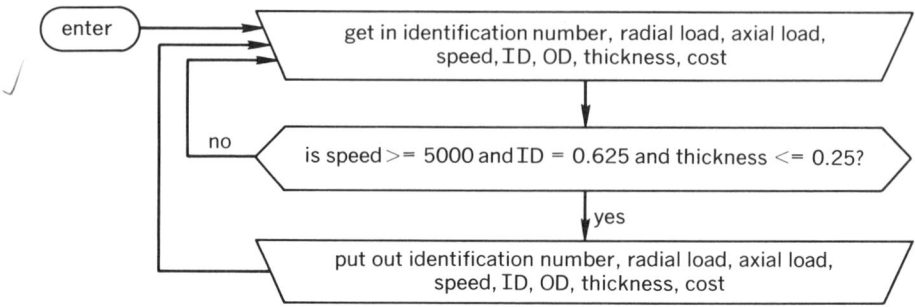

The corresponding IF statement would use the *and operator* "&":

IF (SPEED >= 5000) & (ID = 0.625) & (T <= 0.25) THEN PUT SKIP LIST (#, R_LOAD, A_LOAD, SPEED, ID, OD, T, COST);[4]

The parentheses around the relationships are included only to improve legibility. Actually, the comparison operators have higher priority than the *and* operator, which in turn has higher priority than the *or* operator. Note also that "$4 < X < 9$" and "$X < 9 \ \& > 4$" are incorrect representations of "$(4 < X) \ \& \ (X < 9)$." Relationships may not share a value. For the 48-character set, & and | are encoded as AND and OR respectively. Operators written in alphabetic characters are reserved identifiers.

4.5 NONINTEGER FIXED-POINT VARIABLES[5]

Before writing the program, the designer must consider who will be using the output. The designer would probably encircle his first, second, and third choices, then send the output sheet to a purchasing agent. Being unfamiliar with fraction anomaly and PL/I scientific notation, the purchasing agent would be liable to balk at a number such as 8.49999E−01. He might think it was 8.49999 in. with a tolerance of −0.01 in. If he discovered it was 0.849999 in. he might think the designer foolish to ask for a custom bearing when there is a standard one with a measurement of 0.85 in. Carrying our fantasies even further, a machinist might spend hundreds of dollars trying to machine a bearing to an accuracy of 0.000001 in. At best, the purchasing agent would bother the designer for an explanation, and the agent would be irritated at being forced to decipher floating-point numbers.

[4] The *and* and *or* operators are unsupported by the Model 20; so we must use three separate IF statements or the *nested* IF statement discussed in Section 4.6.
[5] Unsupported by Student PL.

Even programmers should not have to suffer fraction anomaly and floating-point output for applications such as this. Consequently, fixed-point decimal noninteger variables will be established for this application.

The ID, OD, and thickness of bearings are customarily presented in inches and thousandths of inches. Moreover, none of these dimensions ever exceeds 99.999 in.; so a declaration of FIXED(5, 3) DECIMAL is appropriate for these variables. The "5" specifies the total number of digits or *length*, and the "3" specifies the number of fractional digits or *scale factor*. The price is given in dollars and cents, never exceeding $999.99. Consequently, a declaration of FIXED(5, 2) DECIMAL is appropriate for this variable. The axial load, radial load, and speed are customarily represented as whole numbers which never exceed five digits; so the declaration of FIXED(5) DECIMAL is appropriate for them. This is the same as FIXED(5, 0) but if unspecified, a zero scale factor is assumed by default. Similarly, a declaration of FIXED(9) DECIMAL is appropriate for the identification number which is always an integer of up to nine digits. BINARY would be all right for the last four variables because whole numbers do not have any fraction anomaly. However, DECIMAL is faster for this program since the ratio of input and output to calculation is high. The following program incorporates these declarations in factored form:

```
BEARING: PROCEDURE OPTIONS (MAIN);
    DECLARE
        ( #         /* IDENTIFICATION NUMBER      */  FIXED(9),
         (R_LOAD    /* RADIAL LOAD IN POUNDS      */,
          A_LOAD    /* AXIAL LOAD IN POUNDS       */,
          SPEED     /* MAXIMUM RPM                */) FIXED(5),
         (ID        /* INSIDE DIAMETER IN INCHES  */,
          OD        /* OUTSIDE DIAMETER IN INCHES */,
          T         /* THICKNESS IN INCHES        */) FIXED(5,3),
          COST      /* PRICE PER UNIT IN $        */  FIXED(5,2)) DECIMAL;
    PUT PAGE LIST ('BEARING NUMBER', 'RADIAL LOAD', 'AXIAL LOAD',
        'MAXIMUM RPM', 'INSIDE DIAM.', 'OUTSIDE DIAM.', 'THICKNESS',
        'PRICE/UNIT');
    NEXT: GET LIST (#, R_LOAD, A_LOAD, SPEED, ID, OD, T, COST);
        IF (SPEED >= 5000) & (ID = 0.625) & (T <= 0.25) THEN PUT
            SKIP(2) LIST (#, R_LOAD, A_LOAD, SPEED, ID, OD, T, COST);
        GO TO NEXT;
END;
```

Note that there are two levels of factoring—DECIMAL applying to all of the variables. Also, SKIP(2) causes double-spacing. In general, SKIP(N) causes N-spacing or zero-spacing—whichever is larger. The default value of N is 1. The output would look similar to the following for an implementation with six print positions:

| BEARING NUMBER | RADIAL LOAD | AXIAL LOAD | MAXIMUM RPM | INSIDE DIAM. |
OUTSIDE DIAM.	THICKNESS	PRICE/UNIT		
57100	800	100	6500	0.625
0.850	0.235	6.35		
8923498	475	50	9000	0.625
0.755	0.200	7.50		

Each time the designer has a different design problem, he may modify the IF statement and resubmit the program with the same data set.

4.6 NESTED IF STATEMENTS

A THEN clause may be another IF statement. For example, we may write our IF statement in the procedure BEARNG as follows:

```
IF  SPEED  >=  5000  THEN
  IF  ID  >=  0.625  THEN
    IF  T  <=  0.25  THEN  PUT  SKIP(2)  LIST
      (#,  R_LOAD,  A_LOAD,  SPEED,  ID,  OD,  T,  COST);
```

The third IF statement is nested in the second, which in turn is nested in the first. Even more levels of nesting are allowed. For some implementations, nesting is more efficient than the *and* operator.

SUMMARY

The IF statement may have the form IF *relationship* THEN *statement*. The relationship may be a compound relationship consisting of several simple relationships joined by the "&" or " | " operator. A simple relationship consists of two arithmetic expressions joined by one of the comparison operators, $<, =, >, <=, >=, \neg<, \neg=$, and $\neg>$. The THEN clause may be a RETURN, GO TO, GET, PUT, IF, or assignment statement. Except for the RETURN and GO TO case, control proceeds to the statement following the IF statement regardless of whether or not the THEN clause was executed. Examples are:

```
IF  C  >  15.6  THEN  RETURN  (B  +  SIN(X));
IF  C/B  =  SIN(X  +  5)  THEN  GO  TO  S50;
IF  (B  >  0)  |  (B  >  0)  THEN  GET  LIST  (Q);
IF  (B  >  0)  &  (B  <  7)  THEN  PUT  LIST  (P);
IF  (C  <  D)  &  ((C  =  B)  |  (C  =  E))  THEN  B  =  C;
```

Flow Charts

The corresponding flow chart for the above examples, taken in sequence, is at the top of the following page.

Priority of Operators

1st priority	**, prefix −, prefix +	(right to left ←)
2nd priority	*, /	
3rd priority	infix −, infix +	
4th priority	$<, =, >, <=, >=, \neg<, \neg=, \neg>$	(left to right →)
5th priority	&	
6th priority	|	

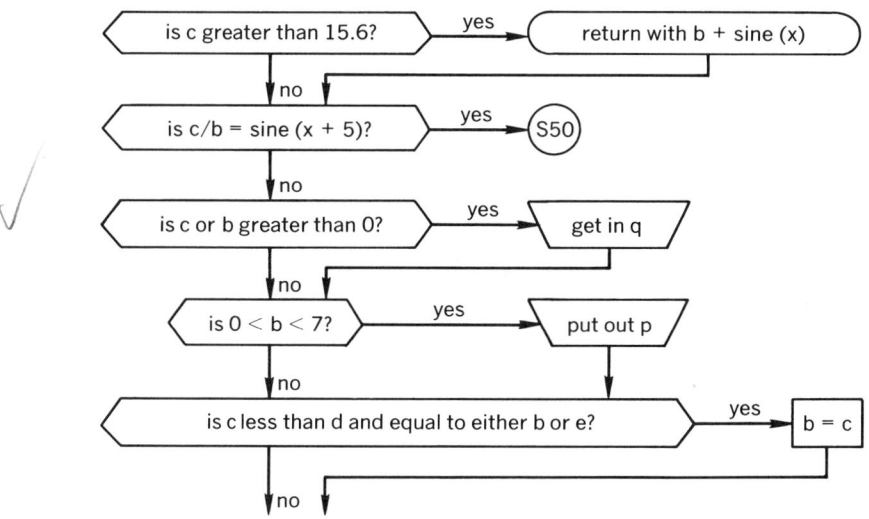

Built-in Arithmetic Functions

The built-in arithmetic functions return a fixed-point value if all of the arguments are fixed-point and a decimal value if all of the arguments are decimal. Examples are:

MAX(any number of arguments)   ABS(X)   CEIL(X)   TRUNC(X)   MOD(A, B)
MIN(any number of arguments)   SIGN(X) FLOOR(X) ROUND(X, N)

Noninteger Fixed-Point Variables

The declaration of FIXED (length, scale factor) DECIMAL is one way to avoid fraction anomaly and achieve fixed-point output without superfluous decimal places.

Skip Option

SKIP(N) will cause the output to start at the beginning of a line with N-spacing. N is one (single spacing) by default. A zero or negative N causes the last line to be overwritten.

EXERCISES

20. Write a flow chart and a bearing selection procedure which will list the specifications for bearings which have a radial load capacity of at least 800 lb, an axial load capacity of at least 500 lb, and a cost of no more than $6.00.
21. Devise a numerical data description of chemical compounds, species of fish, transistors, or whatever is relevant to your discipline. Then, write an example flow

chart and program which will search through your data set, listing the specifications of all items which meet certain restrictions. Problems such as this are elementary examples of *information retrieval*.

22. Develop a date-making program which lists the social security number, telephone number, physical specifications, and coded answers to a question sheet (e.g., 0 for no, 1 for yes, or 1 through 6 for a multiple choice) for all persons who meet particular requirements.

▼23. Write a function procedure to be used in the form NQUAD(X, Y) which returns the quadrant of the point (X, Y) —1, 2, 3, or 4. Consider values of zero to be positive. You will find a preliminary flow chart quite helpful in solving this problem.

24. Write a function procedure to be used in the form NOCT(X, Y, Z) which returns the octant of the point (X, Y, Z). Make use of the function NQUAD defined in exercise 23, letting NOCT(X, Y, Z) = NQUAD(X, Y) if Z is positive and letting NOCT(X, Y, Z) = NQUAD(X, Y) + 4 if Z is negative. Consider zero values to be positive.

▼25. (Difficult) Modify the SINE procedure of Case Study 2 to reduce any angle to the range $-\pi/2$ to $\pi/2$ before applying Hasting's formula. First, let ADUP = A to avoid changing the value of A in case the original value is needed later in the invoking procedure. Then repetitively add or subtract 2π until ADUP is in the range $-\pi$ to π. Next assign ADUP the value π — ADUP if it is in the range $\pi/2$ to π, or $-\pi$ — ADUP if it is in the range $-\pi/2$ to $-\pi$. You will find a preliminary flow chart quite helpful in solving this problem.

26. (Difficult) Same as exercise 25 except make use of the MOD function to reduce any angle to the range 0 to 2π in one step.

27. (Mathematically Advanced) Write a function procedure to be used in the form NGCD(L, M) which returns the greatest common divisor of two integers L and M. The following flow chart shows how to do this by Euclid's method:

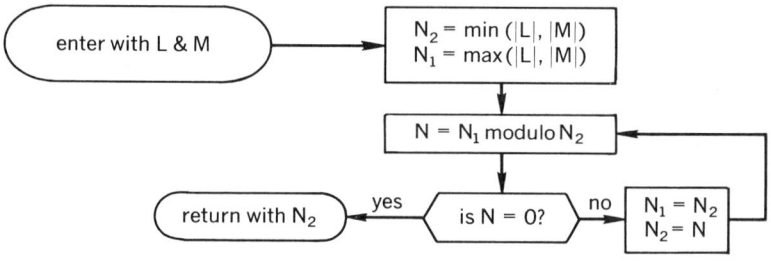

If you have never been exposed to this method, try it out on a few numerical examples to convince yourself that it works.

28. Among other things, SKIP(0) may be used to underline headings and create composite characters. For example, the statements "PUT SKIP LIST ('ANGLE 0'); PUT SKIP(0) LIST ('_____-');" would result in the heading *ANGLE θ*. What other uses and useful composite characters can you think of?

29. (Difficult) Assume that the angular data for a program is given as an integer whose lowest two digits stand for the number of seconds, whose next two digits and for the number of minutes, and whose remaining digits stand for the number of degrees. For example, 1105741 would stand for 110 degrees 57 minutes and 41 seconds. Write a function procedure which converts the data to degrees and decimal fractions thereof.

▼**30.** Write a function procedure IN(X, Y) which returns 1 if the point (X, Y) is in or on the edge of the shaded triangle and returns 0 otherwise.

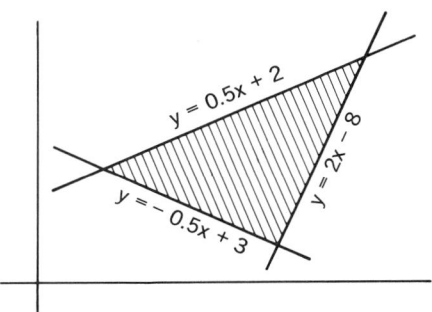

▼**31.** Write a function procedure IN(X, Y) which returns 1 if the point (X, Y) is in or on the edge of the shaded area and returns 0 otherwise.

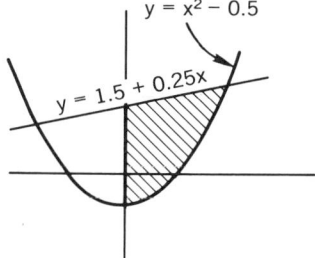

32. Write a function procedure IN(X, Y) which returns 1 if the point (X, Y) is in but not on the edge of the shaded triangle and returns 0 otherwise.

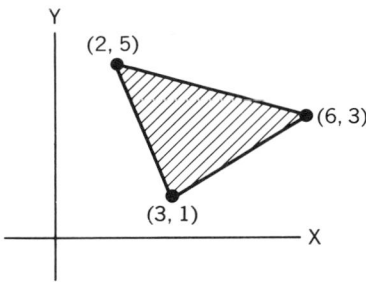

▼**33.** Write a function procedure IN(X, Y) which returns 1 if the point (X, Y) is in or on the edge of the shaded area and returns 0 otherwise.

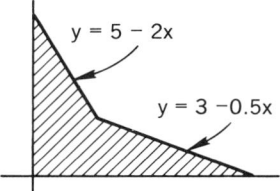

▼34. (Difficult) Write a function procedure IN(X, Y, Z) which returns 1 if the point (X, Y, Z) is inside or on the edge of the shaded tetrahedron.

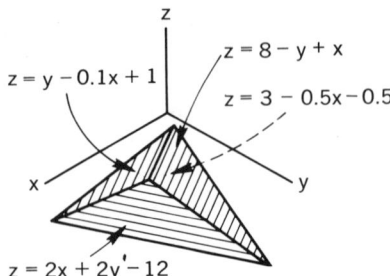

CASE STUDY 5

SOLUTION OF QUADRATIC EQUATIONS

Subroutines

We have already learned that most computer centers maintain a subordinate procedure library. However, most of those procedures are usually of a more general type than a function procedure. Most are *subroutines*. To appreciate the need for this more general type, consider the transformation from spherical to Cartesian coordinates as depicted in Fig. 5-1.

This transformation occurs frequently enough to warrant being programmed as a subordinate procedure. However, there are three results to be returned, and only one may be associated with the name of the procedure via a **RETURN** statement. One solution to this dilemma is to have three separate function procedures for X, Y, and Z. However, this approach requires three separate **PROCEDURE** and **END** statements for the definition as well as three separate assignment statements for each use. To avoid this inefficiency and nuisance, we may take advantage of the fact that a subordinate procedure may *modify* the values of arguments. Until now, our subordinate procedures have merely *used* the values of arguments. For example, we may write a subordinate procedure for our coordinate transformation as follows:

```
CART:   PROCEDURE (R, THETA, PHI, X, Y, Z);
    COEF = R*SIN(THETA);
    X = COEF*COS(PHI);
    Y = COEF*SIN(PHI);
    Z = R*COS(THETA);
END;
```

$X = R \sin \theta \cos \phi$

$Y = R \sin \theta \sin \phi$

$Z = R \cos \theta$

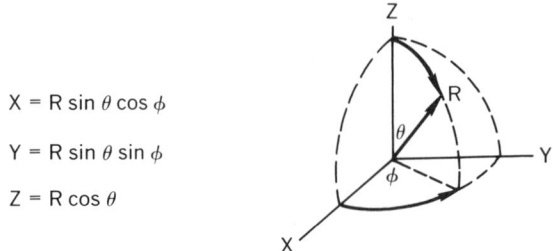

Fig. 5-1 Transformation from spherical to Cartesian coordinates.

This procedure uses the values of its first three parameters and assigns values to its last three parameters. For this reason, the first three may be called *given* parameters and the second three may be called *result* parameters. The order in which the parameters are listed is irrelevant, but the corresponding arguments must be listed in the same order. Some programmers prefer to list the result parameters first, others prefer to scramble them.

This subordinate procedure is said to be of the *subroutine* type (as opposed to function type) because no value is returned in a **RETURN** statement to be associated with the name **CART**. *Control*, but no function value, is returned when execution reaches the **END** statement of **CART**. This procedure may not be invoked by using its name in an expression because there is no value associated with the name. Consequently, we need a new kind of statement to invoke subroutines. It is the **CALL** statement which fills this need.

The following procedure illustrates the use of the **CALL** statement, using it to invoke **CART** to transform the coordinates for a series of points:

```
TRNSFM:   PROCEDURE OPTIONS (MAIN);
    PUT PAGE;
    IN:   GET LIST (RADIUS, COELEVATION, AZIMUTH);
        CALL CART (RADIUS, COELEVATION, AZIMUTH, D_NORTH, D_WEST, D_UP);
        PUT SKIP DATA (RADIUS, COELEVATION, AZIMUTH, D_NORTH, D_WEST,
            D_UP);
        GO TO IN;
END;
```

The proper values of the output arguments are established during execution of the call; so they are available for use any time after the **CALL** statement has been executed. Remember that arguments and their corresponding parameters need not have the same names.[1] **RADIUS** corresponds to **R**, **COELEVATION** to **THETA**, **AZIMUTH** to **PHI**, **D__NORTH** to **X**, **D__WEST** to **Y**, and **D__UP** to **Z**. At first it is disturbing to have dependent variables or results in an argument list, but after a while it becomes quite natural. It may help to consider this **CALL** statement as being similar to

GET   LIST   (D__NORTH,   D__WEST,   D__UP);

[1] For the Model 20, remember that we must also have the declaration "DECLARE CART ENTRY;" in TRNSFM.

5.1 QUADRATIC EQUATIONS EXAMPLE

As another example of subroutines, we will develop one to solve quadratic equations. The general form for these equations is $ax^2 + bx + c = 0$ where x is the unknown and a, b, and c are known coefficients. The theoretical solutions are given by the quadratic formula:

$$x = \frac{-b \pm \sqrt{b^2 - 4ac}}{2a}$$

However, this formula is not very good for computational purposes. First, we can cut down on the computational work by *normalizing* or dividing the quadratic equation by the highest power coefficient. Normalization gives the equation $x^2 + Bx + C = 0$ where $B = b/a$ and $C = c/a$. This does not change the solutions, but it simplifies their formulas to

$$x = -0.5B \pm \sqrt{(0.5B)^2 - C}$$

This solution requires two less multiplications than the standard solution. There are also two less divisions if the equation originates in normalized form, as is often the case.

We must take some precautions before blindly programming this formula. The expression under the square root sign, which is called the *discriminant*, may be negative. If so, the roots are said to be *complex* because they have both a *real* and an *imaginary* part. Complex roots are meaningful and of interest in many physical applications even though the square roots of negative numbers are not real. When the discriminant is negative, the real parts of the roots are $-0.5B$, and the imaginary parts are $\pm\sqrt{|(0.5B)^2 - C|}$ where the straight brackets denote the absolute value. Mathematically, these complex roots are written $-0.5B \pm i\sqrt{|(0.5B)^2 - C|}$ where the i denotes $\sqrt{-1}$. The i serves as a warning that the two terms cannot be added together—they are like apples and oranges. Rather than using an i, in our subroutine we may keep the real and imaginary parts separate by simply returning them as separate result parameters.

We must be sure to determine the sign of the discriminant before attempting to take the square root because an attempt to take the square root of a negative real number leads to termination of execution. Although PL/I has the capability of performing complex arithmetic, it is easier and more efficient in this instance to simply use the absolute value of negative discriminants, treating the real and imaginary parts separately.

Another problem with the quadratic and simplified quadratic formulas is that they may produce a large relative error when both of the solutions are real and one of these is near zero in magnitude. In this case, depending upon the sign of $-0.5B$, either the plus or minus sign in front of the square root will cause subtraction of two nearly equal numbers. If, for example, $-0.5B = -1.59683$ and $\sqrt{(-0.5B)^2 - C} = 1.59688$, with all digits being significant, use of the $+$ sign results in a degeneration from six significant digits to one. This may be serious even for applications concerned with absolute error because, for example, subsequent divisions by this small relatively inaccurate root could cause a large absolute error. Happily, we may use only the sign which is the same as that of $-0.5B$ and evaluate the other root by making use of the fact that C equals the product of the roots. This fact is true because if the roots are r_1 and r_2, then $x = r_1$ or $x = r_2$; and so $(x - r_1)(x - r_2) = 0 = x^2 - (r_1 + r_2)x + r_1 r_2 = x^2 + Bx + C$.

The following flow chart incorporates these ideas:

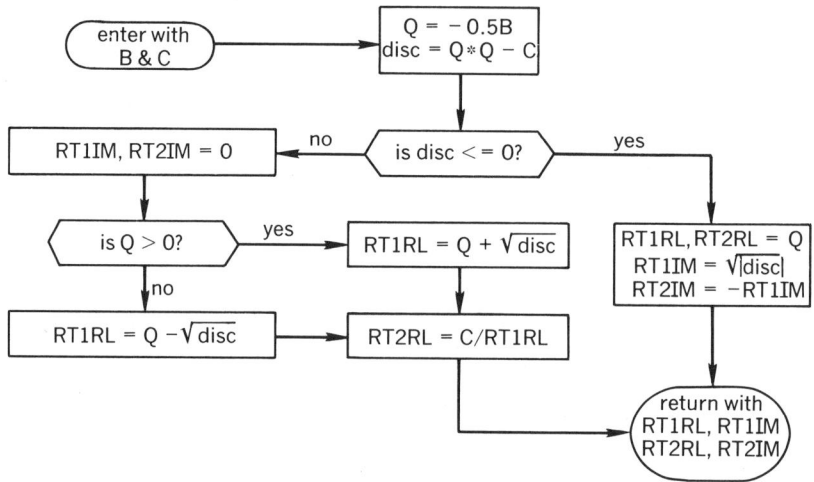

The subroutine below corresponds to this flow chart. However, before you look at it, I suggest that you convince yourself that the flow chart is valid, then attempt to write your own subroutine. It is not easy, but you will learn a great deal about programming regardless of whether or not you succeed.

```
QUAD:   PROCEDURE (B, C, RT1RL, RT1IM, RT2RL, RT2IM);
    /* THIS PROCEDURE FINDS THE REAL AND IMAGINARY PARTS OF THE TWO
       ROOTS OF THE NORMALIXED QUADRATIC EQUATION X*X + B*X + C = 0  */
        Q = -.5*B;    DISC = Q*Q - C;
        IF DISC <= 0   THEN GO TO S10;
        RT1IM, RT2IM = 0;
        IF Q > 0   THEN GO TO S5;
        RT1RL = Q - SQRT(DISC);
        GO TO S7;
    S5:    RT1RL = Q + SQRT(DISC);
    S7:    RT2RL = C/RT1RL;
        RETURN;
    S10:   RT1RL, RT2RL = Q;
        RT1IM = SQRT(ABS(DISC));   RT2IM = -RT1IM;
END;
```

This procedure illustrates the convenient feature of multiple assignment. We may list any number of variables separated by commas on the left of an assignment statement.[2] The value on the right is assigned to all of these variables, beginning with the leftmost variable. Another feature is the simple **RETURN** statement. Like the **END** statement, it accomplishes a return of control without a function value. Alternatively but less efficiently, we could have given the **END** statement a label such as S20, then used "GO TO S20;" instead of "RETURN;." Finally, note how **GO TO** statements have been used to dodge unwanted portions of the procedure.

As an example of how this procedure may be used, the following main procedure uses it to solve any number of unnormalized quadratic equations:

```
QUADS:   PROCEDURE OPTIONS (MAIN);
    PUT PAGE LIST ('A', 'B', 'C', 'ROOT U REAL', 'ROOT 1 IMAG',
        'ROOT 2 REAL', 'ROOT 2 IMAG');
    S10:   GET LIST (A, B, C);
        CALL QUAD (B/A, C/A, RT1RL, RT1IM, RT2RL, RT2IM);
        PUT SKIP(2) LIST (A, B, C, RT1RL, RT1IM, RT2RL, RT2IM);
        GO TO S10;
END;
```

Most computer installations maintain a library of general-purpose subordinate procedures and programs stored on cards, tapes, or other media. If they are stored on a resident auxiliary memory device such as a disc, we need to know only their names and the significance of their parameters in order to invoke them! Most installations will have a manual describing how to use each subordinate procedure. As an example of typical library programs of interest to engineers and scientists, the following section lists the subroutines in the PL/I Scientific Subroutine Package. These subroutines cover a wide range of applications and techniques. The list is extensive enough to be of use in perhaps 75 per cent of engineering and scientific programs. Every time we have a nontrivial new program to write, it is good practice to check the library to see if the program or components for it have already been written. If so, a great deal of time is saved by taking advantage of the previous work of professional programmers and numerical analysts. If not, perhaps the finished program could be contributed to the library! In this way programmers cooperate and each accomplishes far more than he could have starting from scratch.

5.2 SUBROUTINES IN THE IBM PL/I SCIENTIFIC SUBROUTINE PACKAGE[3]

MATRIX OPERATIONS
Elementary Operations

MSCS	Storage conversion—two-dimensional to compressed
MSCG	Storage conversion—compressed to two-dimensional
MAGS	Add–subtract general and symmetric matrices

---

[2] Multiple assignment is unsupported by the D, SL1, and Model 20 compilers.

[3] Reprinted by permission from *H20-0544-0—System/360 Scientific Subroutine Package (PL/I)*. © 1968 by International Business Machines Corporation.

MMGG	Product of two general matrices
MMSS	Product of two symmetric matrices
MMGS	Product of a general matrix and a symmetric matrix
MMGT	Product of a general matrix and its transpose
MPRM	Permutation of rows or columns of a matrix
MTPI	Calculation of permutations from transpositions
MPIT	Calculation of inverse permutation and transpositions

Linear Equations and Related Topics

MFG	Triangular factorization of a general nonsingular matrix
MFS	Triangular factorization of a symmetric positive definite matrix
MFSB	Triangular factorization of a symmetric positive definite band matrix
MFGR	Factorization and rank determination of a general rectangular matrix
MDLS/MDRS	Dividing a matrix by a triangular matrix that has been factored from a symmetric positive definite matrix
MDSB	Dividing a matrix by a triangular matrix that has been factored from a symmetric positive definite band matrix
MDLG	Dividing a matrix by a lower or upper triangular matrix that has been factored from a general nonsingular matrix
MIG	Inverting a general nonsingular matrix that has been factored into upper and lower triangular factors
MIS	Inverting a symmetric positive definite matrix that has been factored into a triangular matrix and its transpose
MINV	Inverting a general square matrix
MLSQ	Solution of a system of linear equations, the least squares solution being obtained in case of an overdetermined system
MGB1/MGB2	Solution of simultaneous linear equations with band matrix of coefficients

Eigenvalues and Related Topics

MATE	Reduction of a real matrix to upper almost-triangular form by elementary transformations
MATU	Reduction of a real matrix to upper almost-triangular form by orthogonal transformations
MSTU	Reduction of a symmetric matrix to tridiagonal form by orthogonal transformations
MEAT	Eigenvalues of a real upper almost-triangular matrix
MEST	Eigenvalues of a real symmetric tridiagonal matrix
MEBS	Bounds for the eigenvalues of a real symmetric matrix
MVST	Eigenvector of a symmetric tridiagonal matrix, corresponding to a given eigenvalue
MSDU	Eigenvalues and eigenvectors of a real symmetric matrix
MGDU	Eigenvalues and eigenvectors of a special real nonsymmetric matrix
MVAT	Eigenvector of a complex almost-triangular matrix, corresponding to a given eigenvalue
MVSU	Eigenvector of a symmetric matrix from the corresponding eigenvector of the associated tridiagonal form
MVUB	Eigenvector of a real matrix from the corresponding eigenvector of the associated almost-triangular matrix, which has been developed using MATU
MVEB	Eigenvector of a real matrix from the corresponding eigenvector of the associated almost-triangular matrix, which has been developed using MATE

POLYNOMIAL OPERATIONS

POV	Values of orthogonal polynomials (Chebyshev, Legendre, Laguerre, and Hermite)
POSV	Values of series expansion in orthogonal polynomials (Chebyshev, Legendre, Laguerre, and Hermite)
PEC/PTC	Economization of a polynomial for symmetric and asymmetric range, transformation of polynomial to expansion in Chebyshev or shifted Chebyshev polynomials
POST	Transformation of orthogonal polynomial expansion to a polynomial
PRTC	Roots of a complex polynomial by Nickel's method based on a method of Newton

NUMERICAL QUADRATURE

Quadrature of Tabulated Functions

QTFG/QTFE	Integration of monotonically or equidistantly tabulated function by trapezoidal rule
QSF	Integration of equidistantly tabulated function by Simpson's rule
QHFG/QHSG/ QHFE/QHSE	Integration of monotonically or equidistantly tabulated function with first (and second) derivatives by Hermitian formula of the first (and second) order

Quadrature of Nontabulated Functions

QATR	Integration of a given function by the trapezoidal rule together with Romberg's extrapolation method
QG2, QG4, QG8, QG16, QG24, QG32, QG48	Integration of a given function by Gaussian quadrature formulas
QL2, QL4, QL8, QL12, QL16, QL24	Integration of a given function by Gaussian–Laguerre quadrature formulas
QH2, QH4, QH8, QH16, QH24, QH32, QH48	Integration of a given function by Gaussian–Hermite quadrature formulas
QA2, QA4, QA8, QA12, QA16, QA24	Integration of a given function by associated Gaussian–Laguerre quadrature formulas

NUMERICAL DIFFERENTIATION

Differentiation of Tabulated Functions

DGT3	Differentiation of a tabulated function by Lagrangian interpolation
DET3	Differentiation of an equidistantly tabulated function using Lagrangian interpolation formulas
DET5	Differentiation of an equidistantly tabulated function using Lagrangian interpolation formulas

Differentiation of Nontabulated Functions

DFEC	Derivative of a function at the center of an interval by Richardson's and Romberg's extrapolation method
DFEO	Derivative of a function at the end of an interval by Richardson's and Romberg's extrapolation method

INTERPOLATION OF TABULATED FUNCTIONS

ALIM/ALIE	Aitken–Lagrange interpolation, monotonic, and equidistant tables
AHIM/AHIE	Aitken–Hermite interpolation, monotonic, and equidistant tables
ACFM/ACFE	Continued fraction interpolation, monotonic, and equidistant tables

APPROXIMATION OF TABULATED FUNCTIONS

FFT	Fast Fourier transform for real or complex one-dimensional array
FFTM	Fast Fourier transform for real or complex multidimensional array
APLL	Setting up normal equations for least-squares polynomial approximation
APC1/APC2	Setting up normal equations for least-squares Chebyshev polynomial approximation
ASN	Solving normal equations for least-squares fit

SMOOTHING OF TABULATED FUNCTIONS

SG13/SE13	Local least-squares smoothing of a tabulated function using a linear fit relative to three points
SE15	Local least-squares smoothing of an equidistantly tabulated function using a linear fit relative to five points
SE35	Local least-squares smoothing of an equidistantly tabulated function using a cubic fit relative to five points
EXSM	Triple exponential smoothing of a given series

ROOTS AND EXTREMA OF FUNCTIONS

FMFP	Minimization of a function of several variables without constraints
RTF	Root of a function using linear, quadratic, or hyperbolic interpolation
RTFD	Root of a function with given derivatives, by linear, inverse, quadratic, or hyperbolic interpolation

SYSTEMS OF ORDINARY DIFFERENTIAL EQUATIONS

DERE	Performing one integration step on a system of first-order ordinary differential equations

SPECIAL MATHEMATICAL FUNCTIONS

CEL1/CEL2	Complete elliptic integral of first and second kind
ELI1/ELI2	Incomplete elliptic integral of first and second kind
JELF	Jacobian elliptic functions
LGAM	Log of the gamma function

DATA SCREENING AND ANALYSIS

TALY	Totals, means, standard deviations, minima, and maxima
BOUN	Selection of observations over, under, and within bounds
ABST	Detection of missing data
SBST	Subset selection from observation matrix satisfying certain conditions
TAB1	Tabulation of data (one variable) including frequencies, over class intervals, mean, standard deviation, minimum, and maximum
TAB2	Tabulation of data (two variables)

SUBM	Copying a subset matrix that satisfies certain conditions from an observation matrix

ELEMENTARY STATISTICS

MOMN	First four moments for grouped data on equal class intervals
TTST	Certain t-statistics on the means of populations

CORRELATION AND REGRESSION ANALYSIS

CORR	Means, standard deviations, and correlation coefficients
ORDR	Selection of submatrix from matrix of correlation coefficients for multiple linear regression analysis
MLTR	Multiple linear regression analysis
STRG	Stepwise multiple linear regression analysis
CANC	Canonical correlation between two sets of variables

ANALYSIS OF VARIANCE

AVAR	Analysis of variance for a complete factorial design

DISCRIMINANT ANALYSIS

DMTX	Means and dispersion matrix for all groups
DSCR	Discriminant functions

PRINCIPAL COMPONENTS ANALYSIS

TRAC	Cumulative percentage of eigenvalues
LOAD	Factor loading
VRMX	Varimax rotation

NONPARAMETRIC STATISTICS

KLMO	Kolmogorov–Smirnov one-sample test
KLM2	Kolmogorov–Smirnov two-sample test
SMIR	Kolmogorov–Smirnov limiting distribution values
CHSQ	Chi-square test for contingency tables
KRNK	Kendall rank correlation
QTST	Cochran Q-test
RANK	Rank observation
SRNK	Spearman rank correlation
TIE	Calculation of correction factor due to ties
TWAV	Friedmann two-way analysis of variance statistic
UTST	Mann-Whitney U-test
WTST	Kendall coefficient of concordance
HTES	Kruskal-Wallis H-test

DISTRIBUTION FUNCTIONS

NDRT	Normal distribution function
BDTR	Beta distribution function
CDTR	Chi-square distribution function
NDTI	Inverse of normal distribution function

5.3 OPTIONAL LANGUAGE FEATURE: INTERNAL PROCEDURES[4]

So far we have considered only external procedures. Internal procedures may be nested within procedures. For example, the following program solves any number of triangles, given three sides:

```
SSS:    PROCEDURE OPTIONS (MAIN);
   ACOSD:   PROCEDURE (U);
               RETURN (ATAND(SQRT(1 - U*U), U));
            END;
   S10:     GET LIST (X, Y, Z);
            XSQ = X*X;   YSQ = Y*Y;   ZSQ = Z*Z;
            A = ACOSD((XSQ - ZSQ)/(2*Y*Z));
            B = ACOSD((YSQ - ZSQ - XSQ)/(2*Z*X));
            PUT SKIP LIST (X, Y, Z, A, B, C);
            GO TO S10;
   END;
```

Execution begins at statement S10. Like external procedures, internal procedures are entered only when invoked or called, rather than sequentially. Unlike external procedures, the names in an internal procedure are not necessarily independent of those in the invoking procedure. To determine where a name has a particular meaning, we must make some formal definitions. These definitions may seem rather complex and arbitrary at first; but they maximize the interaction flexibility while they minimize the possibility of conflict; and they cover nesting to any level.

A name is said to be *explicitly* declared if it appears—

1. in a **DECLARE** statement,
2. in a parameter list,
3. as a statement label, or
4. as a procedure label.

A name which is not explicitly declared is said to be *contextually* or *implicitly* declared. (The distinction between the two does not concern us here.) In our example, **SSS, ACOSD, U,** and **S10** are explicitly declared whereas **X, Y, Z, XSQ, YSQ, ZSQ, A, B,** and **C** are contextually or implicitly declared.

All of the text from the keyword **PROCEDURE** through the corresponding **END** statement is said to be *contained* in the procedure. In our example, the text from "**PROCEDURE OPTIONS (MAIN);**" through the second **END** statement is contained in the procedure **SSS**, and the text from "**PROCEDURE (U);**" through the first **END** statement is contained in the procedure **ACOSD**.

The *scope* of a declaration is the text in which it applies. *The scope of an explicit*

---

[4] Unsupported by the Model 20.

declaration is the contents of the innermost procedure which contains it, excluding any nested procedures which have another explicit declaration for the same identifier. In our example, the scope of the name SSS is throughout the entire program; the scope of the names ACOSD and S10 is within the procedures SSS and ACOSD; and the scope of the name U is within the procedure ACOSD.

The scope of a contextual or implicit declaration is the contents of the external procedure which contains it, excluding any nested procedures which have another explicit declaration for the same identifier. In our example, the scope of X, Y, Z, XSQ, YSQ, ZSQ, A, B, and C is within SSS and ACOSD.

The important consequences of these rules are:

1. We do not need to use an internal procedure parameter if the corresponding argument is the same for every invocation. For example, suppose that we had a main procedure which required a transformation from spherical to Cartesian coordinates at two different places in the procedure, but the argument names were the same in each case. We could write the procedure as follows:

```
COORD: PROCEDURE OPTIONS (MAIN);
   CART: PROCEDURE;
      COEF = R * SIN(THETA);
         X = COEF * COS(PHI);
         Y = COEF * SIN(PHI);
         Z = R * COS(THETA);
   END;
      . . .
      CALL  CART;
      . . .
      CALL  CART;
      . . .
   END;
```

We must establish the proper values of R, THETA, and PHI before each invocation, and the corresponding values of X, Y, and Z are available for our use after each invocation.

2. We may have a GO TO statement transfer control from an internal procedure to any procedures which contain it—for example:

```
A: PROCEDURE (X, Y);
   B: PROCEDURE (Z);
      . . .
      GO TO S10;
      . . .
   END;
   . . .
   S10:
   . . .
END;
```

3. We may invoke a procedure from within itself. However this is called a *recursive* invocation, and it requires other considerations which are not discussed in this text.

4. An internal procedure may not be invoked from a procedure which does not contain it.

5. We may make a name in an internal procedure independent of the same name outside of it by explicitly declaring the name in the internal procedure. As a precaution, it is recommended that each name in an internal procedure be explicitly declared in that internal procedure unless it is specifically desired that the scope extend outside the internal procedure. For example, suppose we had written our internal ACOSD procedure as follows:

```
ACOSD: PROCEDURE (U);
    X = ATAND(SQRT(1−U∗U), U);
    RETURN (X);
END;
```

This X would then be the same as that in SSS, causing incorrect calculation of B. This error is avoided by including the statement "DECLARE X;" in this version of ACOSD.

Internal Procedures generally execute faster than external procedures, but the advantage of independent compilation and carefree use of names is lost.

5.4 OPTIONAL LANGUAGE FEATURE: THE EXTERNAL ATTRIBUTE[5]

Another way to communicate between two procedures is by declaring the same names to have the EXTERNAL attribute in both procedures. For example, we could write CART and TRANSFM as follows:

```
CART: PROCEDURE;
    DECLARE (R, THETA, PHI, X, Y, Z) EXTERNAL;
    .
    .
    .
END;
TRNSFM: PROCEDURE OPTIONS (MAIN);
    DECLARE (R, THETA, PHI, X, Y, Z) EXTERNAL;
    .
    .
    .
END;
```

[5] Unsupported by SL1 and Student PL.

A variable will occupy the same storage location for every procedure in which it is declared EXTERNAL.. This technique generally executes faster than a parameter list, and the procedures may still be compiled separately. However, the advantage of independent argument-parameter names is lost. Consequently this form of communication is generally used for a group of subroutines that are intended to be used only with each other. This technique is often used when such a group is too large to fit into memory simultaneously. The first procedure is loaded, then replaced by the second procedure, which inherits their mutual EXTERNAL data; and so on.

External names are limited to seven characters for the F compiler and to six characters for the D and Model 20 compilers.

SUMMARY

Typical Subroutine

    subroutine name: PROCEDURE (list of input and output parameters);
    . . .
    output parameter = expression;
END;

Return Statement

A simple RETURN statement is used to return control from a point other than the END.

Call Statement

No value is associated with a subroutine name; so a subroutine is invoked with the statement:

    CALL subroutine name (list of input and output arguments);

EXERCISES

1. Write a subroutine SPHERE which transforms Cartesian coordinates into spherical, with the angles in degrees, according to the formulas $r = \sqrt{x^2 + y^2 + z^2}$, $\theta = \arctan(\sqrt{x^2 + y^2}, z)$, $\phi = \arctan(y, x)$.
2. Write a main procedure which repetitively gets the x, y, and z coordinates of a satellite, and uses SPHERE from the previous exercise to calculate and put out its longitude, latitude, and distance from the center of the earth.

▼3. The solution of two simultaneous linear equations is required so frequently that it would be helpful to have a special subroutine for it. In general we have $ax + by = c$ together with $px + qy = r$ where x and y are unknowns and $a, b, c, p, q,$ and r are known constants. By Cramer's rule, the solution is $x = (cq - rb)/d$ and $y = (ar - pc)/d$ where d is the *determinant* of the coefficients, $aq - pb$. Write a corresponding subroutine to be used in the form: TWOEQN (A, B, C, P, Q, R, X, Y).

4. Same as exercise 3 except use Gaussian Elimination, which is faster. Multiplying the first equation by $f = p/a$ and subtracting from the second equation to eliminate x gives $y = (r - fc)/(q - fb)$. y may then be back-substituted into the first equation, giving $x = (c - by)/a$. (The time savings becomes increasingly dramatic over Cramer's rule as the number of simultaneous equations increases.) To minimize the loss of accuracy due to subtraction of two nearly equal numbers, we would like the absolute value of fc to be small relative to that of r, and we would like the absolute value of fb to be small relative to that of q. Since $f = p/a$, we should order the equations and unknowns so that the absolute value of a is as large as possible compared to that of the other coefficients. In practice, an adequate strategy called *scaled partial pivoting* is simply to choose as the first equation that which has the larger absolute ratio of first to second coefficient. For example we should let -5 be "a" for the following two simultaneous equations: $6x - 7800y = 15, -5x + 12y = 450$.

▼5. Write a subroutine to be used in the form AAS(A, B, X, C, Y, Z) which calculates the other angle and sides of a triangle, given two angles in degrees and the side opposite the first—A, B, and X.

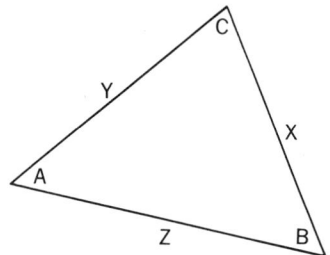

▼6. Write a main procedure which repetitively gets two angles of a triangle and the side opposite the first, uses AAS from the previous exercise to calculate the other angle and sides, and puts out all six quantities.

7. Write a subroutine to be used in the form ASA(A, Z, B, X, C, Y) which calculates the other angle and sides of a triangle, given two angles in degrees and the included side—A, B, and Z.

8. Write a main procedure which repetitively gets two angles and the included side of a triangle, uses ASA from the previous exercise to calculate the other angle and sides, and puts out all six quantities.

▼9. Write a subroutine to be used in the form SSS(X, Y, Z, A, B, C) which calculates the three angles, in degrees, of a triangle, given the three sides. You may wish to make use of the ACOSD procedure written in exercise 12 in Case Study 2.

▼10. Write a main procedure which repetitively gets three sides of a triangle, uses SSS from the previous exercise to calculate the three angles, and puts out all quantities.

11. Write a subroutine to be used in the form SAS(X, C, Y, A, Z, B) which calculates the other side and angles of a triangle, given two sides and their included angle in degrees—X, Y, and C. (Avoid the temptation to use the law of sines together with the identity $\arcsin(q) = \arctan(q, \sqrt{1-q^2})$ because the arcsin function cannot distinguish between angles in the first and second quadrants.)

12. Write a main procedure which repetitively gets two sides and the included angle of a triangle, uses SAS from the previous exercise to calculate the sides and angles, and puts out all six quantities.

▼13. (Difficult) Write a subroutine to be used in the form SSA(X, Y, A, B1, C1, Z1, B2, C2, Z2) which calculates the other side and angles of a triangle, given two sides and the angle opposite the first in degrees—X, Y, and A. There are two solutions if X is less than or equal to Y. Otherwise, make the second solution equal to the first. You may wish to make use of the procedure QUAD from Sec. 5.1 or the identity $\arcsin(q) = \arctan(q, \sqrt{1-q^2})$.

▼14. (Difficult) Write a main procedure which uses AAS, ASA, SSS, SAS, and SSA from the previous exercises to solve any number of triangles of any type. The data for each triangle will consist of an integer 1, 2, 3, 4, or 5, indicating which type of triangle it is, followed by the three known quantities for the triangle.

15. The normalized cubic equation $x^3 + Dx^2 + Ex + F = 0$ has the roots $x_1 = G + H - D/3$, $x_2 = -(G+H)/2 - D/3 + i(G-H)\sqrt{3}/2$, and $x_3 = -(G+H)/2 - D/3 - i(G-H)\sqrt{3}/2$ where $G = (R + \sqrt{Q^3 + R^2})^{1/3}$ and $H = (R - \sqrt{Q^3 + R^2})^{1/3}$ with $R = (ED/3 - F)/2 - (D/3)^3$ and $Q = E/3 - (D/3)^2$. Write a subroutine to be used in the form CUBIC(D, E, F, RT1, RT2RL, RT2IM, RT3RL, RT3IM). (This method will not work when $Q^3 + R^2$ is negative, which corresponds to the case of three distinct real roots; so iterative methods such as those to be discussed in Case Study 7 are generally used for cubic equations.)

16. (Difficult) The normalized quartic equation $x^4 + Px^3 + Sx^2 + Tx + W = 0$ has roots equal to those of the two quadratic equations $v^2 + [P/2 \pm \sqrt{(P/2)^2 + u - S}]v + u/2 \pm [(u/2)^2 - W] = 0$ where u is the real root of the equation $u^3 - Su^2 + (PT - 4W)u - (T^2 + WP^2 - 4WS) = 0$. Write a subroutine to be used in the form QUARTC(P, S, T, W, RT1RL, RT1IM, RT2RL, RT2IM, RT3RL, RT3IM, RT4RL, RT4IM). Use CUBIC from the previous exercise and QUAD from Sec. 5.1 to solve the auxiliary equations. (This method will not work if the auxiliary cubic has three distinct real roots or if $[(P/2)^2 + u - S]$ is negative; so iterative methods are generally used for quartic equations. Iterative methods *must* be used for fifth and higher-order polynomials unless some of the terms are missing.)

17. The sixth-order polynomial $x^6 + 5.783x^3 - 13.75 = 0$ may be solved by QUAD from Sec. 5.1 because it is quadratic in x^3. Write a quickie-type program which uses QUAD to solve this equation.

▼18. The translated product and moments of inertia of an area A with its centroid located at coordinates (x, y) are given by the formulas $F'_{xy} = F_{xy} + Axy$, $F'_{xx} = F_{xx} + Ax^2$, and $F'_{yy} = F_{yy} + Ay^2$ where F_{xy}, F_{xx}, and F_{yy} are the centroidal product and moments of inertia. Write a subroutine to be used in the form TRNSLT(X, Y, A, FXY, FXX, FYY) where FXY, FXX, and FYY are given-result parameters which TRNSLT modifies from the centroidal to the translated product and moments of inertia.

19. The rotated product and moments of inertia of an area which is rotated counterclockwise through an angle α are given by the formulas $F'_{xy} = D\sin(2\alpha) + F_{xy}\cos(2\alpha)$, $F'_{xx} = S + D\cos(2\alpha) - F_{xy}\sin(2\alpha)$, and $F'_{yy} = S - D\cos(2\alpha) + F_{xy}\sin(2\alpha)$ where $S = (F_{xx} + F_{yy})/2$ and $D = (F_{xx} - F_{yy})/2$ with F_{xy}, F_{xx}, and F_{yy} being the unrotated product and moments of inertia. Write a subroutine to be used in the form ROTATE(ALPHA, FXY, FXX, FYY) where FXY, FXX, and FYY are given-result parameters which ROTATE modifies from the unrotated to the rotated product and moments of inertia.

▼20. (Difficult) Write a main procedure which uses ROTATE and TRNSLT from exercises 18 and 19 to find the translated, rotated product and moments of inertia of any number of areas. The data for each area will consist of the area, the orientation angle, the coordinates of the centroid, the product of inertia, and the moments of inertia.

21. The principal angle and principal stresses for a state of plane stress are given by the formulas

$$\theta = \arctan[2\tau_{xy}, (\sigma_x - \sigma_y)]$$

and

$$\sigma_1, \sigma_2 = \frac{\sigma_x + \sigma_y}{2} \pm \sqrt{\left(\frac{\sigma_y - \sigma_x}{2}\right)^2 + \tau_{xy}^2}$$

where τ_{xy}, σ_x, and σ_y are the shear stress and the two normal stresses respectively. Write a subroutine PRNCPL which calculates θ, σ_1, and σ_2 from τ_{xy}, σ_x, and σ_y.

22. Write a main procedure which repetitively gets τ_{xy}, σ_x, and σ_y; uses PRNCPL from the previous exercise to calculate θ, σ_1 and σ_2; then puts out all six quantities.

▼23. Instead of the fact that the product of the roots is C, why not make use of the fact in QUAD of Sec. 5.1 that the sum of the roots is −B, since addition and subtraction are generally faster than division? Also, is it possible for QUAD to raise the ZERODIVIDE condition?

24. There is still room for improvement in our subroutine QUAD. Read Forsythe's article, then write a better version of QUAD.[6]

---

[6] George E. Forsythe, "Pitfalls in Computation, or Why a Math Book Isn't Enough," *American Math Monthly*, vol. 77, 1970, pp. 931–56.

CASE STUDY 6

DATA REDUCTION

The ON ENDFILE Statement, Double Precision, and the ADD, MULTIPLY, DIVIDE, and PRECISION Functions

Experiments frequently generate so much data that they are difficult to interpret. An objective of data reduction or statistics is to reduce these data to a few simple numbers that reveal the essence of the experimental results. For example, a psychologist might have to measure the test response of a dozen individuals in order to calculate a statistically accurate enough average. Very few people would be interested in the 12 individual responses. Statistics such as the average are more meaningful because the 12 individual numbers contain too much information to comprehend at one time.

The arithmetic mean or average of n values is

$$\bar{X} = (X_1 + X_2 + X_3 + \cdots X_n)/N = \left(\sum_{K=1}^{N} X_K\right)/N$$

where the Greek capital sigma is shorthand for "sum for K equals 1 to N." The following flow chart and procedure indicate the *brute force* approach to programming this formula for the specific case when $N = 12$:

```
get in x₁, x₂, x₃, x₄, x₅, x₆, x₇, x₈, x₉, x₁₀, x₁₁, x₁₂
                        ↓
x̄ = (x₁ + x₂ + x₃ + x₄ + x₅ + x₆ + x₇ + x₈ + x₉ + x₁₀ + x₁₁ + x₁₂)/12
                        ↓
                   put out x̄
```

```
MEAN:   PROCEDURE OPTIONS (MAIN);
   GET LIST (X1, X2, X3, X4, X5, X6, X7, X8, X9, X10, X11, X12);
   AVG = (X1+X2+X3+X4+X5+X6+X7+X8+X9+X10+X11+X12)/12;
   PUT PAGE DATA (AVG);
END;
```

The following flow chart and procedure indicate a better approach that accommodates any value of **N** and avoids the tedium resulting from using a distinct name for each value in the average:

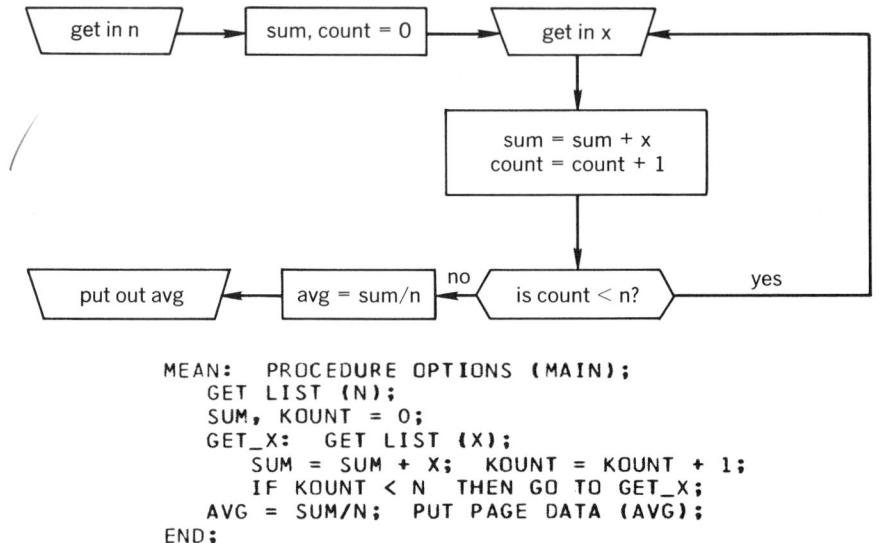

```
MEAN:   PROCEDURE OPTIONS (MAIN);
   GET LIST (N);
   SUM, KOUNT = 0;
GET_X:   GET LIST (X);
      SUM = SUM + X;   KOUNT = KOUNT + 1;
      IF KOUNT < N THEN GO TO GET_X;
   AVG = SUM/N;   PUT PAGE DATA (AVG);
END;
```

The trick is to build up the numerator of the average term-by-term in a loop. Each time through the loop, the value of the most recently gotten value of **X** is added to the numerator. It is important to note that **SUM** must be initialized to zero before entering the loop. Otherwise, **SUM** would have an undetermined value when needed for the first evaluation of SUM + X. The computer does not automatically initialize variables to zero. The initial value is whatever is left in the storage location by the previous program.

In this program we cannot use the trick of terminating by the standard **END-FILE** action because there is more to be done after the looping is completed. Consequently we have set up a counter, increased it by one each time through the loop, and checked its value against **N**. This *counter* technique allows us to terminate a loop without terminating execution. In the program, counter is spelled with a K in order to make it fixed-point the lazy way. It would be inefficient to use floating-point for the counter since counting involves only integers.

The difference in the amount of typing between the counter and the brute force techniques becomes more dramatic as the number of values increases. More important is the fact that the brute force approach is inflexible—it must be rewritten each time a different number of values is to be averaged.

6.1 THE ON ENDFILE STATEMENT[1]

If there are a great number of values, human fallibility makes it easy to miscount them and thus provide an incorrect value for N. The following flow chart and procedure use the ON statement to make the computer calculate N, thus avoiding the possibility of human miscount:

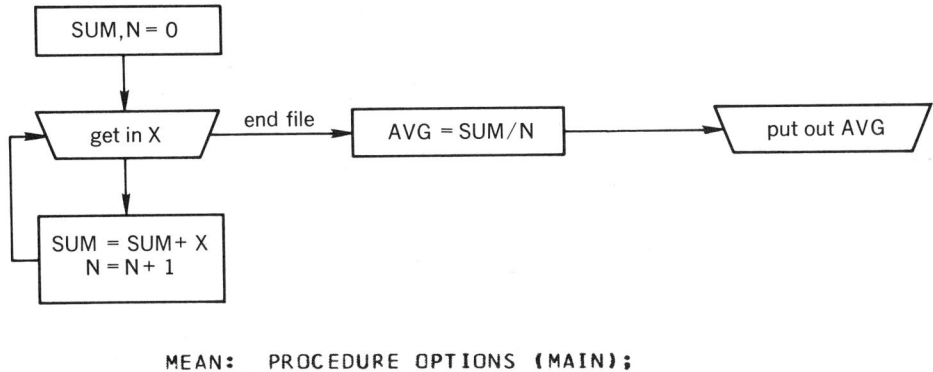

```
MEAN:  PROCEDURE OPTIONS (MAIN);
       SUM, N = 0;
       ON ENDFILE(SYSIN)  GO TO S10;
  S5:  GET LIST (X);
       SUM = SUM + X;  N = N + 1;
       GO TO S5;
 S10:  AVG = SUM/N;  PUT PAGE DATA (AVG);
       END;
```

In general there may be many files residing on magnetic tape, cards, or elsewhere. SYSIN stands for the standard *SYStem INput* file.[2] The ON statement is merely *enabled* when it is reached in the execution sequence. Its *on-unit* is executed only when the *condition* arises. In our example, the on-unit is "GO TO S10" and the condition is "ENDFILE (SYSIN)." More generally, the ON statement has the form

ON condition on-unit

Of the statements discussed so far, the on-unit may be a GO TO, GET, PUT, or assignment statement. The condition may be ENDFILE (SYSIN), ZERODIVIDE, OVERFLOW, UNDERFLOW, FIXEDOVERFLOW, or others that have not yet been discussed. In this case study, we will use only "ON ENDFILE (SYSIN)

---

[1] Unsupported by Student PL. A similar effect may be achieved by using an extra terminal *sentinel* data value with an absurd value such as −1E−70, then testing for it with an IF statement.

[2] This file must be explicitly declared for the Model 20 and D compilers with a statement such as DECLARE SYSIN FILE INPUT ENVIRONMENT (MEDIUM (SYSIPT, 2540) F(80)); When this is done, the GET statements should be written GET FILE (SYSIN) LIST (list of variables separated by commas) ;.

GO TO statement label;." Once enabled, an ON statement remains in effect until termination of the procedure in which it appears or until overridden by another ON statement with the same condition and a different on-unit. Consequently, one ON ENDFILE statement may apply to any number of subsequent GET statements.

6.2 LEAST-SQUARES EXAMPLE

As another example of the use of this ON ENDFILE technique, we will consider a program that finds the least-squares fit of a straight line to experimental data. Frequently, an experimental relation between two variables may be approximated as linear: Y = A · X + B. Naturally, the measured values of Y(X) will not usually satisfy this approximation exactly. However, the values of A and B for which the sum of the squared discrepancies is minimal are:

$$A = \frac{N \sum XY - \sum X \sum Y}{N \sum X^2 - (\sum X)^2} \qquad B = \frac{\sum Y - A \sum X}{N}$$

All of the sums run from 1 to N where there are N pairs of values (X, Y). The following procedure evaluates these formulas using the loop and ON ENDFILE technique to simultaneously form all of the sums:

```
LINE:    PROCEDURE OPTIONS (MAIN);
     ON ENDFILE(SYSIN)    GO TO FINISH;
     SUMXY, SUMX, SUMY, SUMXSQ, N = 0;
     IN:   GET LIST (X, Y);
        SUMXY = SUMXY + X*Y;
        SUMX = SUMX + X;
        SUMY = SUMY + Y;
        SUMXSQ = SUMXSQ + X*X;
        N = N + 1;
     GO TO IN;
     FINISH:   A = (N*SUMXY - SUMX*SUMY)/(N*SUMXSQ - SUMX*SUMX);
        B = (SUMY - A*SUMX)/N;
        PUT DATA (A, B);
END;
```

EXERCISES

1. The harmonic mean of n quantities is defined as

$$H = \frac{n}{\frac{1}{x_1} + \frac{1}{x_2} + \frac{1}{x_3} \cdots \frac{1}{x_n}} = \frac{n}{\sum_{k=1}^{n} \frac{1}{x_k}}$$

Exercises

Write a procedure using a counter that will get n followed by that many values of x, calculate H, and put out H.

▼2. Same as exercise 1 except use the ON ENDFILE technique.

3. The geometric mean is defined as

$$G = (a_1 a_2 a_3 \cdots a_n)^{1/n} = \left[\prod_{k=1}^{n} a_k\right]^{1/n}$$

(Big π is used for product notation just like Σ is used for summation notation.) Write a procedure using the ON ENDFILE technique to calculate the geometric mean.

▼4. (Mathematically Advanced). For a function $f(x)$, the "f mean" (e.g., fifth power mean or exponential mean) is defined as

$$f^{-1}\left[\frac{1}{n}\sum_{k=1}^{n} f(x_k)\right]$$

where f^{-1} denotes the inverse of the f function. Write a program using the ON ENDFILE technique to evaluate the arctangential mean,

$$\text{TAN}\left[\frac{1}{n}\sum_{k=1}^{n} \text{ATAN}(x_k)\right]$$

From this perspective, what are the arithmetic, harmonic, and geometric means?

5. If a body with weight W_j has its center of gravity at a point given by the coordinates x_j, y_j, and z_j, then the combined center of gravity of n such bodies in given by the formulas

$$\bar{x} = \frac{\sum_{j=1}^{n} W_j x_j}{W_{\text{total}}} \qquad \bar{y} = \frac{\sum_{j=1}^{n} W_j y_j}{W_{\text{total}}} \qquad \bar{z} = \frac{\sum_{j=1}^{n} W_j z_j}{W_{\text{total}}}$$

where

$$W_{\text{total}} = \sum_{j=1}^{n} W_j$$

Write a procedure similar to LINE that calculates $\bar{x}$, $\bar{y}$, and $\bar{z}$ from data consisting of x, y, z, and W for each body in turn.

▼6. The area beneath an empirical curve may be approximated by a sum of rectangular strips as is shown below. Write a program that will calculate the total area under a curve from data which consist of the height and width of the first strip followed by those of the second, and so on. This is called *numerical integration*.

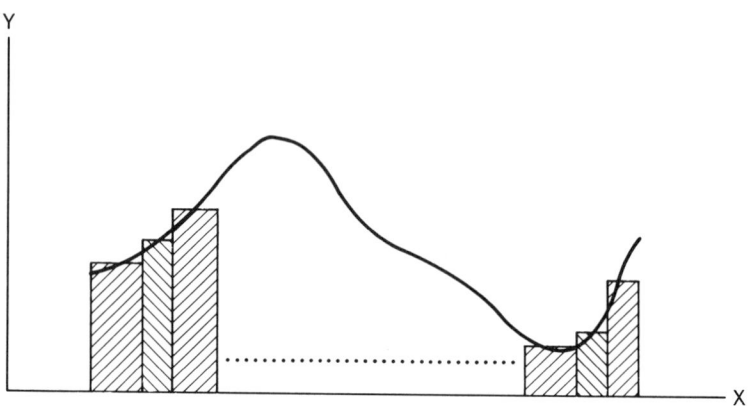

6.3 CHOPOFF ERROR

Let us now consider the question of accuracy in floating-point arithmetic. Throughout this discussion, we will presume the computer stores its floating-point numbers with a decimal mantissa of length two. This will make it easy for you to verify the results—which you *must* do to achieve full understanding. The results in this section must be regarded as qualitative only, but similar results pertain for mantissas of any length and base. Most computers *chopoff* rather than roundoff numbers. For example, the result of **22E1 + 66E0** will be **28E1** rather than **29E1**.[3] It is more economical to design hardware which supplies a few extra digits than it is to design hardware which rounds. Also, both chopoff and roundoff are less serious with low base number systems such as binary.

We have already seen how arithmetic subtraction of two nearly equal numbers can cause a large magnification of the relative experimental error.[4] We have also seen how such subtractions may sometimes be avoided by use of a different formula. Sometimes a theoretically less exact formula is computationally more exact. For example, the rise of a pendulum is proportional to $1 - \cos \theta$ where θ is the

---

[3] Many authors call this *truncation*, but we will avoid this terminology because PL/I makes special use of it for chopping to an integer and because it is also used for the error due to finite termination of a theoretically infinite process. Other authors use the term *roundoff* in a generalized sense which includes chopoff.

[4] In this section we mean *arithmetic* subtraction unless stated otherwise. The algebraic addition $A + B$ is an arithmetic subtraction if A and B have opposite signs.

angular displacement from the center. However, for small angles $\cos\theta$ is nearly 1, causing a bad subtraction. Consequently, the small-angle approximation to this expression, $\theta^2/2$, turns out to be computationally more accurate for angles as large as 0.025 radians (1.5 degrees) with six-digit mantissas.

Bad subtractions may often be eliminated or at least reduced in severity by a mere rearrangement of a formula. For example, 68E0 **2 − 66E0 **2 is more accurately evaluated as (68E0 − 66E0)*(68E0 + 66E0). The first form gives 30E1 with chopoff and 20E1 with roundoff. The factored form gives 26E1 which is much closer to the exact value of 26.8E1. The factored form is generally faster, too, because most computers add faster than they multiply. The better accuracy is due to performing the bad subtraction before the accuracy has been degraded by intermediate calculations. The general rule is to *perform unavoidable bad subtractions early*. As another example of this rule, (72E0 + 49E0) − 43E0 gives 70E0 whereas 72E0 + (49E0 − 43E0) gives the exact value of 78E0. As yet another example, 1/80E0 − 1/85E0 is more accurately evaluated as (85E0 − 80E0)/(80E0 * 85E0). The first form given 10E−4 whereas the second gives 73E−5 with chopoff or 74E−5 with roundoff—both of which are much closer to the exact value of 73.5⋯E−5. The second form is faster, too, for computers which multiply faster than they divide.

When a bad subtraction is unavoidable and we have arranged the computation to minimize the loss of accuracy, we may take comfort in the fact that the remaining error is often indicative of how erratic the physical system will be. In other words, unavoidable computational inaccuracy often reflects a physically inaccurate system.

The other arithmetic operations are not so problematic as subtraction. A loss of significance with addition is usually harmless unless it results from a poor choice of reference as is discussed in Case Study 3. For example, 48E2 + 36E0 gives 48E2. The smaller term has no effect on the result, but this fact usually properly reflects the fact that the second term is physically negligible. The sum still has essentially as many significant digits as did the constituent 48E2.

Addition usually results in an error of no more that ± 1 in the last digit. The same is true of multiplication and division. However, a series of additions, multiplications, or divisions may cause a large cumulative error even though a single operation is rarely troublesome. Since this case study deals with summation, we will use that as an example. With summation, the least significant digits of the constituents cease to participate as the subtotal grows in comparison to the constituents. Eventually, the subtotal may become relatively so large that none of the constituent's digits participate. The subtotal may remain unchanged from that point on. The second two columns in Fig. 6-1 illustrate this with a series of

ADDITION OF UNIFORM RANDOM NUMBERS BETWEEN 00 AND 99			ADDITION OF SAME NUMBERS REARRANGED INTO ASCENDING ORDER		
exact	roundoff	chopoff	exact	roundoff	chopoff
68E0	68E0	68E0	03E0	03E0	03E0
+ 18E0	+ 18E0	+ 18E0	+ 04E0	+ 04E0	+ 04E0
86E0	86E0	86E0	07E0	07E0	07E0
+ 14E0	+ 14E0	+ 14E0	+ 11E0	+ 11E0	+ 11E0
100E0 →	10 E1 →	10 E1	18E0	18E0	18E0
+ 36E0	36E0	36E0	+ 14E0	+ 14E0	+ 14E0
136E0	14 E1	13 E1	32E0	32E0	32E0
+ 88E0	+ 88E0	+ 88E0	+ 18E0	+ 18E0	+ 18E0
224E0	23 E1	21 E1	50E0	50E0	50E0
+ 11E0	+ 11E0	+ 11E0	+ 21E0	+ 21E0	+ 21E0
235E0	24 E1	22 E1	71E0	71E0	71E0
+ 87E0	+ 87E0	+ 87E0	+ 28E0	+ 28E0	+ 28E0
322E0	33 E1	30 E1	99E0	99E0	99E0
+ 92E0	+ 92E0	+ 92E0	+ 32E0	+ 32E0	+ 32E0
414E0	42 E1	39 E1	131E0 →	13 E1 →	13 E1
+ 92E0	+ 92E0	+ 92E0	+ 33E0	+ 33E0	+ 33E0
506E0	51 E1	48 E1	164E0	16 E1	16 E1
+ 88E0	+ 88E0	+ 88E0	+ 36E0	+ 36E0	+ 36E0
594E0	60 E1	56 E1	200E0	20 E1	19 E1
+ 32E0	+ 32E0	+ 32E0	+ 40E0	+ 40E0	+ 40E0
626E0	63 E1	59 E1	240E0	24 E1	23 E1
+ 99E0	+ 99E0	+ 99E0	+ 45E0	+ 45E0	+ 45E0
725E0	73 E1	68 E1	285E0	29 E1	27 E1
+ 04E0	+ 04E0	+ 04E0	+ 56E0	+ 56E0	+ 56E0
729E0	73 E1	68 E1	341E0	35 E1	32 E1
+ 21E0	+ 21E0	+ 21E0	+ 62E0	+ 62E0	+ 62E0
750E0	75 E1	70 E1	403E0	41 E1	38 E1
+ 66E0	+ 56E0	+ 56E0	+ 68E0	+ 68E0	+ 68E0
806E0	81 E1	75 E1	471E0	48 E1	44 E1
+ 62E0	+ 62E0	+ 62E0	+ 73E0	+ 73E0	+ 73E0
868E0	87 E1	81 E1	544E0	55 E0	51 E0
+ 45E0	+ 45E0	+ 45E0	+ 87E0	+ 87E0	+ 87E0
913E0	92 E0	85 E1	631E0	64 E1	59 E1
+ 28E0	+ 28E0	+ 28E0	+ 88E0	+ 88E0	+ 88E0
941E0	95 E1	87 E1	719E0	73 E1	67 E1
+ 33E0	+ 33E0	+ 33E0	+ 88E0	+ 88E0	+ 88E0
974E0	98 E1	90 E1	807E0	82 E1	75 E1
+ 03E0	+ 03E0	+ 03E0	+ 92E0	+ 92E0	+ 92E0
977E0	98 E1	90 E1	899E0	91 E1	84 E1
+ 73E0	+ 73E0	+ 73E0	+ 92E0	+ 92E0	+ 92E0
1050E0 →	11 E2	97 E1	991E0 →	10 E2	93 E1
+ 40E0	+ 40E0	+ 40E0	+ 99E0	+ 99E0	+ 99E0
1090E0	11 E2 →	10 E2	1090E0	11 E2 →	10 E2

Figure 6-1 Chopoff and roundoff errors for addition

randomly chosen two-digit floating-point numbers in the interval 00 to 99. (The constituents are directly above the lines and the subtotals are directly below them.) It is not necessary to check all of these additions, but notice the following features: With roundoff, the sum is sometimes too large and sometimes too small. With chopoff the sum is consistently too low after about two additions, and it is usually less accurate. With both techniques, one digit ceases to participate after the addition of about two numbers, but with roundoff it still participates in a statistical sense according to whether it is greater than four or not. Neither digit participates after the addition of about 20 numbers although with roundoff the second digit continues to participate in a statistical sense until the addition of about 200 numbers. These results hold true for whole numbers even if the data are converted internally to binary. This means that for M-digit arithmetic, we may add up an average of $2 \times 10^{N-M}$ uniformly distributed N-digit numbers of one sign before incurring any absolute chopoff or roundoff error. We may add considerably more than these without incurring appreciable *relative* error. If all of the mantissa's digits are generally nonzero, it turns out that the relative chopoff error for the addition of N numbers with a decimal mantissa of length L averages about $0.664 \times 10^{-L}N$. For binary mantissa of length equivalent to L decimal digits, the relative chopoff error averages about $0.17 \times 10^{-L}N$.[5] Similar formulas apply for roundoff but with N replaced by $\sqrt{N}$. Whichever formula is applicable should be compared with the experimental errors of the constituents to see if they are tolerable.

For example, with a mantissa of binary length equivalent to six decimal digits, the addition of 100 uniformly distributed positive numbers will give an average relative chopoff error of $0.17 \times 10^{-6} \times 100$. If the mantissa is true decimal, the average relative chopoff error would be $0.664 \times 10^{-6} \times 100$. Either way, this would certainly be acceptable if the constituents had a relative experimental error of 10^{-4}. In fact, it might be acceptable even if constituents had a greater relative experimental error. We may not require an answer as accurate as our data. Since the chopoff error is of predictable sign when the constituents are all of one sign, for utmost accuracy we could consider adjusting for its average value after completing the summation. Theoretically, this would reduce the average chopoff error to the values given for roundoff. However, these formulas are just crude approximations; so it would be wise to make experiments with a compiler before using them in this way.

With numbers equally distributed between positive and negative, the "random walk" algebraic sum *grows more slowly* on the average. It turns out that on the average it takes four such additions before the first digit ceases to participate, 400 such additions before the second digit ceases to participate, 40,000 such additions before the third digit ceases to participate, and so on. The average relative chopoff or roundoff error is similar to that given for roundoff with numbers of one

---

[5] The D, F, and Student PL compilers use hexadecimal, which is less accurate than binary or decimal. An appropriate formula is approximately $10^{-L}N$.

sign. Chopoff compares more favorably with roundoff for constituents of both signs. Most situations will lie between the extremes of all one sign and both signs equi-probably. However, because of the subtractions when constituents have both signs, the relative error in the algebraic sum due to *experimental* errors will be quite large whenever the sum is near zero.

The situation is not as severe when the numbers with smallest magnitude are combined first. This postpones the growth of the subtotal and thus enables more small numbers to participate. The numbers from the previous example have been rearranged in ascending order in columns 4, 5, and 6 of Fig. 6-1 to illustrate this principle. Both digits participate until after the seventh addition rather than merely the second. Of course, it will not always be possible or practical to know the proper order in advance. When practical, it is also helpful to combine the constituents into several subtotals, combining the subtotals at the end.

6.4 DOUBLE PRECISION[6]

When default precision is not sufficient for our variables—either immediately, by virtue of a bad subtraction, or by virtue of cumulative chopoff error, we may declare greater precision for these variables—e.g., FLOAT(16). Most implementations have only two lengths for floating-point numbers—*single precision* and *double precision*. All constants not exceeding single precision and all variables with declarations not exceeding single precision are stored as single precision. Longer constants and variables with longer declarations are stored as double precision. When this is the case, it is sensible to standardize on default length and whatever is double precision. For example, single precision is the equivalent of six decimal digits and double precision is the equivalent of 16 decimal digits for the D and F compilers; so in this text we shall standardize on the two floating-point declarations FLOAT and FLOAT(16).[7] The reason double precision is more than twice single precision is because no part of the second word of storage need be devoted to a sign or exponent.

As an example of how to use double precision, assume that we are building up a sum, SUM = SUM + X; and default precision adequately represents the data, but some extra digits are needed due to the great number of additions. We may DECLARE SUM FLOAT(16). There is no need to also declare X to be double precision because when floating-point quantities of differing precisions are combined, the precision of the result is the greater of the two precisions. However, it saves the time of a conversion to also declare X to be double precision.

[6] Unsupported by Student PL.
[7] SL1 uses 3, 7, 11, or 15 digits with 7 being default. The Model 20 uses 6 or 15 digits.

6.5 THE ADD, MULTIPLY, DIVIDE, AND PRECISION BUILT-IN FUNCTIONS[8]

The built-in arithmetic functions ADD, MULTIPLY, DIVIDE, and PRECISION may be used to control precision within an expression. For example, if we have a bad subtraction "X = A*B − C*D ;," we may improve the accuracy without making A, B, C, or D double precision by using "X = MULTIPLY(A, B, 16) − MULTIPLY(C, D, 16) ;." The result of each product is 16 digits rather than default precision. However, it is more efficient to declare double precision for A, B, C, and D if a high proportion of the expressions involving them require double precision.

We may use ADD(A, −B, N) to subtract floating-point variable B from A, chopping the result to N digits. However, note that a double-precision subtraction alone does not usually improve accuracy: ADD(1.23456E0, − 1.23452E0, 16) is 4.000000000000000E−5 with only one accurate digit. The same is true even if A and B are declared to be double precision unless the two constituents are known or can be computed to more than single precision. Double precision is usually of help only when other operations *precede* a bad subtraction. Double precision is no panacea. Before blindly using it, we should carefully consider whether or not it will help and whether or not it is needed.

We may use DIVIDE(A, B, N) to divide floating-point variable A by B, chopping the result to N digits. For example, we may wish to evaluate the continued fraction

$$X - \frac{5}{\dfrac{4}{X+3}}$$

as 5/(X − DIVIDE(4, (X + 3), 16)) if it is to be used in the neighborhood of X = 4/(X + 3).

We may use PRECISION(X, N) to cause a copy of X with length N to be used. For example, if single precision is generally adequate for A and B, but at one point we have the bad subtraction SIN(A) − SIN(B), we can cause the double-precision versions of these mathematical functions to be used by the expression SIN(PRECISION(A, 16)) − SIN(PRECISION(B, 16)).

If ADD, MULTIPLY, DIVIDE, or PRECISION are used with fixed-point arguments, an extra *scale factor* argument must be listed. For example, if K and L are default fixed-point binary integers, DIVIDE(K, L, 15, 0) may be used to get a result which is truncated to an integer. (Implementations which do not have non-integer fixed-point capability may always perform fixed-point division this way.)

---

[8] Student PL, the Model 20, and SL1 do not support these functions.

Some other practical advice about accuracy is to avoid decimal fractions which cannot be represented with a finite number of digits. For example, it is more accurate to use Q/10 than .1*Q unless the implementation has true decimal floating-point. Q may, after all, be evenly divisible by 10. Q/3 is even more preferable to .333333*Q because the latter involves an additional error in the *decimal* representation. (Incidentally, with a six-digit single precision computer, 0.333333*Q would unintentionally slow things down by causing double precision.) However, .5*Q is usually preferable to A/2 because it is faster on most computers, and .5 has an exact representation in both decimal and binary. Halves, fourths, eighths, sixteenths, thirty-seconds, etc. have exact representations in both decimal and binary. This is about the only good thing that can be said about the fractional inch system.

6.6 OPTIONAL MATHEMATICALLY ADVANCED APPLICATION: NUMERICAL INTEGRATION

Many physical problems reduce to that of finding the area under a mathematical curve. Integral calculus is directed toward finding finite expressions for these areas in terms of the mathematical functions which are built into PL/I. However, there are no such formulas for many curves—for example, $\sin(x)/x$, $x/\tan(x)$, e^x/x, $x^3/(e^x - 1)$, $\log(x)/(x - 1)$, and $\log(\sin(x))$. In such cases we may use the concept introduced in exercise 6 of dividing the area into numerous narrow rectangular strips. We may take the height of the curve at the middle of each strip as the height of each rectangle. More specifically, to approximate the area under the function $y(x)$ between $x = a$ and $x = b$, we may divide the area into n strips of equal width $\Delta x = (b - a)/n$; we may calculate the heights of the rectangles $y(a + 0.5\Delta x)$, $y(a + 1.5\Delta x)$, $y(a + 2.5\Delta x)$, ..., $y[a + (n - 0.5)\Delta x]$; and we may sum up the areas of the rectangles:

$$\text{area} = \Delta x \cdot \{y(a + 0.5\Delta x) + y(a + 1.5\Delta x) + \ldots y[a + (n - 0.5)\Delta x]\}$$

This technique is called the *midvalue* rule for numerical integration. The following function procedure uses it to find the area between A and B under any curve given by a function procedure Y(X):

```
AREA:   PROCEDURE OPTIONS (MAIN);
        DX = (B - A)/N;    FIRST = A + .5*DX;
        SUM, K = 0;
ADD:    SUM = SUM + Y(FIRST + K*DX);
        K = K + 1;
        IF K < N   THEN GO TO ADD;
        RETURN (SUM*DX);
END;
```

Although it might be faster to calculate X by adding DX each time, terminating with a test against B, this method might not loop the proper number of times due

Sec. 6.7   Optional Mathematically Advanced Application: Differential Equations   107

to fraction anomaly or cumulative chopoff error. There can be no such error with the test of integer K against integer N.

There will usually be some *discretization* error, often called *truncation* error, due to approximating the area under a smooth curve by that of a discrete number of rectangles. This error may be made arbitrarily small by taking enough rectangles, but the chopoff error (and the computation time) increases with the number of rectangles. It turns out that for best accuracy we should compromise at the number of strips for which the two kinds of error are about equal. If $y'(a)$ is the slope of the curve at a, and $y'(b)$ is the slope of the curve at b, an estimate for the absolute discretization error is

$$\frac{(b-a)^2[y'(b)-y'(a)]}{24n^2}$$

We may set this equal to whatever formula is applicable for the absolute chopoff or roundoff error and solve for n. However, these formulas are based on so many assumptions that it is better to choose n on the basis of experience with a particular compiler. This experience is worth a great many fancy formulas, and it may be gained by trying out various numbers of strips on a variety of functions with known integrals.

As an example of numerical integration, the following two procedures may be used with AREA and input data to evaluate the area under the curve e^x/x between $x = 0.135$ and $x = 3.76$ using 200 strips:

```
Y:     PROCEDURE (X);
          RETURN (EXP(X)/X);
       END;

INTEG: PROCEDURE OPTIONS (MAIN);
          GET LIST (A, B, N);
          PUT LIST (AREA(A, B, N));
       END;
```

We may even use AREA for empirical functions by using

```
Y:     PROCEDURE (X);
          GET LIST (VALUE);
          RETURN (VALUE);
       END;
```

6.7 OPTIONAL MATHEMATICALLY ADVANCED APPLICATION: DIFFERENTIAL EQUATIONS

When a first-order ordinary differential equation with the initial condition $y(t_0) = y_0$ can be put in the form $g(y)\,dy = h(t)\,dt$, we may solve it by the techniques of the previous section if g and h are not analytically integrable. However, often the best we can do is to put the equation in the form

$$\frac{dy}{dt} = f(y, t)$$

In this case we may use *Euler's method* to approximate the solution numerically. Provided t is small enough, we may extrapolate to say that, approximately, $y(t_0 + \Delta t) = y_0 + f(y_0, t_0) \cdot \Delta t$. We may continue in this fashion, getting the solution at each time step from the solution at the previous time step according to the general relationship: $y(t_{\text{next}}) = y(t_{\text{last}}) + f(y_{\text{last}}, t_{\text{last}}) \cdot \Delta t$.

The following procedure uses this method to solve any differential equation up to $t = $ TFINAL where $f(y, t)$ is given by a function procedure F(Y, T):

```
EULER:  PROCEDURE OPTIONS (MAIN);
        /* THIS PROCEDURE SOLVES THE DIFFERENTIAL EQUATION DY/DX = F(Y,T)
           BY EULER'S METHOD, PUTTING OUT T AND Y UNTIL T EXCEEDS TFINAL.
           FOR DATA, GIVE THE INITIAL VALUE OF T AND Y FOLLOWED BY THE TIME
           INCREMENT AND TFINAL.  */
        GET LIST (T, Y, DT, TFINAL);
        PUT PAGE DATA (T, Y);
STEP:   Y = Y + F(Y, T)*DT;
        T = T + DT;
        PUT SKIP LIST (T, Y);
        IF T < TFINAL  THEN GO TO STEP;
END;
```

As an example, the following function procedure may be used with EULER and input data to solve the differential equation

$$\frac{dx}{dt} = \frac{\cos(t-x)}{\log(t+x)}$$

for the initial condition $x(1.32) = 7.85$ with steps of $t = 0.1$ until $t \geq 3.8$:

```
        F:  PROCEDURE (T, X);
            RETURN (COS(T-X)/LOG(T+X));
            END;
```

SUMMARY

Typical Procedure:

```
            procedure name:   PROCEDURE OPTIONS (MAIN);
                ON ENDFILE (SYSIN) GO TO FINISH;
                SUM = 0;
            AGAIN:   GET LIST (list of variables separated by commas);
                SUM = SUM + expression;
                GO TO AGAIN;
            FINISH:  variable = expression;
                PUT DATA (list of variables separated by commas);
            END;
```

On Statement

ON condition on-unit is *enabled* when reached in the execution sequence. It is *executed* when the condition arises. Among the conditions are ENDFILE (SYSIN), UNDERFLOW, ZERODIVIDE, OVERFLOW, and FIXEDOVERFLOW. Among the on-units are a GO TO, GET, PUT, or assignment statement. Once enabled, an ON statement remains in effect throughout the procedure (and any subsequently invoked procedure) unless overridden by an ON statement with the same condition and a different on-unit.

Chopoff Error

Chopoff error may be minimized by

1. avoiding formulas with bad subtractions
2. performing unavoidable subtractions early
3. combining small numbers first
4. avoiding fractions which have inexact binary or decimal representations

When extra precision is necessary we may declare extra precision with the FLOAT (length) attribute and/or use the function ADD(a, b, length), MULTIPLY(a, b, length), DIVIDE(a, b, length), or PRECISION(a, length).

EXERCISES

7. Rearrange the following expressions to minimize the error:
    (a) $x^4 - y^4$
    ▼(b) $cu - du - dv + cv$
    (c) $\dfrac{1}{\dfrac{1}{a} - \dfrac{1}{b}}$
    (d) $8.7652E0 + 3.7956E4 - 9.11151E1$
    (e) $c(a + b) - a(c + b)$
    ▼(f) $0.1 \sin(90° - a)$
    (g) $a\left(\dfrac{p}{q} - 1\right) - a\left(\dfrac{p}{q} + 1\right)$
    (h) $\log(p) - \log(q)$

▼8. Is the subtraction in the discriminant of QUAD in Case Study 5 ever a serious problem? Why?

▼9. Rewrite the subroutine in exercise 4 in Case Study 5, using the built-in functions to perform the critical operations in double precision.

10. Rewrite the subroutine in exercise 4 in Case Study 5, declaring all of the variables to be double precision.

11. Check your solutions to the other exercises in Case Study 5 to see if their accuracy might be improved by a different choice of formulas.

12. Using default-length floating-point variables on your implementation, how many positive four-digit numbers would you expect to find the sum of, without any absolute chopoff error?

13. On your implementation, how many positive default-length floating variables would you expect to be able to add before accumulating a relative error of 10^{-3}?

14. (Mathematically Advanced) The relativistic kinetic energy, as a function of rest mass m, speed of light c, and momentum p, is given by the formula $mc^2[\sqrt{1 - p^2/(mc)^2} - 1]$. Use the binomial expansion

$$(1 - x)^n = 1 - nx + \frac{n(n-1)x^2}{2} - \frac{n(n-1)(n-2)x^3}{3 \cdot 2} + \cdots$$

to derive a computationally more accurate formula for p^2 small relative to $(mc)^2$.

▼15. Derive a formula for $1 - \cos(\theta)$ which is theoretically exact as well as computationally more accurate for small angles.

▼16. (Difficult) The area of the shaded triangle is given by the formula $0.5(x_1 y_2 - x_2 y_1)$, which is negative if point 2 is clockwise from point 1. Write a procedure which finds the area of a polygon given data consisting of the x and y coordinates of its first point, followed by those of its second point, then the third point, and so on in counterclockwise order, finishing with relisting of the coordinates of its first point. Will the procedure work for polygons which do not enclose the origin and/or have holes in them?

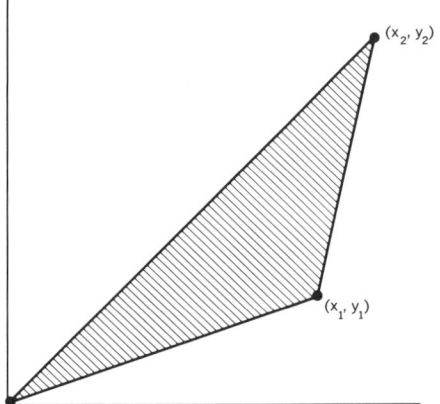

17. (Difficult) The angle subtended from point 1 to point 2 at point 0 is given by the formula ATAN(U1*V2 − U2*V1, U1*U2 + V1*V2), being negative if point 2 is clockwise of point 1, where U1 = X1 − X0, V1 = Y1 − Y0, U2 = X2 − X0, and V2 = Y2 − X0. Write a program which gets in X0 and Y0 followed by the x and y coordinates of the vertices of a polygon, in counterclockwise order, then determines whether or not the point X0, Y0 is in the polygon, and prints out the message INSIDE or OUTSIDE.

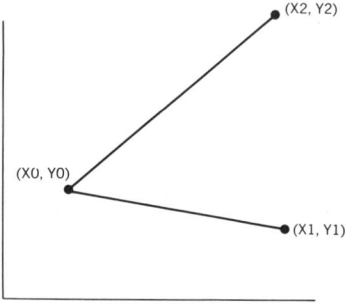

18. (Difficult) Same as exercise 17 except use the faster approximate formula for the angle:

$$\frac{U1*V2 - U2*V1}{(U1 + U2)^2 + (V1 + V2)^2}$$

19. (Mathematically Advanced) The *trapezoidal rule* is another numerical integration formula, and it has an error estimate equal to −2 times that of the midvalue rule.

The trapezoidal rule derives from approximating the area by n equal-width trapezoidal strips. The area of these strips is simply the width times the average height, and the average height is simply one-half the sum of the left and right ordinates. However, each ordinate with the exception of the very first and very last will participate in the areas of the two trapezoids; so the total area will be $\Delta x \cdot \{0.5y(a) + y(a+\Delta x) + y(a+2\Delta x) + \cdots y[a + (n-1)\Delta x] + 0.5y(b)\}$. Write a function procedure for the trapezoidal rule.

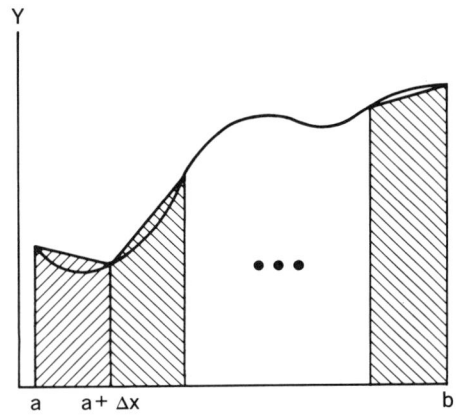

▼20. (Mathematically Advanced) Same as exercise 19 except use Simpson's rule which is derived from having equal-width parabolic-topped strips:

area = $(\Delta x/3)\{y(a) + 4y(a + \Delta x) + 2y(a + 2\Delta x) + 4y(a + 3\Delta x)$
$+ 2y(a + 4\Delta x) + \cdots 4y[a + (n-1)\Delta x] + y(b)\}$

Note that n must be odd and greater than or equal to 3. Simpson's rule is usually more accurate than the midvalue or trapezoidal rule for smooth functions. A bound for the discretization error is given by

$$\frac{-(b-a)^5 q}{180 n^4}$$

where q is the maximum value of the fourth derivative of y between a and b.

21. (Mathematically Advanced) The period of a pendulum of length l which swings through an angular amplitude θ is given by the formula

$$T = 4\sqrt{\frac{l}{g}} \int_0^{\pi/2} \frac{d\alpha}{\sqrt{1 - \sin^2\left(\frac{\theta}{2}\right) \sin^2 \alpha}}$$

where α is a dummy variable of integration in radians and g is the acceleration of gravity. This integral is called the complete elliptic integral of the first kind, and it cannot be symbolically integrated by even the cleverest manipulative tricks. (Try it and see!) Most physics textbooks give only the approximate formula

$$\tilde{T} = 2\pi \sqrt{\frac{l}{g}}$$

Write a program which evaluates and puts out the ratio of $T/\tilde{T}$ for $\theta = \pi/20$, $2\pi/20$, $3\pi/20 \cdots 19\pi/20$ using 200 steps with the midvalue rule. Note that the integrand is a function of two variables, and therefore the procedure from the text must be modified.

22. (Mathematically Advanced) Euler's method is the simplest; so it is a natural point to begin studying the numerical solution of differential equations. However, its

discretization error is too large for practical purposes, and the method is often unstable with small errors becoming greatly magnified. To overcome these disadvantages we may use the *improved* Euler method which uses the average of the slopes at the two end points of each interval rather than just the slope at the left end:

$$f_1 = f(y_{last}, t_{last})$$
$$f_2 = f(y_{last} + f_1 \cdot \Delta t, t_{last} + \Delta t)$$
$$y_{next} = y_{last} + 0.5(f_1 + f_2) \cdot \Delta t$$

Write a corresponding procedure for this improved Euler method.

▼23. Write a function procedure FACTRL (N) that evaluates

$$\text{factorial (N)} = (N)! = 1 \cdot 2 \cdot 3 \ldots N$$
$$\text{with } 0! = 1$$

For the F compiler, 8! exceeds the capacity of default-length fixed-binary variables; but for $N > 7$, we may use the approximation

$$N! \cong 2.50663 \exp((N + 0.5) \log_e (N) - N)\left(1 + \frac{1}{12N} + \frac{1}{288N^2} - \frac{139}{51840N^3}\right)$$

which has an error of less than one part per million.

▼24. (Mathematically Advanced) Hastings gives the following approximation for the gamma function:[9]

$$\Gamma(x) = 1 - 0.574865a + 0.951236a^2 - 0.699859a^3 + 0.42455a^4 - 0.101068a^5$$
$$\text{for } 0 \leq a = x - 1 \leq 1$$

The formula may be used for values of x outside this range by taking advantage of the fact that $\Gamma(x + 1) = x\Gamma(x)$. Write a corresponding function procedure GAMMA(X).

---

[9] Cecil Hastings, Jr., *Approximations for Digital Computers* (Princeton N.J.: Princeton University Press, 1955), p. 188.

CASE STUDY 7

SOLUTION OF ALGEBRAIC EQUATIONS

The DO Group, the ELSE Clause, and the STOP Statement

A great many problems ultimately reduce to an equation of the form $f(x) = 0$. Unless $f(x)$ is a polynomial of degree less than five, or else a simple relation involving only logarithms and constants, exponentials and constants, or trigonometric functions and constants, it is usually impossible to solve the equations in terms of a finite number of elementary algebraic or transcendental functions. For example, the following equations can be solved by manipulation:

1. $x^2 - 6x + 3 = 0$ $(x = 3 \pm \sqrt{9 - 3})$
2. $(\log_e(x))^2 - 16 = 0$ $(x = e^4)$
3. $10^x - 13.5 = 0$ $(x = \log_{10}(13.5))$
4. $\sin^2(x) + 4\sin(x) - 1.6 = 0$ $(x = \arcsin(-2 \pm \sqrt{4 + 1.6}))$

However, this kind of solution is usually impossible when the types of function become mixed, such as in the following two examples:

5. $x^2 \log_e(x) - 5.8 = 0$
6. $\sin(x)/\log_e(x) - 8.0 = 0$

In such cases, the equations are said to be transcendental, and graphical or approximate methods must be used to solve them. Actually, the solutions to equations 1 through 4 are approximate too, insofar as $\sqrt{6}$ and the transcendental functions can only be approximated. Nevertheless, their solutions can at least be explicitly *stated* in terms of finite combinations of familiar functions whereas the solutions to examples 5 and 6 cannot.

Solving the equation $f(x) = 0$ is equivalent to finding where the function $y = f(x)$ crosses the x axis because at those points, $y = 0$.

7.1 THE ELSE CLAUSE

A root can be isolated by evaluating y for different values of x until the sign of y changes. For continuous functions, there must be an axis crossing; hence, a root, within the interval between sign changes. Once a root has been isolated in this manner the root may be found to any desired finite accuracy by the iterative *bisection technique*, also called the *Bolzano's method*: y is evaluated at the midpoint of the interval. The sign of y at the midpoint will indicate which half-interval contains the root—the side with the sign change. That half-interval is then bisected and so on until the interval width is less than an acceptable amount of error.

It might appear that this method is terribly time consuming, but every three bisections reduces the original interval by a factor of eight, which is equivalent to a factor of ten every 3.32 bisections.

As is plotted in Fig. 7-1, a root for equation 5 has been isolated by a rough slide rule calculation.

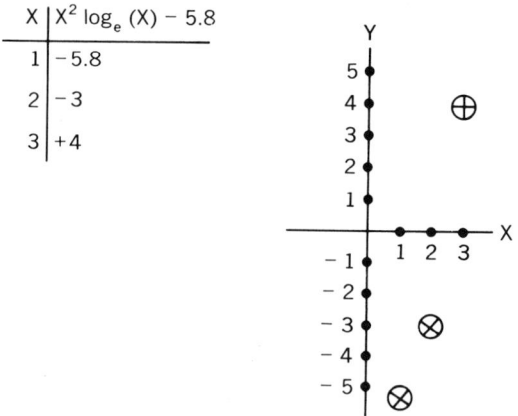

Fig. 7-1 Root isolation for the equation: $x^2 \log_e(x) = 5.8$.

Rather than write a program which solves just this equation, we will write one that solves any equation given by a function procedure F(X). Otherwise, the program would be of use only once, making it hardly worth the effort.

We must decide how much to refine the root. Perhaps with our example we would be satisfied when the interval of uncertainty, which is a bound on the absolute error, has been reduced to 0.0001. The required accuracy depends upon the application. 0.0001 per cent accuracy is often required for physics and chemistry,

Sec. 7.1 The Else Clause

but one per cent is often adequate for engineering, the life sciences, and the behavioral sciences. Note, however, that the important consideration is the per cent or relative accuracy. An absolute accuracy specification would be unsuitable for a general-purpose program such as this because it would be entirely too demanding for large roots and entirely too lenient for small roots. For instance, although an absolute accuracy of 0.0001 may be all right for equation 5, it would be unattainable with default precision for a root on the order of 10^8, and it might be insufficient for a root on the order of 10^{-12}. Consequently, we may use a relative tolerance as input data and refine until the interval of uncertainty divided by the root is less than this relative tolerance. In terms of the variables to be used in the program, this relationship is "DIFF/X < RELTOL." However, to make the program work for roots of value zero, we may express this as "DIFF <= RELTOL*X." Otherwise, an attempted division by zero would raise the ZERODIVIDE condition, which causes the computer to put out a message, and then terminate execution. Also, to account for the possibilities that the root or the interval is negative, we may take the absolute value of both sides.

$$ABS(DIFF) <= ABS(RELTOL*X)$$

This, now, is a foolproof completion relationship, and we will use it for many of our programs.

Besides the relative tolerance, we may list as input the initial bounds between which the root is known to lie. In general the function may cross the axis with either a positive or negative slope; so as input data we should also give the sign of the function ($+1$ of -1) at one of the initial bounds. The flow chart in Fig. 7-2 incorporates these ideas.

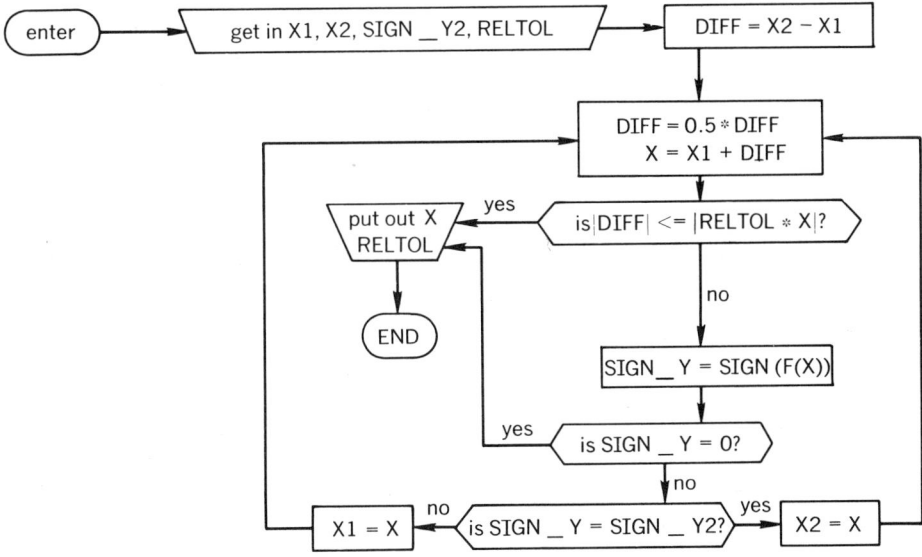

Fig. 7-2 Flow chart for Bolzano's method.

It is helpful to follow the instructions in this flow chart using a slide rule with $F(X) = X^2 \log_e(X) - 5.8$ starting with $X1 = 2$, $X2 = 3$, $SIGN\_Y2 = +1$, and $RELTOL = 0.05$.

Does it work? Correct it if necessary, then consider the corresponding program:

```
ROOT:   PROCEDURE OPTIONS (MAIN);
    /* THIS PROCEDURE USES BOLZANO'S METHOD TO FIND THE ROOT OF
       A FUNCTION NAMED 'F' TO WITHIN A RELATIVE TOLERANCE OF
       'RELTOL'.  INPUT CONSISTS OF THE INITIAL BOUNDS ON THE ROOT,
       THE SIGN OF 'F' AT THE SECOND BOUND, AND THE RELATIVE
       TOLERANCE.   */
    GET LIST (X1, X2, SIGN_Y2, RETOL);
    DIFF = X2 - X1;
    BISECT:  DIFF = 0.5*DIFF;   X = X1 + DIFF;
        IF ABS(DIFF) <= ABS(RETOL*X)  THEN GO TO OUTPUT;
        SIGN_Y = SIGN(F(X));
        IF SIGN_Y = 0   THEN GO TO OUTPUT;
        IF SIGN_Y = SIGN_Y2  THEN X2 = X;
            ELSE X1 = X;
        GO TO BISECT;
    OUTPUT:  PUT PAGE DATA (X, RETOL);
END;
```

This procedure introduces the **ELSE** clause. The **THEN** and **ELSE** clauses are mutually exclusive. The **ELSE** clause is skipped if the **THEN** clause is executed. If the keyword **ELSE** was removed from this procedure, X2 and X1 would *both* end up with the value X if SIGN__Y = SIGN__Y2. Without the **ELSE** clause, we would have to program this portion as follows:

```
        IF SIGN_Y = SIGN_Y2 THEN GO TO LEFT_HALF;
        X1 = X;  GO TO BISECT;
LEFT_HALF:  X2 = X;
        GO TO BISECT;
```

We could do without the **ELSE** clause, but it is a nice convenience. However, it is inefficient to use the **ELSE** clause unnecessarily. There is never any need to use it when the **THEN** clause is a **RETURN** or **GO TO** statement, and as we have seen, there is often no need to use it when the **THEN** clause is a **GET**, **PUT** or assignment statement.

7.2 THE NULL ON STATEMENT

This procedure is quite likely to result in **UNDERFLOW** since it is looking for function values near zero, and many equations have a root of zero. In this program the standard **UNDERFLOW** message is annoying since it clutters the output without giving any vital information. We may suppress the standard message by using the *null* **ON** statement.

```
            ON UNDERFLOW;
```

This statement says to do nothing whenever UNDERFLOW occurs—i.e., return to the point following the underflow, using the zero result of the underflow. Grammatically, this statement is an incomplete sentence, but what more elegant way is there to say "do nothing" than to say nothing! This ON statement will be in effect for every subsequently executed statement in ROOT or in F, which is invoked by ROOT.

Similarly, we could have "ON OVERFLOW;" and "ON ZERODIVIDE," statements, but if any of the corresponding conditions should arise in ROOT, it means that we have been quite careless in making our preliminary sketch to isolate the root. In these cases the standard result of a message followed by termination is appropriate.

7.3 FUNCTIONS AS PARAMETERS

Refining the root of an equation is often required as part of a larger program—perhaps there is more than one root to be found or perhaps the root is to be substituted into a formula. Consequently, we should make ROOT a subordinate procedure. A function procedure would be appropriate. Using the name BLZNO to distinguish it from the previous version, it could be used in the form BLZNO(X1, X2, SIGN_Y2, RELTOL). However as a subordinate procedure, BLZNO can overcome another shortcoming of its predecessor. ROOT has the disadvantage of finding the root only of functions named F. This is annoying if we have already written the function procedure with another name. It is an imposition to have to rewrite the procedure statement and recompile. Worse yet, this would preclude finding the roots of two different functions in one program. As a subordinate procedure with a parameter list, BLZNO may find the root of functions with any name if F is listed as a parameter:

```
BLZNO:   PROCEDURE (X1, X2, SIGN_Y2, RELTOL, F);
    DECLARE   F   ENTRY;
    ON UNDERFLOW;
    DIFF = X2 - X1;
    BISECT:   DIFF = 0.5*DIFF; X = X1 + DIFF;
        IF ABS(DIFF) <= ABS(RELTOL*X)   THEN RETURN(X);
        SIGN_Y = SIGN(F(X));
        IF SIGN_Y = 0   THEN RETURN (X);
        IF SIGN_Y = SIGN_Y2   THEN X2 = X;
            ELSE X1 = X;
        GO TO BISECT;
END;
```

Procedure names are called *entries*, and we have declared F accordingly so that it will not be taken by default as a floating-point variable.[1] Undeclared parameters are defaulted according to the first-letter convention regardless of how they are used within a subordinate procedure.

[1] Entry names may not be used as parameters for the Model 20, SL1, or Student PL.

When we use this procedure, we may use any function name in place of F just as we may use any names in place of X1, X2, SIGN__Y2, and RELTOL. For example, if we had a function procedure BESSEL (X), and we knew it was positive at X = 2 and negative at X = 3, we could use the following main procedure together with BESSEL and BLZNO to refine the root to within a relative error of 10^{-4}.

```
B_RT:   PROCEDURE OPTIONS (MAIN);
        DECLARE BESSEL ENTRY;
        PUT LIST (BLZNO(2E0, 3E0, -1E0, 1E-4, BESSEL));
        END;
```

Note that we list the function name without its argument list because we want to pass the *name* rather than a specific value. Consequently, we have had to explicitly declare that BESSEL is an ENTRY name; otherwise the compiler would assume that BESSEL was a variable because it does not appear anywhere in B__RT with an argument list. One of the features of subordinate procedures is that they may be compiled separately from a main procedure. Consequently, when compiling the invoking procedure, the compiler has no way of peeking inside the invoked procedure to see what kind of arguments are expected.

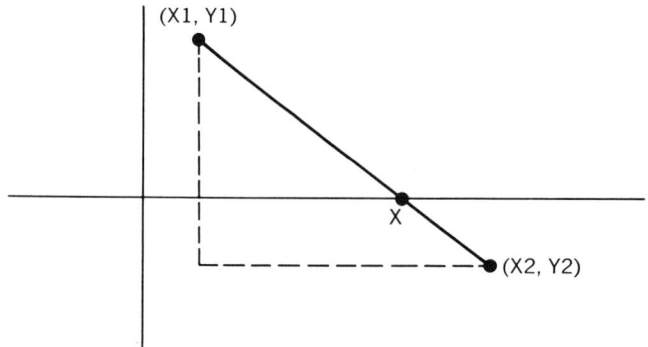

Fig. 7-3 Geometrical basis for *regula falsi*.

7.4 THE DO GROUP

A more sophisticated method for refining the root is the *regula falsi*. This is Latin, not slang, and it stands for "rule of false position." It is based on linear interpolation for each successive trial rather than mere bisection. Given two function values of different sign, Y1 = F(X1) and Y2 = F(X2), similar triangles give X = X1 + Y1*(X2 − X1)/(Y1 − Y2) for the interpolated root as is shown in Fig. 7-3. Using this formula, the flow chart in Fig. 7-4, and the procedure RGLFLS, show how the root may be refined by the *regula falsi*.

Sec. 7.4 The DO Group

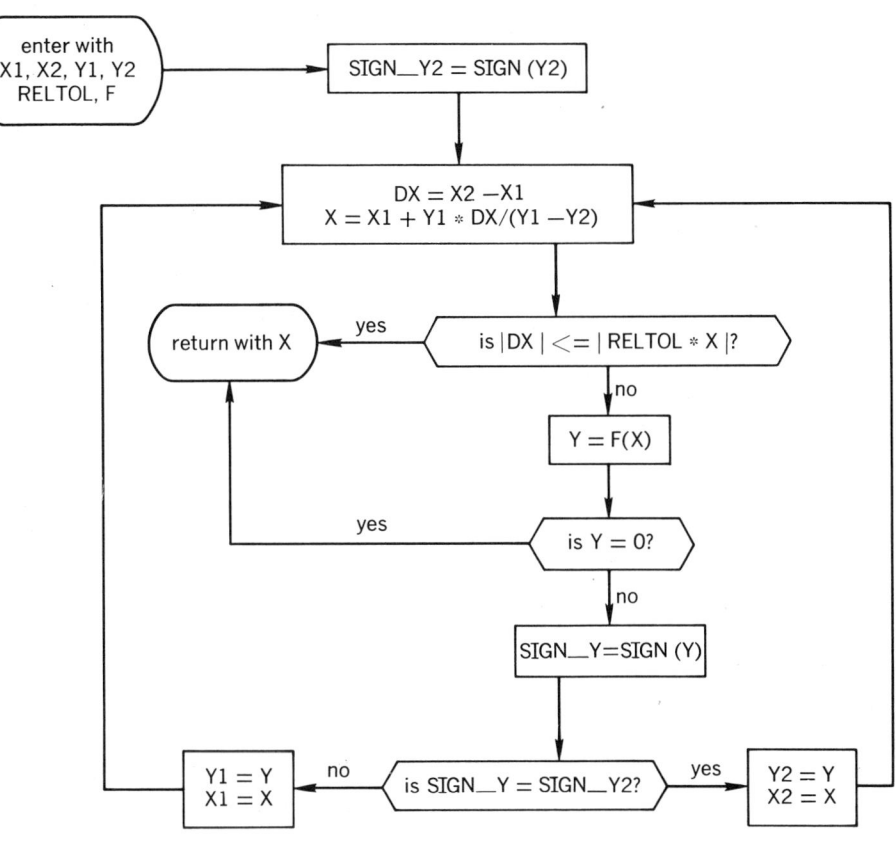

Fig. 7-4 Flow chart for *regula falsi*.

```
RGLFS:   PROCEDURE (X1, X2, Y1, Y2, RELTOL, F);
         /* THIS FUNCTION PROCEDURE USES REGULA FALSI TO REFINE THE ROOT OF
            ANY FUNCTION.  */
         DECLARE
            (X1, Y1)  /* GIVEN:  ONE BOUND ON THE ROOT AND THE FUNCTION
                                 VALUE AT THAT BOUND.  */,
            (X2, Y2)  /* GIVEN:  THE OTHER BOUND ON THE ROOT AND THE
                                 FUNCTION VALUE AT THAT BOUND.  Y1 AND Y2 MUST
                                 HAVE OPPOSITE SIGNS.  */,
            RELTOL    /* GIVEN:  THE BOUND ON THE RELATIVE ERROR IN THE
                                 ROOT.  */,
            F ENTRY   /* GIVEN:  NAME OF THE FUNCTION PROCEDURE WHOSE
                                 ROOT IS TO BE FOUND.  */;
         ON UNDERFLOW;
         SIGN_Y2 = SIGN(Y2);
   S10:  DX = X2 - X1;
         X = X1 + Y1*DX/(Y1 - Y2);
         IF ABS(DX) <= ABS(RELTOL*X)   THEN RETURN (X);
         Y = F(X);
         IF Y = 0   THEN RETURN (X);
         SIGN_Y = SIGN(Y);
         IF SIGN_Y = SIGN_Y2   THEN DO;   Y2 = Y;   X2 = X;   GO TO S10;   END;
         Y1 = Y;   X1 = X;   GO TO S10;
   END;
```

This procedure introduces the DO *group* whose purpose is to let us get around the restriction to only one statement for a THEN or ELSE clause. All of the statements between "DO;" and "END;" are treated as one clause. The first END will not end the procedure because there is a DO group for it to end. Without the DO group we would have to write this as:

```
IF SIGN_Y = SIGN_Y2 THEN GO TO S20;
   Y1 = Y; X1 = X; GO TO S10;
S20: Y2 = Y; X2 = X; GO TO S10;
```

We could do without the DO group, but like the ELSE clause, it is a useful feature. On the other hand, it is inefficient to use the DO group unnecessarily. For example, there was no need to use "ELSE DO; Y1 = Y; X1 = X; GO TO S10; END;" in RGLFLS because the THEN clause includes a GO TO statement. However, both DO groups are necessary in the following version:

```
IF SIGN_Y = SIGN_Y2 THEN DO; Y2 = Y; X2 = X; END;
   ELSE DO; Y1 = Y; X1 = X; END;
GO TO S10;
```

7.5 THE STOP STATEMENT[2]

Regula falsi and Bolzano's method both require that a root be isolated between function values of differing sign; so they do not work for a root which just touches the axis. Also, although faster on the average, *regula falsi* may be slower than Bolzano's method in some instances, as is shown in Fig. 7-5. The *secant method* is generally considerably faster than both, and it works for roots which just touch the axis. This method uses the same formula as *regula falsi*, but always with the two most recently calculated estimates of the root—even if they are on the same side of the axis. When the values are on the same side, the formula is effectively being used for extrapolation rather than interpolation as is shown in Fig. 7-6. However, in exchange for the greater speed, there is the possibility that the method

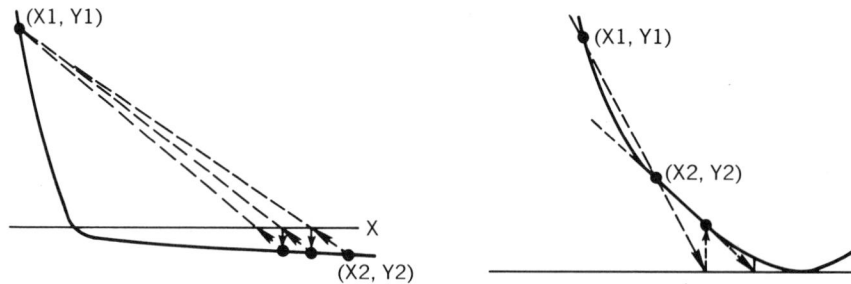

Fig. 7-5 A case of slow convergence for *regula falsi*.

Fig. 7-6 An example of the secant method.

[2] Unsupported by the Model 20.

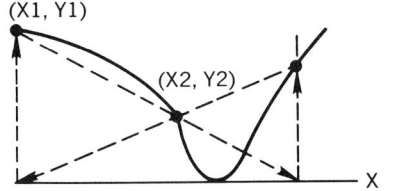
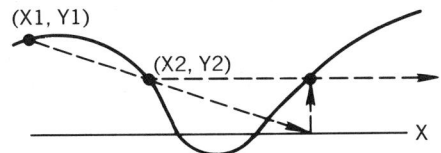

Fig. 7-7 A case of a limit cycle for the secant method.

Fig. 7-8 A case of divergence for the secant method.

may go unstable if the initial guesses for the root are not close enough. This behavior is depicted in Figs. 7-7 and 7-8. Because of this possibility, we should include a counter to stop the process if it has not converged within a reasonable number of iterations—say 20. In this instance, we could also put out a message to tell the user why he received no answer. Otherwise he may spend a lot of unnecessary time trying to figure out why his program did not run. These ideas and a few others are incorporated in the flow chart in Fig. 7-9 and in the program that follows it.

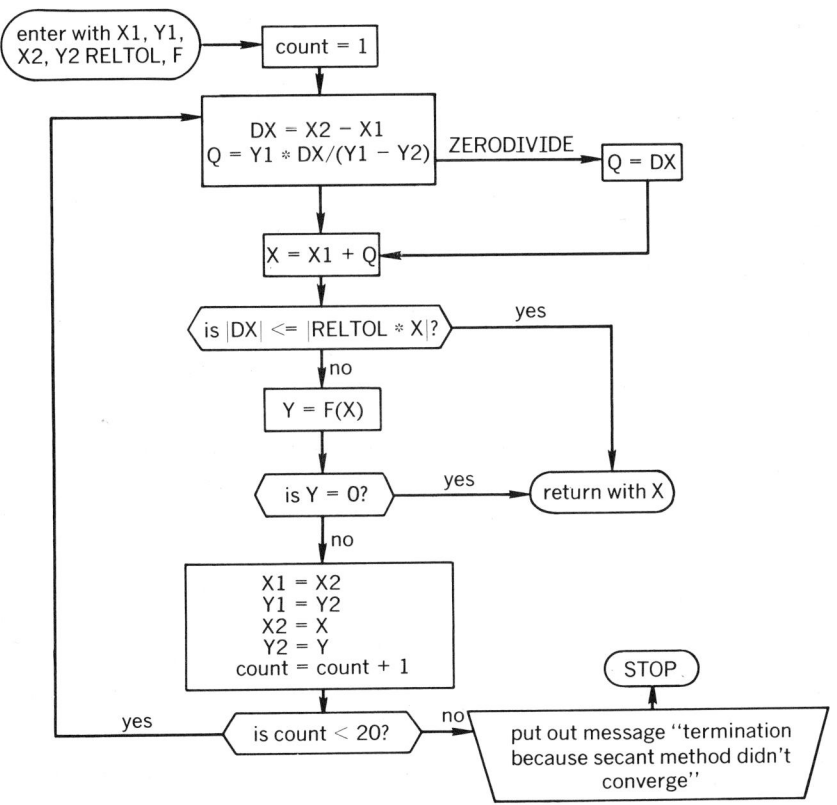

Fig. 7-9 A flow chart for the secant method.

```
SCNT:   PROCEDURE (X1, Y1, X2, Y2, RELTOL, F);
   DECLARE  F   ENTRY;
   KOUNT = 0;
   ON UNDERFLOW;
      ON ZERODIVIDE   GO TO S3;
   S1:   DX = X2 - X1;   Q = Y1*DX/(Y1 - Y2);
   S2:   X = X1 + Q;
      IF ABS(DX) <= ABS(RELTOL*X)   THEN RETURN (X);
      Y = F(X);
      IF Y = 0   THEN RETURN (X);
      X1 = X2;   Y1 = Y2;   X2 = X;   Y2 = Y;
      KOUNT = KOUNT + 1;
      IF KOUNT < 20   THEN GO TO S1;
   PUT LIST ('TERMINATION BECAUSE SECANT METHOD DIDN''T CONVERGE');
   STOP;
   S3:   Q = DX;   GO TO S2;
END;
```

As before, it is a good idea to include the null "ON UNDERFLOW;" statement. However, underflow could occur in the divisor Y1 − Y2 since the function values may be of the same sign. The consequent termination due to ZERODIVIDE would be undesirable if underflow is caused by convergence to a root which just touches the axis as in Fig. 7-6. On the other hand, a null ON ZERODIVIDE statement would be equally undesirable because it would allow X to have an unpredictable value. A reasonable policy in this instance is to change X by the amount it changed the time before. This policy gives a good chance for Y to attain a zero value before the iteration limit is reached.

Unless the on-unit contains a GO TO statement, the *normal return* upon completion of the on-unit is to the statement following the one which causes an ENDFILE interrupt. In contrast, the normal return is to the point following the *operation* which causes an OVERFLOW, UNDERFLOW, FIXEDOVERFLOW, or ZERODIVIDE interrupt. Thus, writing "ON ZERODIVIDE Q = DX;" would be futile in SCNT because Q would then be reassigned an undefined value upon normal return. Note also that DO groups may not be used for on-units; so we may not write

   ON ZERODIVIDE DO; Q = DX; GO TO S2; END;

Another feature of this procedure is the introduction of the STOP statement. It terminates execution, as does executing the END statement of a main procedure. It is not a good idea to merely allow a return to the invoking procedure from the end of SCNT because the invoking procedure will blindly use the most recent value of SCNT which is invalid. This would be a poor policy in spite of the message because the invoking procedure might go on performing a lot of futile calculations; and if there is more than one invocation, the user will not necessarily know which result is erroneous.

Admittedly, there may be instances where the user would prefer to proceed with other calculations rather than terminate with an error message. For maximum flexibility, this decision is best left up to the user. We may do this by having a result parameter IER which is set to one if the method diverges and set to zero

Sec. 7.6 Dummy Arguments

otherwise. The user may test IER and, if IER is one, he may put out a message and stop, suppress the use of that root and continue, or do whatever else he chooses. He may even ignore IER if he is overconfident. Since there are now two results and since the user should test IER before using the results, we may as well rewrite this procedure as a subroutine. This has been done below, using the name SECANT to distinguish it from the former version:

```
    SECANT:  PROCEDURE (X1, Y1, X2, Y2, F, X, IER);
        DECLARE  F  ENTRY;
        KOUNT, IER = 0;
        ON UNDERFLOW;
            ON ZERODIVIDE  GO TO S3;
    S1:    DX = X2 - X1;    Q = Y1*DX/(Y1 - Y2);
    S2:    X = X1 + Q;
            IF ABS(DX) <= ABS(RELTOL*X)   THEN RETURN;
            Y = F(X);
            IF Y = 0  THEN RETURN;
            X1 = X2;   Y1 = Y2;   X2 = X;   Y2 = Y;
            KOUNT = KOUNT + 1;
            IF KOUNT < 20   THEN GO TO S1;
            IER = 1;   RETURN;
    S3:    Q = DX;   GO TO S2;
    END;
```

7.6 DUMMY ARGUMENTS

Our example function of $X^2 \log_e (X) - 5.8$ is undefined for negative X, and a sketch quickly reveals that it has only one positive root. However, in general we would like a program which could isolate as well as refine all of the roots of a function. For example, a sketch of the two terms of $0.01x = \sin x$ reveals that this equation has several roots. The equation $x^5 + 3x^4 - 8x^3 + 3x^2 + 4x - 9 = 0$ may have as many as five real roots. There are many special iterative techniques for finding all the real and complex roots of polynomials, and most libraries have a subroutine for this. Consequently, we will consider how to find all of the real roots of a more general function.

To isolate the roots, we may have the computer start at one extreme of its floating-point range and systematically step toward the other extreme. Each time a sign change in the function is detected, a root may be refined; then the stepping may be resumed. In fact, to detect roots which just touch the axis or cross it and return within one step, we may use the secant method every time there is a positive local minimum or a negative local maximum as is depicted in Fig. 7-10.

We must increase the step size as we get further from the origin. Otherwise the stepping would come to a halt when X becomes too large to be influenced by the step size; and long before that, the *relatively* small step size would represent a great waste of time and money. What we want is a step size which is a constant percentage of X. We may accomplish this by multiplying successive values by a

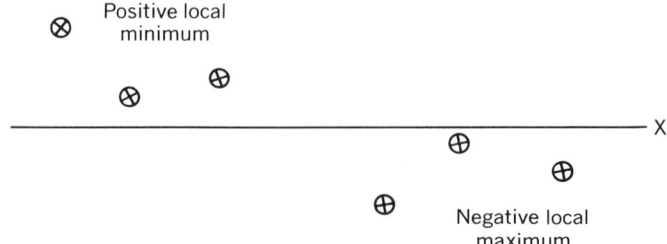

Fig. 7-10 Examples of possible roots.

constant factor. For example, if we start at 10^{-10}, we may successively multiply by 1.125, chosen because of its exact binary representation. To reach 10^{+10}, this requires only about $[10 - (-10)]/\log_{10}(1.125) = 400$ steps. An additional 400 steps would be required to cover the negative numbers. Even though this is a rather modest number of steps, they are about eight times finer than were used to isolate the root of $X^2 \log_e(X) - 5.8$ by slide rule, and roots outside the range -10^{10} to $+10^{10}$ generally correspond to problems stated in physically inappropriate units. Also, the likelihood of ZERODIVIDE or OVERFLOW in the evaluation of F(X) generally increases with the magnitude of X. In fact, these conditions are quite likely to arise even within the range -10^{10} to $+10^{10}$; so we had better decide what to do about it.

Rather than terminate, we would like the computer to simply shrug and go on to the next step. We may do this by including the null ON ZERODIVIDE and null ON OVERFLOW statements. If the incorrect result causes an unnecessary invocation of SECANT, all that is lost is a bit of time because SECANT will terminate without a result after 20 iterations. On the other hand, the incorrect result of a ZERODIVIDE or OVERFLOW is unlikely to cause omission of a root because these conditions are associated with *singularities* or infinite values which are the opposite of roots. It must be realized, however, that there are circumstances where roots will be overlooked. There is no substitute for a preliminary sketch and familiarity with the behavior of mathematical functions. Appropriately, then, we will name our procedure MOSTRT rather than ALLRT. Moreover, we will search only for positive roots, leaving it as an exercise to write a procedure which searches for both positive and negative roots:

```
MOSTRT:   PROCEDURE (RELTOL, F);
   DECLARE   F   ENTRY;
   ON ZERODIVIDE;   ON OVERFLOW;   ON UNDERFLOW;
   PUT PAGE LIST ('ROOTS:');
   X1 = 1E-10;   Y1 = F(X1);
   X2 = 1.125*X1;   Y2 = F(X2);
   S10:   IF SIGN(Y2) ¬= SIGN(Y1)   THEN DO;
      CALL SECANT((X1), (Y1), (X2), (Y2), RELTOL, F, X, IER);
      IF IER = 0   THEN PUT SKIP LIST (X);
   END;
   X3 = 1.125*X2;   Y3 = F(X3);
   IF (Y2>0 & Y2<=Y1 & Y2<Y3) | (Y2<0 & Y2>Y1 & Y2>=Y3)   THEN DO;
      CALL SECANT((X1), (Y1), (X2), (Y2), RELTOL, F, X, IER);
         IF IER = 0   THEN PUT SKIP LIST (X);
   END;
```

```
        X1 = X2;   Y1 = Y2;   X2 = X3;   Y2 = Y3;
        IF X3 < 1E10   THEN GO TO S10;
    END;
```

The first four parameters are modified by **SECANT** even though they are essentially *given* parameters. To keep this from fouling up our isolation scheme, we have used *copies* of **X1**, **Y1**, **X2**, and **Y2** as arguments. The parentheses around them makes them expressions, and the result of an expression is stored in a different location than the constituents of the expression. It is these *dummy arguments* which are modified by **SECANT**. An equivalent alternative would have been to write "X1COPY = X1; Y1COPY = Y1; X2COPY = X2; Y2COPY = Y2; CALL SECANT (X1COPY, Y1COPY, X2COPY, Y2COPY, RELTOL, F, X, IER);."

It is unusual for general-purpose subordinate procedures to have **GET** or **PUT** statements because they cramp the style of the user. Perhaps the user wants to use the results in further calculations rather than put them out. Perhaps the user would prefer a different style of output. However, we have deviated from this practice with **MOSTRT** because until we discuss subscripted variables, there is no convenient way to pass an indefinite number of roots back to the invoking procedure.

7.7 DO GROUPS AND THE ELSE CLAUSE IN NESTED IF STATEMENTS

ELSE clauses and/or **DO** groups may be used in nested **IF** statements. Most programmers are better off avoiding such elaborate nesting schemes because the result is liable to be confusing. Nevertheless, this optional section is included for the sake of completeness. As an example, we may write the following function procedure to return the quadrant of the point (X, Y) as defined in Fig. 2-3:

```
        NQUAD:  PROCEDURE (X, Y);
            IF X >= 0
                THEN IF Y >= 0
                    THEN N = 1;
                    ELSE N = 4;
                ELSE IF Y >= 0
                    THEN N = 2;
                    ELSE N = 3;
            RETURN (N);
        END;
```

Figure 7-11 shows the flow chart for this example.

The answer in the Appendix to exercise 4.23 shows an equivalent procedure which does not require nested **IF** statements.

As with unnested **IF** statements, each **THEN** or **ELSE** clause may be a **DO**

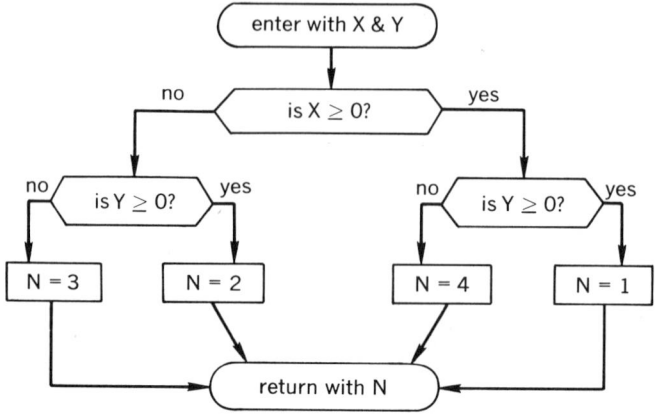

Fig. 7-11 Flow chart for quadrant determination.

group. In general, each ELSE clause is paired with the innermost unmatched THEN clause in the same DO group. This rule occasionally necessitates a null THEN or ELSE clause to achieve the desired result. Suppose, for example, that we wish to put out the value of the following function:

$$Y(X) = \begin{cases} 0.63 + 0.8X^2 + \sin(X) & \text{if } X \leq 0 \\ 0.63 + 0.8X^2 & \text{if } 0 < X \leq 1 \\ 0.63 + 0.8X^2 + \log(X) & \text{if } X > 1 \end{cases}$$

We may do this as follows:

```
Y = .63 + .8*X*X;
IF  X > 0
    THEN IF X > 1
        THEN Y = Y + LOG(X);
        ELSE;
    ELSE Y = Y + SIN(X);
PUT LIST (Y);
```

The null clause simply says to do nothing. It is used merely to make subsequent clauses properly paired. However, nested IF statements are confusing enough without null clauses, and it is always possible to rearrange the logic to avoid them. For example, the following statements are equivalent to those above:

```
Y = .63 + .8*X*X;
IF  X <= 0
    THEN Y = Y + SIN(X);
    ELSE IF X > 1 THEN Y = Y + LOG(X);
PUT LIST (Y);
```

7.8 OPTIONAL MATHEMATICALLY ADVANCED SECTION: INFINITE SERIES

Many mathematical and empirical functions may be approximated by a finite number of terms from an infinite series. For example, the cosine function for radian arguments is given by the *Taylor* series

$$\cos(x) = 1 - \frac{x^2}{2} + \frac{x^4}{4\cdot 3\cdot 2} - \frac{x^6}{6\cdot 5\cdot 4\cdot 3\cdot 2} + \frac{x^8}{8\cdot 7\cdot 6\cdot 5\cdot 4\cdot 3\cdot 2} - \cdots$$

For Taylor series the absolute value of the error due to termination of the series is always less than the last term included. With the cosine series, this *discretization* error can be made arbitrarily small for any value of x because $k!$ eventually grows faster than x^k. However, as a practical matter x is best reduced to an equivalent angle in the range $-\pi \leq x \leq \pi$ before using this series because otherwise the large, canceling middle terms will lead to large chopoff errors.

Although there is already a cosine function built into PL/I, the cosine Taylor series would be useful for writing a *haversine* function. The haversine is defined as $\text{hav}(x) = 0.5(1 - \cos(x))$. However, this definition is computationally inaccurate for small angles where $\cos x$ is nearly equal to one. To avoid this inaccuracy, we may substitute the cosine Taylor series into this formula, giving

$$\text{hav}(x) = \frac{x^2}{2\cdot 2} - \frac{x^4}{4\cdot 3\cdot 2\cdot 2} + \frac{x^6}{6\cdot 5\cdot 4\cdot 3\cdot 2\cdot 2} - \cdots$$

We may successively evaluate and accumulate terms of this series until the last term is so small relative to the total that it no longer affects the total. In evaluating the terms, we may save a great deal of time by noting that each term may be evaluated from the previous one—for example

$$\frac{+x^6}{6\cdot 5\cdot 4\cdot 3\cdot 2\cdot 2} = \left(\frac{-x^2}{6\cdot 5}\right)\left(\frac{-x^4}{4\cdot 3\cdot 2\cdot 2}\right)$$

In our subordinate procedure for the haversine function, we must provide for the possibility that an attempt is made to use it for angles with magnitudes which are too large. Otherwise a great deal of computer time may be wasted, and the user may put his trust in an inaccurate result. We do not generally wish to use more than perhaps a dozen terms of a series with alternately positive and negative terms because large, nearly-canceling intermediate terms may cause a large relative error. We may safely use more terms when they all have the same sign—

perhaps two or three dozen. When the approximation is to be used many times, computation time may become a limiting factor. Accordingly, we have written our haversine function as a subroutine with an error parameter which is set to one if convergence is not achieved in nine terms.

```
HAV:    PROCEDURE (X, HAVNEW, IER);
    /* THIS PROCEDURE CALCULATES HAVNEW = HAVERSINE (X).  IER IS SET
       TO 1 IF TOO LARGE A VALUE OF X IS ATTEMPTED.  OTHERWISE IER IS
       SET TO 0.  */
    IER = 0;   XSQ = X*X;   K = 4;   HAVOLD, TERM = .25*XSQ;
NEXT:   TERM = -TERM*XSQ/(K*(K - 1));
        HAVNEW = HAVOLD + TERM;
        IF HAVNEW = HAVOLD THEN RETURN;
        HAVOLD = HAVENEW;   K = K + 2;
        IF K < 20   THEN GO TO NEXT;
    IER = 1;
END;
```

A better way of limiting the chopoff error is to conduct a comparison with a double precision version to see at what point the chopoff error becomes intolerable. This was done with the haversine, and it was found that for the D and F compilers, the Taylor series is less accurate than $0.5(1 - \cos x)$ for $|x| > 0.5$. Accordingly, we could write the haversine procedure as follows:

```
HAV:    PROCEDURE (X);   /* THIS FUNCTION PROCEDURE RETURNS HAVERSINE(X).
    WARNING:  ABS(X) > .5  WILL CAUSE TERMINATION.  */
    IF ABS(X) > .5  THEN DO;
        PUT LIST ('THE FOLLOWING ARGUMENT IS TOO LARGE FOR THE HAV FUNCTI
ON:', X);   STOP;   END;
        XSQ = X*X;   K = 4; HAVOLD, TERM = .25*XSQ;
NEXT:   TERM = -TERM*XSQ/(K*(K-1));
        HAVNEW = HAVOLD + TERM;
        IF HAVNEW = HAVOLD   THEN RETURN (HAVNEW);
        HAVOLD = HAVNEW;   K = K + 2;
        GO TO NEXT;
END;
```

SUMMARY

The ELSE Clause and DO Groups

The IF statement may have the form:

> IF relationship THEN DO; any number of statements END;
> ELSE DO; any number of statements END;

The THEN and ELSE clauses are mutually exclusive, and the DO groups may be used to group any number of statements into one clause.

ENTRY Names as Arguments

Procedure names may be listed as parameters and arguments, but in order to be recognized as procedure names they must be declared to have the ENTRY attribute.

The ON Statement

The ON statement may have the form:

ON condition on-unit

The conditions discussed thus far are ENDFILE(SYSIN), UNDERFLOW, ZERODIVIDE, OVERFLOW, and FIXEDOVERFLOW.

The on-unit may be the null statement ";" which means "do nothing" or a GET, PUT, assignment, or GO TO statement.

With the exception of the GO TO on-unit, control generally returns to the operation following the interrrupt after completion of the on-unit. However, with ENDFILE(SYSIN) it returns to the *statement* following the interrupt.

The result of UNDERFLOW is zero, and standard system action is a message followed by continuation.

The results of ZERODIVIDE, OVERFLOW, and FIXEDOVERFLOW are unpredictable, and standard system action is a message followed by termination.

EXERCISES

1. Sketch the function $y = x^3 - .5\sqrt{x} - 0.6$ to isolate a root, then write a function procedure for this function and a main procedure which uses it together with BLZNO to refine the root to a relative error of no more than 10^{-3}.

▼2. Isolate a root of the equation $\theta \cos \theta = 0.5$, then write an appropriate function procedure and a main procedure which uses it together with REGFLS to refine the root to a relative error of no more than 10^{-4}.

3. Isolate a root of the equation $xe^x = 1.67$, then write an appropriate function procedure and a main procedure which uses it together with SECANT to refine the root to a relative error of no more 5×10^{-4}. Put out an appropriate message if SECANT does not converge.

4. The equation $\cos(\log_e(x)) = 1/x$ has an infinite number of positive roots. Write an appropriate function procedure and a main procedure which uses it together with MOSTRT to find some of them.

▼5. (Difficult) Rewrite MOSTRT to be used in the form MOSTRT(XLEFT, XRIGHT, RELTOL, F) so that it searches in the interval from XLEFT to XRIGHT which may be partly or totally negative.

6. In some applications, it may be reasonable to terminate when the absolute value of the *function* is less than some absolute tolerance ABTOL. Rewrite BLZNO accordingly.

7. In some applications, it may be reasonable to terminate when the absolute value of the function is less than ABTOL *and* the relative error in x is less than RELTOL. Rewrite RGLFLS accordingly.

8. Same as exercise 7 except for SECANT.

▼9. If an equation can be rearranged in the form $x = g(x)$, the following resubstitution technique will refine a root, provided the root and the initial estimate of it are in a neighborhood where the absolute value of the slope of g is less than one:
  (a) Evaluate $g(x)$ for the first estimate of the root x.
  (b) Assign that value to x and reevaluate x.
  (c) Keep repeating until the relative difference between two successive values of x is negligible.

The accompanying figure portrays graphically how the process works when the absolute value of the slope is less than one and how it fails when the slope is outside this range. Write a corresponding subroutine RESUB(X1, RELTOL, G, X, IER) where IER is zero if the process converges within 100 iterations and one otherwise.

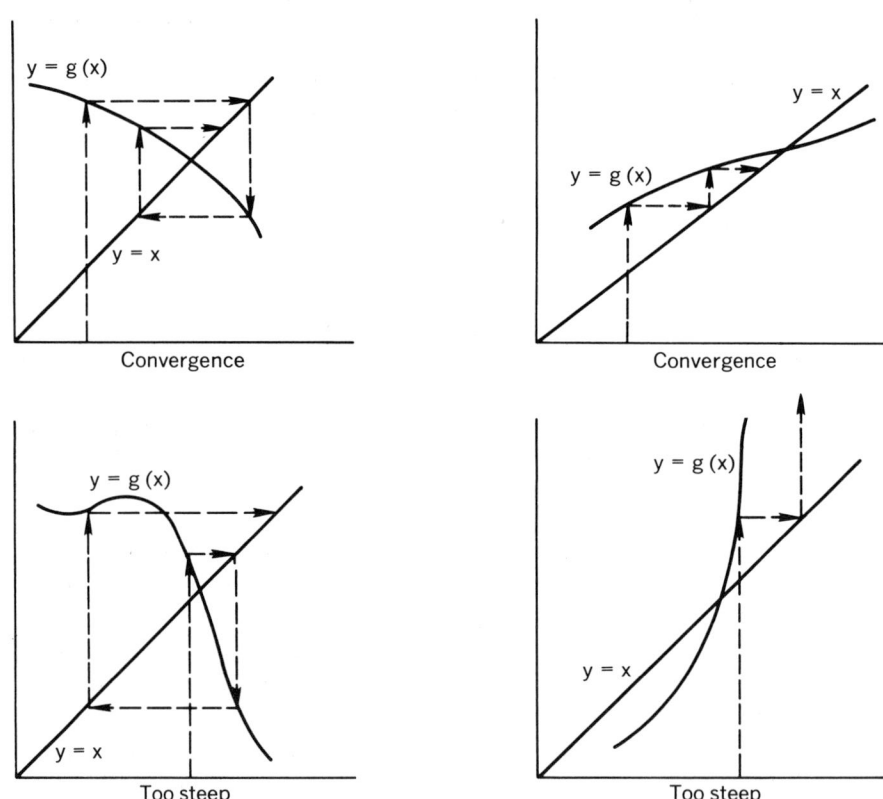

10. Same as exercise 9 except write it in the form RESUB(RELTOL, G, X, IER) where X is both the initial guess and the final answer.

11. (Mathematically Advanced) The Newton-Rapheson method depends upon being able to derive a formula for the slope of the curve, $y'(x)$. This is usually a straightforward application of differential calculus. As is illustrated in the accompanying figure, this slope may be used as a basis for extrapolating from an initial guess for the root:

$$x_{new} = x_{old} - y(x_{old})/y'(x_{old})$$

This process can be repeated until the relative difference between two successive values of x is less than RELTOL. However, the method may not converge if the initial guess is not close enough. Write a corresponding subroutine NEWTON (RELTOL, FCN, X, IER) where FCN is the function $y(x)/y'(x)$, X is the initial guess and the final result, and IER is zero if the process converges within 20 iterations and one otherwise.

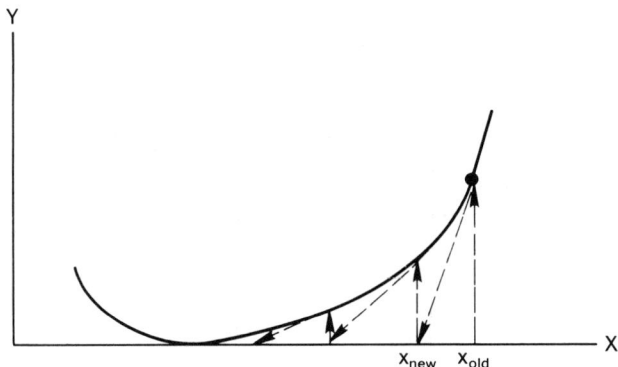

▼**12.** The maximum of a function $y = f(x)$ within an interval $x_1 \leq x \leq x_4$ can be found by a *dichotomous search* as indicated in the accompanying illustration. Two function evaluations are made near the midpoint of the interval—for example, at 15/32 and 17/32 of the interval. The larger value indicates which "half" contains the maximum. The indicated subinterval is then subdivided again and so on until the relative size of the interval is less than some tolerance. This technique is summarized in the following flow chart. Write a corresponding subroutine MAXDCT(X1, X4, RELTOL, F, X3, Y3) where Y3 is the maximum and X3 is the X value at which it occurs, to within a relative error of RELTOL.

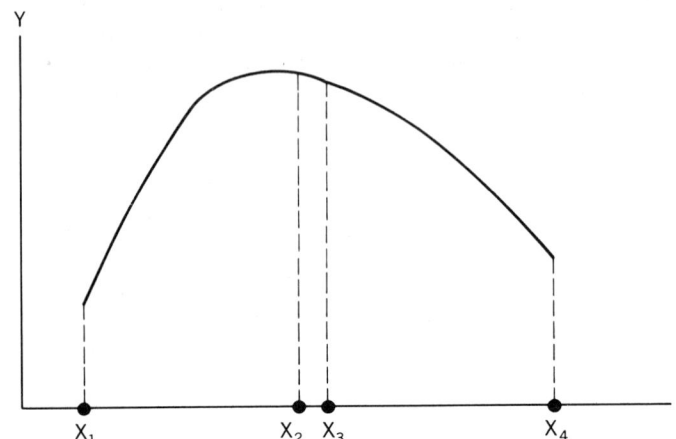

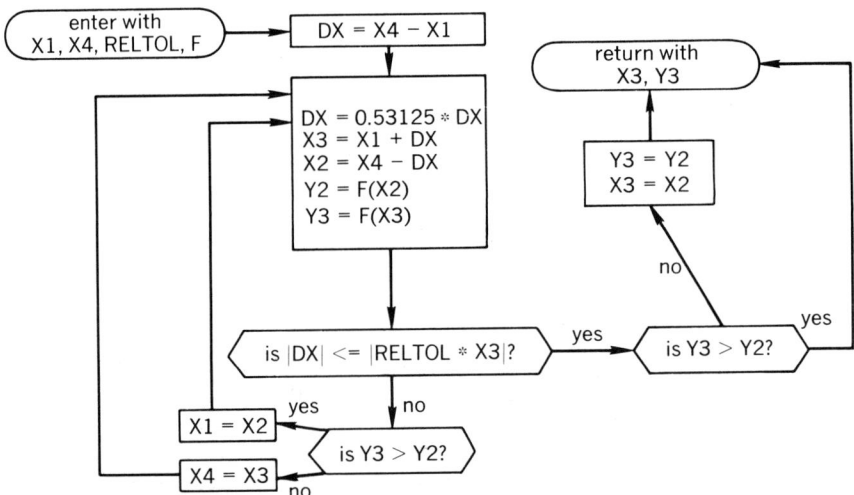

13. Same as exercise 12 except for the *minimum* of a function.
14. The dichotomous search of exercise 12 reduces the interval by a factor of almost 0.5 for every two function evaluations. The *golden mean search* is more efficient because it reduces the interval by a factor of about 0.38 for every *one* function evaluation. This search starts like the dichotomous search except the proportions are chosen so that after each subdivision, one of the last two midpoints is properly positioned to be one of the next two midpoints. This technique is summarized in the following flow chart. Write a corresponding subroutine MAXGDN (X1, X4, RELTOL, F, X3, Y3).

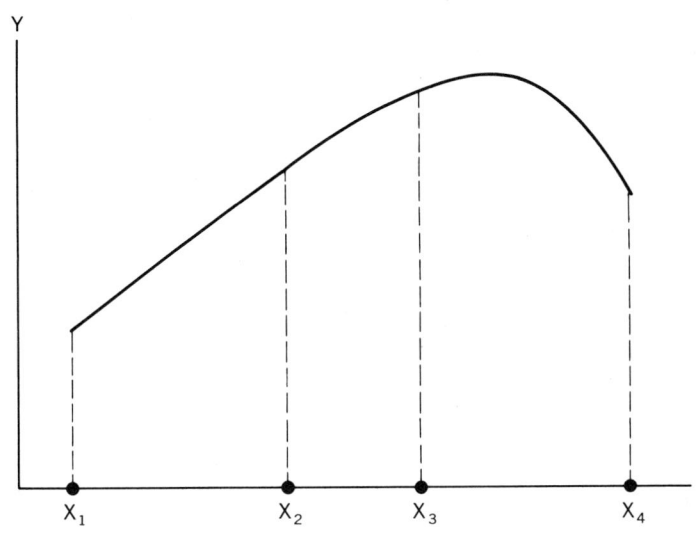
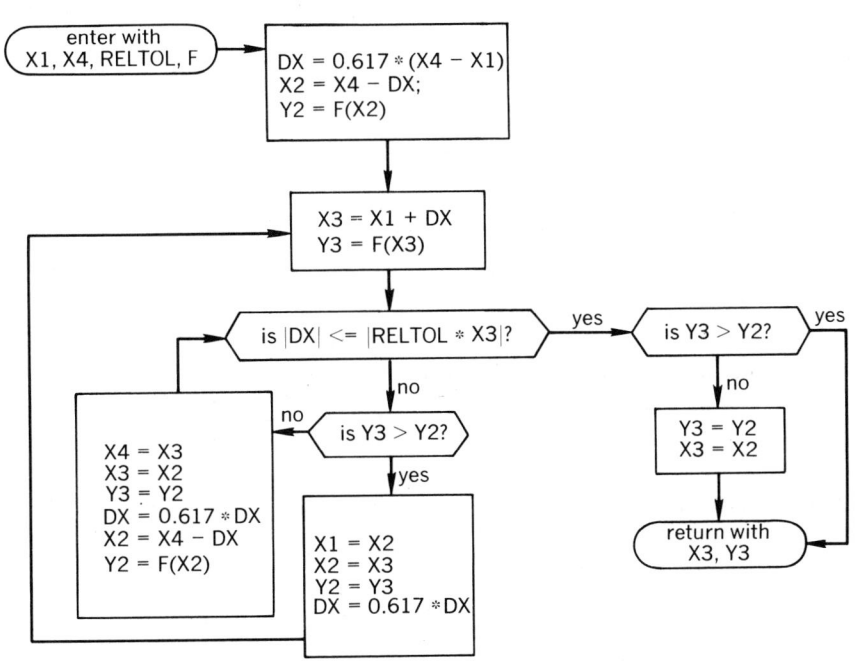

▼**15.** (Difficult) Same as exercise 14 except generalize it to find either a maximum or a minimum according to whether RELTOL is positive or negative, respectively.

16. (Mathematically Advanced) The *Fibonacci search* is slightly more efficient than the golden mean search.[3] Rather than specify the relative tolerance, with the Fibonacci search it turns out to be more convenient to specify the number of significant-digits improvement desired in the interval of uncertainty. The following flow chart presents a close approximation to the Fibonacci search. Write a corresponding subroutine MAXFIB(X1, X4, DIGITS_IMPRV, F, X3, Y3).

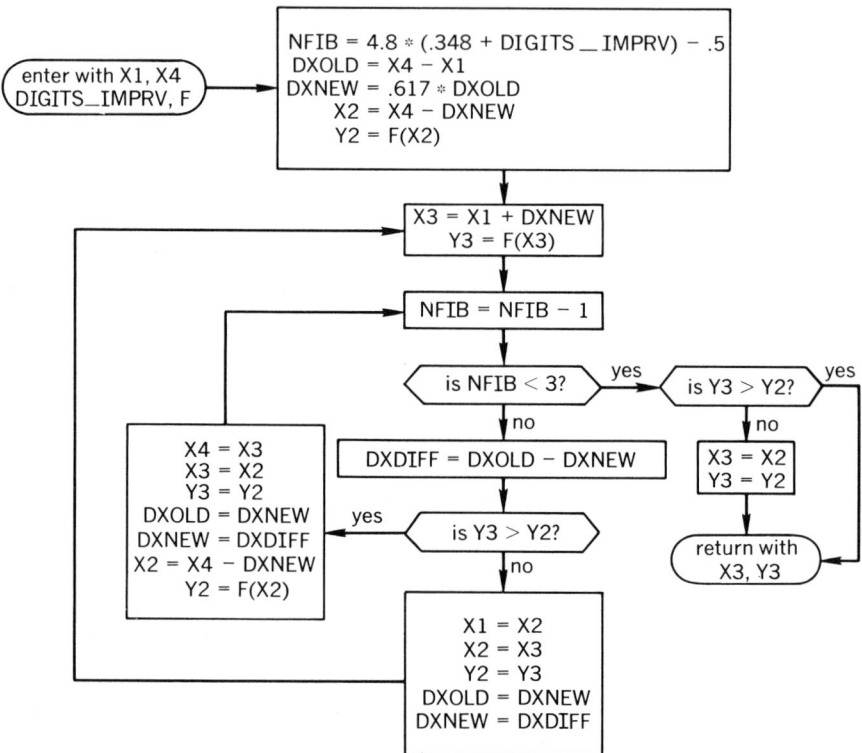

17. (Mathematically Advanced) Write a subordinate procedure that evaluates the zeroth order Bessel function using the Taylor series

$$J_0(x) = 1 - \frac{x^2}{4(2!)^2} + \frac{x^4}{4^2(4!)^2} - \frac{x^6}{4^3(6!)^2} + \frac{x^8}{4^4(8!)^2} \cdots$$

---

[3] For a thorough discussion of both of them, and a discussion of a particularly efficient root refining technique, see Bellman and Dreyfus, *Applied Dynamic Programming* (Princeton, N.J.: Princeton University Press, 1962), Chap. 4.

Exercises

▼18. (Mathematically Advanced) Write a subordinate procedure for the dilogarithm function that is defined for $0 \leq x \leq 2$ by the series

$$\text{dilog}(x) = (1-x) + \frac{(1-x)^2}{2^2} + \frac{(1-x)^3}{3^2} + \frac{(1-x)^4}{4^2} + \cdots$$

Note that only the numerator of each term can be calculated from its predecessor by multiplication.

19. (Mathematically Advanced) Write a subordinate procedure for the hypergeometric function that is defined for $x < 1$ by the series

$$F(a, b, c, x) = 1 + \frac{abx}{c} + \frac{a(a+1)b(b+1)x^2}{c(c+1)\cdot 2!}$$
$$+ \frac{a(a+1)(a+2)b(b+1)(b+2)x^2}{c(c+1)(c+2)\cdot 3!} + \cdots$$

Many elementary and not-so-elementary transcendental functions turn out to be related to the hypergeometric function.

20. (Mathematically Advanced) Write a subordinate procedure for the Riemann Zeta function which is defined for $x > 1$ by the series

$$\zeta(x) = 1 + 2^{-x} + 3^{-x} + 4^{-x} + 5^{-x} + \cdots$$

Note that this is not a power series; so each term must be calculated from scratch.

▼21. (Mathematically Advanced) Write a subordinate procedure for the function defined by the infinite product

$$x^2 \left[1 - \left(\frac{x}{\pi}\right)^4\right]\left[1 - \left(\frac{x}{2\pi}\right)^4\right]\left[1 - \left(\frac{x}{3\pi}\right)^4\right]\cdots$$

(This example is primarily illustrative because this function is the frequently occurring combination $\sin(x) \cdot \sinh(x)$, and both factors are built-in functions. However, the infinite product is probably faster for small x.)

22. (Mathematically Advanced) Write a subordinate procedure for the cosecant function defined by the infinite partial fraction expansion

$$\csc(x) = \frac{1}{x^2} + \frac{1}{(x-\pi)^2} + \frac{1}{(x+\pi)^2} + \frac{1}{(x-2\pi)^2} + \frac{1}{(x+2\pi)^2} + \frac{1}{(x-3\pi)^2} + \cdots$$

(This example is illustrative only, because it would be more reasonable simply to use the reciprocal of the sine function.)

▼23. (Mathematically Advanced) The nth order *exponential integral function* has the asymptotic expansion

$$E_n(x) \cong \frac{e^{-x}}{x}\left[1 - \frac{n}{x} + \frac{n(n+1)}{x^2} - \frac{n(n+1)(n+2)}{x^3} + \cdots\right]$$

However, asymptotic series have the characteristic that rather than decreasing to zero, the magnitude of successive terms generally decreases to a nonzero minimum, then increases, with the error always being less than the first term *ignored*. Consequently, the discretization error cannot theoretically be reduced to zero, and the summation should be terminated just short of the term with smallest magnitude. Write a corresponding subroutine EXPINT(N, X, EXPOLD, ABERROR) which calculates the Nth order exponential integral function of X together with a bound on its absolute discretization error.

▼24. (Mathematically Advanced) The nth order exponential integral function may be represented by the infinite continued fraction

$$E_n(x) = e^{-x}\left(\frac{1}{x+}\ \frac{n}{1+}\ \frac{1}{x+}\ \frac{n+1}{1+}\ \frac{2}{x+}\ \frac{n+2}{1+}\ \frac{3}{x+}\ \cdots\right)$$

This notation is explained in exercise 15 of Case Study 2. In general notation, the kth *convergent* of an infinite continued fraction is denoted by the expression

$$\left(b_0 + \frac{a_1}{b_1+}\ \frac{a_2}{b_2+}\ \frac{a_3}{b_3+}\ \cdots\ \frac{a_k}{b_k}\right)$$

Successive convergents may be evaluated until their relative difference is less than a specified relative tolerance. Since successive convergents require complete reevaluation of the fractions and since the pattern for E_n extends over two successive terms, a good strategy would be to start with $k = 2$ and increase it by two each iteration. (There is a general method of calculating a convergent from the preceding ones, but it is of academic interest only because it requires four multiplications, a division, and two additions making it slower than complete reevaluation for all k less than about 12.) Write a subroutine EXPINT(N, X, RELTOL, CNVGNTNEW) where CNVGNTNEW is the answer and RELTOL the desired relative tolerance which is modified to the *achieved* relative tolerance if the desired relative tolerance is not achieved by the 16th convergent.

25. (Mathematically Advanced) Modify the procedure AREA discussed in the integration section of Case Study 6 so that it will work for functions with any name, write a function EXPONX(X) = EXP(X)/X, and write an appropriate main procedure to find its area between X = 0.132 and X = 3.76.

26. (Mathematically Advanced) Same as exercise 19 in Case Study 6 except write the procedure to work for functions with any name.

27. (Mathematically Advanced) Same as exercise 20 in Case Study 6 except write the procedure to work for functions with any name.

28. (Mathematically Advanced) Same as exercise 22 in Case Study 6 except make the procedure a subroutine that works for functions with any name.

▼29. (Addition Language Feature) (unsupported by Student PL) Frequently, an alternative to the null ON statement is to disable the corresponding condition with a *condition prefix*. For example, if we have a function procedure FCN(X) which is quite likely to cause underflow and we are satisfied with the resultant zero, we may attach the NOUNDERFLOW condition prefix to the procedure statement:

(NOUNDERFLOW): FCN: PROCEDURE(X);

Some other disabling condition prefixes are NOZERODIVIDE, NOOVERFLOW, NOFIXEDOVERFLOW, and NOCONVERSION. (The CONVERSION condition, which leads to termination, is most often raised in scientific programming when the letter O is typed instead of the digit zero for numerical data.) There is no interruption with a disabled condition so it is generally faster than a null ON statement, but an on-unit remains in effect during the execution of an invoked procedure whereas a condition prefix does not. Check the procedures in this case study to determine which, if any, could have a null ON statement replaced with a disabling condition prefix.

More than one prefix may be listed, separated by commas within the parentheses. Prefixes may also be attached to individual statements, in which case they are effective only during execution of the clause to which they are attached. Also, there are ordinarily disabled conditions which may be *enabled* by a condition prefix. Among these is the SIZE condition which, if enabled by the prefix SIZE, is raised when too large a whole part is assigned to a fixed-point variable. When raised, the standard system action is to put out a message and terminate.

▼30. (Additional Language Feature) Conditions which cause termination do so indirectly by raising the ERROR condition. The ERROR condition provides a common condition so that one ON statement may be used for many conditions. The ERROR condition cannot be disabled, and the standard system action is taken upon normal return from an ERROR on-unit. For the D and Model 20 compilers, the standard system action is to terminate execution. For the F compiler, the standard system action is to raise the FINISH condition, which in turn causes termination. The FINISH condition is also raised by execution of a STOP statement or a main procedure END statement. Check the procedures in this case study to see which, if any, could make good use of an ON ERROR statement.

▼31. (Additional Language Feature) For this exercise, do exercise 9 using the EXTERNAL attribute for IER, as discussed in Sec. 5.4.

▼32. (Additional Language Feature—unsupported by the Model 20) As an alternative to an error parameter, we may have a statement label parameter giving a statement label in the main procedure to which a GO TO statement is to send control in the event of an error. For example, we may write SECANT and MOSTRT as follows:

```
SECANT:   PROCEDURE(X1, Y1, X2, Y2, RELTOL, F, X, LOC);
   DECLARE LOC LABEL, F ENTRY;
         .
         .
         .
   IF KOUNT < 20 THEN GO TO S1;
   GO TO LOC;
END;
```

```
MOSTRT:  PROCECURE (RELTOL, F);
   DECLARE F ENTRY;
   ON ZERODIVIDE;   ON OVERFLOW;   ON UNDERFLOW;
   PUT PAGE LIST ('ROOTS:');
   X1 = 1E-10;    Y1 = F(X1);
   X2 = 1.125*X1; Y2 = F(X2);
S10:  IF SIGN(Y2) ¬= SIGN(Y1)   THEN DO;
         CALL SECANT ((X1), (Y1), (X2), (Y2), RELTOL, F, X, S20);
         PUT SKIP LIST (X);   END;
S20:  X3 = 1.125*X2;   Y3 = F(X3);
      IF (Y2>0 & Y2<=Y1 & Y2<Y3) | (Y2<0 & Y2>Y1 & Y2>= Y3)   THEN DO;
         CALL SECANT ((X1), (Y1), (X2), (Y2), RELTOL, F, X, S30);
         PUT SKIP LIST (X);   END;
S30:  X1 = X2;   Y1 = Y2;   X2 = X3;   Y2 = Y3;
   IF X3 < 1E10·  THEN GO TO S10;
END;
```

For this exercise, modify the procedure HAV of Sec. 7.8 in this fashion. Then write a procedure which gets in any number of angles, invokes HAV, and puts out the haversine or an appropriate error message.

33. For integer N, $A^{1/N}$ may be evaluated more rapidly by Newton's iterative method than by writing A**(1/N), which requires both a log and an exponential evaluation. Starting with $X_{old} = A/N$, we may iteratively apply the formula

$$X_{new} = \frac{A}{N \cdot X_{old}^{N-1}} + (1 - 1/N) \cdot X_{old}$$

until successive iterates are the same with respect to the limited precision of the computer. Write a corresponding function procedure ROOT(A, N).

▼34. (Mathematically Advanced) Numerical integration may be performed iteratively. Referring to exercise 19 in Case Study 6, the number of steps may be successively doubled until two successive integrals agree to within some specified tolerance. Only the sum of the new ordinates need be evaluated each time. Write a corresponing subordinate integration procedure.

CASE STUDY 8

TABLE PRODUCTION

The DO Loop

Several times now we have used the IF statement to check a counter. This may be done more conveniently by attaching a repetitive specification to the DO group, making it a DO loop. For example, the counter version of MEAN from Case Study 6 may be written:

```
MEAN:   PROCEDURE OPTIONS (MAIN);
   GET LIST (N);
   SUM = 0;
   DO KOUNT = 1 TO N;
      GET LIST (X);
      SUM = SUM + X;
   END;
   AVG = SUM/N;   PUT PAGE DATA (AVG);
END;
```

8.1 THE DO LOOP

The statements between the DO and END statements will be repeated N times. This DO loop is equivalent to the statements:

```
NPRIME = N;
KOUNT = 1;
```

```
            GO  TO  S20;
        S10:
              .
              .
              .
        KOUNT  =  KOUNT  +  1;
        S20:  IF  KOUNT  <=  NPRIME  THEN  GO  TO  S10;
```

Note that;

1. The group of statements will be executed exactly once if $N = 1$.
2. The group of statements will be skipped if N is less than 1.
3. KOUNT will have the value MAX(1, N + 1) after completion of the DO loop.
4. A change of N within the loop would have no effect on the number of repetitions.

We may use the iteration variable within the loop. For example, to form the sum $\sqrt{2} + \sqrt{3} + \sqrt{4} + \cdots \sqrt{15}$ we may use the statements:

```
        SUM  =  0;
        DO  M  =  2  TO  15;
            SUM  =  SUM  +  SQRT(M);
        END;
```

We are liable to get into trouble if we enter a DO loop by a route other than the DO statement, but we may transfer out of a loop prematurely. For example, SECANT from Case Study 7 may be written:

```
        SECANT:  PROCEDURE  (X1, Y1, X2, Y2, F, X, IER);
            DECLARE   F    ENTRY;
            IER = 0;
            ON UNDERFLOW;  ON ZERODIVIDE;
            DO K = 1 TO 20;
                DX = X2 - X1;    X = X1 + DX/((Y1-Y2)/Y1);
                IF ABS(DX) <= ABS(RELTOL*X)    THEN RETURN;
                Y = F(X);
                IF Y = 0    THEN RETURN;
                X1 = X2;    Y1 = Y2;    X2 = X;    Y2 = Y;
            END;
            IER = 1;
        END;
```

Like the ELSE clause and the DO groups, the DO loop is a convenience rather than a necessity. More generally, the DO group may have the form:

        DO   variable = start TO finish BY increment;

The optional BY phrase enables us to use increments other than one. In fact, the increment, as well as *start* and *finish*, may be negative and/or fractional. For example, we may have:

        DO X = 6.7 TO −8.2 BY −3.1;

The loop will be executed until X becomes less than -8.2. Actually, the increment and the limits may be expressions. If so, they are evaluated once before the looping is begun. For example, we may have

$$\text{DO} \quad Z = \text{SIN}(A + B) \quad \text{TO} \quad C/D \quad \text{BY} - .5 * B;$$

The TO phrase is also optional. When both the BY and the TO phrase are omitted, the statements within the loop are executed only once, with the indicated value of the DO variable:

```
         DO  variable  =  start;
                  .
                  .
                  .
         END;
```

When only the TO phrase is omitted, looping will continue indefinitely until terminated by a condition or an IF statement within the loop. For example, we may rewrite the procedure MEAN of Case Study 6 as follows:

```
MEAN:   PROCEDURE OPTIONS (MAIN);
   SUM = 0;
   ON ENDFILE (SYSIN)   GO TO S10;
   DO N = 0 BY 1;
      GET LIST (X);
      SUM = SUM + X;
   END;
S10:   AVG = SUM/N;   PUT DATA (AVG);
END;
```

8.2 FLOW CHART FOR DO LOOPS

In this text we will use the flow chart symbol in Fig. 8-1 for a DO loop, the dashed line indicating the extent of the loop:

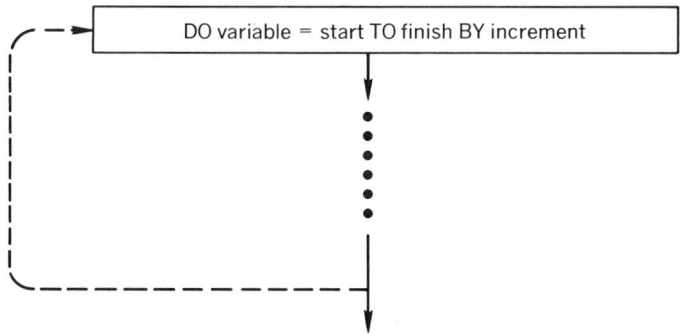

Fig. 8-1   Flow chart for the DO loop.

8.3 OPTICS EXAMPLE

As specialists, engineers and scientists often require frequent use of formulas particular to their specialty. When this is true, it is often worthwhile to have the computer prepare a table from the formula. Thereafter, it is a simple matter of looking up the answer in the table or perhaps interpolating from the table. For example, a specialist in optics who customarily uses glass with an index of refraction 1.52 would find it convenient to have a table summarizing the angle of refraction as a function of the angle of incidence for this index. The relevant formula is Snell's Law: $\theta_r = \arcsin(\sin(\theta_i)/R)$ where R is the index of refraction. The flow chart in Fig. 8-2 and following program use a **DO** loop to produce a corresponding table for $\theta_i = 0.5°$ in steps of $0.5°$ for the specific case when R = 1.52.

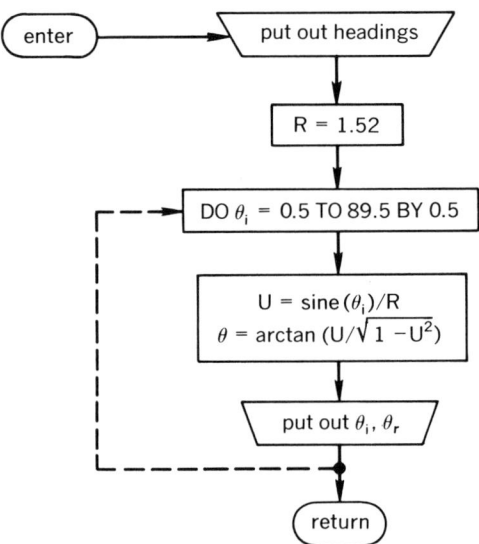

Fig. 8-2 Flow chart for producing an optics table.

```
TABLE:   PROCEDURE OPTIONS (MAIN);
    PUT PAGE LIST('REFRACTION TABLE IN DEGREES FOR AN INDEX OF 1.52:');
    PUT SKIP(2) LIST ('INCIDENT ANGLE', 'REFRACTED ANGLE');
    R = 1.52;
    DO THETA_I = .5 TO 89.5 BY .5;
        U = SIND(THETA_I)/R;
        THETA_R = ATAND(U/SQRT(1-U*U));
        PUT SKIP LIST (THETA_I, THETA_R);
    END;
END;
```

Note that we have used a trigonometric identity to express the arcsin function in terms of the arctan function. The output for this program would look something

Sec. 8.4 *Multiple Specifications* 143

like the following:

```
REFRACTION TABLE IN DEGREES FOR AN INDEX OF 1.52:

INCIDENT ANGLE    REFRACTED ANGLE
  5.00000E-01       3.28944E-01
  1.00000E+00       6.57875E-01
  1.50000E+00       9.86778E-01
  2.00000E+00       1.31563E+00
  2.50000E+00       1.64444E+00
  3.00000E+00       1.97317E+00
  3.50000E+00       2.30181E+00
  4.00000E+00       2.63036E+00
  4.50000E+00       2.95879E+00
  5.00000E+00       3.28710E+00
  5.50000E+00       3.61525E+00
  6.00000E+00       3.94326E+00
  6.50000E+00       4.27109E+00
```

In retrospect, we can see that fixed-point output would improve the legibility because the angles would not require extra zeros to place the decimal point.

EXERCISES

▼1. Operating with air, the theoretical efficiency of a gas turbine as a function of pressure ratio r is $1 - 1/r^{0.285}$. Write a program that will prepare a table of this function for $r = 0.1, 0.2, 0.3, \cdots, 6.0$.

2. The Fermi-Dirac distribution for a normalized energy u is given by the formula $1/(e^u - 1)$. Write a program that will prepare a table of this function for $u = 0.05, 0.1, 0.15, \cdots, 10$.

3. Rewrite exercise 1 of Case Study 6 using a DO loop.

4. Rewrite exercise 9 of Case Study 7 using a DO loop.

▼5. (Mathematically Advanced) Rewrite exercise 19 of Case Study 6 using a DO loop.

6. (Mathematically Advanced) Rewrite exercise 20 of Case Study 6 using a DO loop.

▼7. (Mathematically Advanced) Rewrite exercise 11 of Case Study 7 using a DO loop.

▼8. (Mathematically Advanced). Rewrite exercise 17 of Case Study 7 using a DO loop.

8.4 MULTIPLE SPECIFICATIONS[1]

Most lenses operate with relatively small angles of incidence. Consequently, an optical specialist would probably want a finer table for small angles than for large angles. For example, he might want increments of 0.1° from 0.1° to 10° and

---
[1] Unsupported by the Model 20.

increments of 1° from 10° to 89°. The following DO loop accomplishes this using a DO loop with two specifications:

```
TABLE2:  PROCEDURE OPTIONS (MAIN);
   DECLARE (THETA_I, THETA_R) FIXED(6,4) DECIMAL;
   PUT PAGE LIST ('REFRACTION TABLE IN DEGREES FOR AN INDEX OF 1.52:');
   PUT SKIP(2) LIST ('INCIDENT ANGLE', 'REFRACTED ANGLE');
   R = 1.52;
   DO THETA_I = .1 TO 10 BY .1, 11 TO 89 BY 1;
      U = SIND(THETA_I)/R;
      THETA_R = ATAND(U/SQRT(1-U*U));
      PUT SKIP LIST (THETA_I, THETA_R);
   END;
END;
```

After completing the specification "1 TO 10 BY 1" the computer goes on to complete the specification "11 TO 89 BY 1." Incidentally, the latter BY phrase is superfluous because the increment is one by default. In general, there may be any number of specifications separated by commas. As a degenerate case, a specification may simply be a single number. For example, the indices of the three most commonly used glasses are 1.46, 1.52, and 1.66; so to put out THETA__R for these three indices at a specific angle—say 4.5 degrees—we could use the following statements:

```
         SINE = SIND(4.5);
         DO R = 1.46, 1.52, 1.66;
            U = SINE/R;
            THETA__R = ATAND(U/SQRT(1-U*U));
            PUT SKIP LIST (R, THETA__R);
         END;
```

Note that SIND (4.5) has been calculated outside the loop to avoid having the same calculation done three times. In general, it is good to do as much as possible outside the loops. Some compilers are sophisticated enough to recognize and relocate outside a DO loop any subexpressions which do not change within the loop. However, it is better not to rely on this.

8.5 NESTED LOOPS

Perhaps the specialist would like a table that lists the refracted angle for a whole set of incident angles together with each of the above three values of R. We may do this by changing 4.5 in the above statements to THETA__I, then nesting these statements in another DO loop which increments THETA__I through its range:

```
         DO THETA__I = .1 TO 10 BY .1, 11 TO 89;
            SINE = SIND(THETA__I);
            DO R = 1.46, 1.52, 1.66;
               U = SINE/R;
```

```
        THETA__R = ATAND(U/SQRT(1-U*U));
        PUT SKIP LIST (THETA__I, R, THETA__R);
      END;
    END;
```

Figure 8-3 shows the corresponding flow chart.

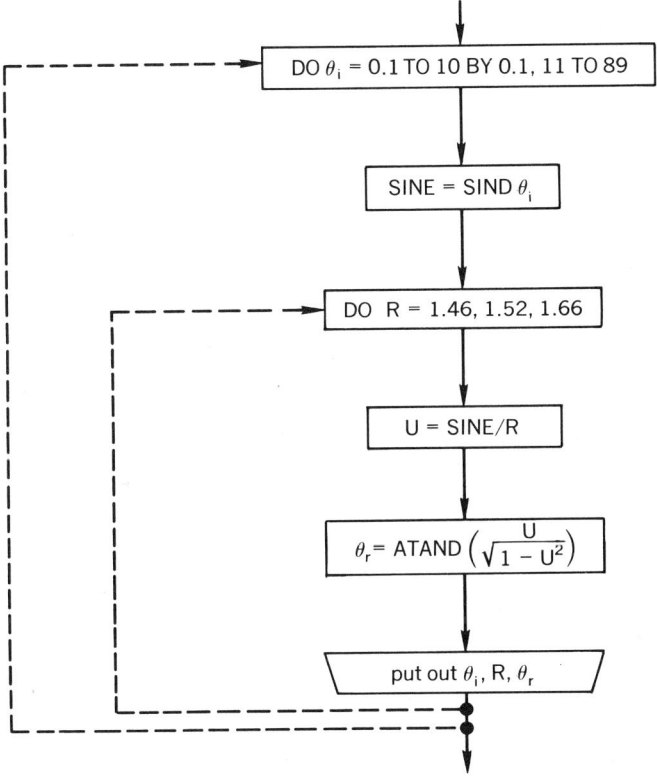

Figure 8-3 Flow chart for a nest of DO loops.

For each value of THETA__I, the inner loop is completely reexecuted for all three values of R. Consequently, all combinations of THETA__I with R will be covered. There are 100 values in the first specification of THETA__I and 79 values in the second specification, both with three values of R; so there will be 179 $\times$ 3 = 537 values of THETA__R produced. The action is rather like that of an odometer. The inner loop cycles more rapidly than the outer loop, just as the lower-value odometer wheels cycles more rapidly than the higher-valued wheels.

The above statements would result in a long slender table. The following compact table is more convenient:

```
ANGLE OF REFRACTION IN DEGREES

INDEX:                    1.46             1.52             1.66
INCIDENT ANGLE
    0.1000              0.0684           0.0657           0.0602
    0.2000              0.1369           0.1315           0.1204
    0.3000              0.2054           0.1973           0.1807
    0.4000              0.2739           0.2631           0.2409
    0.5000              0.3424           0.3289           0.3012
    0.6000              0.4109           0.3947           0.3614
    0.7000              0.4794           0.4605           0.4216
    0.8000              0.5479           0.5263           0.4819
    0.9000              0.6164           0.5920           0.5421
    1.0000              0.6849           0.6578           0.6023
    1.1000              0.7534           0.7236           0.6626
    1.2000              0.8218           0.7894           0.7228
    1.3000              0.8903           0.8552           0.7830
    1.4000              0.9588           0.9210           0.8433
    1.5000              1.0273           0.9867           0.9035
```

This table may be produced by the following program:

```
TABLE3:  PROCEDURE OPTIONS (MAIN);
    DECLARE ((THETA_I, THETA_R) FIXED(6,4), R FIXED(3,2)) DECIMAL;
    PUT PAGE LIST ('ANGLE OF REFRACTION IN DEGREES');
    PUT SKIP(2) LIST ('INDEX:', '1.46', '1.52', '1.66');
    PUT SKIP LIST ('INCIDENT ANGLE');
    DO THETA_I = .1 TO 10 BY .1, 11 TO 89;
        PUT SKIP LIST (THETA_I);
        SINE = SIND(THETA_I);
        DO R = 1.46, 1.52, 1.66;
            U = SINE/R;   THETA_R = ATAND(U/SQRT(1-U*U));
            PUT LIST (THETA_R);
        END;
    END;
END;
```

With nested DO loops, we often have applications which call for the limits or increment of an inner loop to depend upon the DO variable of an outer loop. For example, the wavelengths of the hydrogen spectrum in angstroms are given by the formula

$$\lambda = \frac{911.8}{\left(\dfrac{1}{N^2} - \dfrac{1}{M^2}\right)} \quad \text{where } M = 2, 3, 4, \cdots \text{ together with } N = 1, 2, \cdots, M-1$$

For implementations with six or more print positions, a table similar to the following one may be produced by the program below it.

```
PARTIAL TABLE OF HYDROGEN WAVELENGTHS IN ANGSTROMS
    1215
    4102        6564
    8752       19450       18756
   15195       39075       51285       40522
   23445       65646       98469      105033       74597
   33506       99279      160832      194948      186148      123717
```

```
HYDRGN:  PROCEDURE OPTIONS (MAIN);
   DECLARE LAMBDA FIXED(7) DECIMAL;
   PUT PAGE LIST('PARTIAL TABLE OF HYDROGEN WAVELENGTHS IN ANGSTROMS');
   DO M = 2 TO 7;
      PUT SKIP;  MSQ = M*M;
      DO N = 1 TO M-1;
         LAMBDA = (911.8*MSQ*N*N)/(M + N)*(M - N);
         PUT LIST (LAMBDA);
      END;
   END;
END;
```

Note that the formula is evaluated in fixed-point; so we have rearranged it to make the division come last.

For tables which require more columns than the number of print positions, the output may be broken into several sheets as is discussed in exercise 22. For more compact tables, we may also use the PUT EDIT statement, which is not discussed in this text.

8.6 OPTIONAL LANGUAGE FEATURE: THE WHILE CLAUSE[2]

We may also include a WHILE clause with a DO loop—for example:

```
DO WHILE (ABS(DIFF) <= ABS(RELTOL*X));
   ...
END;
```

Note that the relationship after the keyword WHILE must be enclosed in parentheses. This example is equivalent to:

```
S10:   IF ABS(DIFF) <= ABS(RELTOL*X) THEN GO TO S20;
       ...
       GO TO S10;
S20:   ...
```

The WHILE clause should be used with great caution because many iterative schemes require termination from a point other than the beginning of the loop. For example, the following version of BLZNO from Case Study 7 might easily cause premature termination because X and SIGN__Y are not initialized before the WHILE clause:

```
BLZNO:  PROCEDURE (X1, X2, SIGN_Y2, RELTOL, F);
   DECLARE F ENTRY;
   ON UNDERFLOW;
   DIFF = X2 - X1;
   DO WHILE (ABS(DIFF) <= ABS(RELTOL*X) & SIGN_Y ¬= 0);
      DIFF = .5*DIFF;  X = X1 + DIFF;
      SIGN_Y = SIGN(F(X));
      IF SIGN_Y = SIGN_Y2  THEN X2 = X;
                           ELSE X1 = X;
   END;
END;
```

[2] Unsupported by the Model 20.

BLZNO could be modified to correctly use the WHILE clause only by introducing some duplicate statements.

The WHILE clause may be attached to an iterative specification. For example, we could have the statement:

DO K = 1 TO 20 WHILE (ABS(DX) <= ABS(RELTOL*X));

The loop will be executed until either the iteration limit is exceeded or the WHILE relationship is false.

The WHILE clause usually causes more trouble than it saves, but it is mentioned here for the sake of completeness.

SUMMARY

A general form for the DO loop is

DO variable = list of specifications separated by commas;
.
.
.
LABEL: END;

with LABEL being optional.

Each specification may have one of the forms:

expression$_1$ TO expression$_2$ BY expression$_3$
expression$_1$ TO expression$_2$
expression$_1$ BY expression$_3$
expression$_1$

In the first case the DO variable begins with the value of expression$_1$, changing each cycle by expression$_3$ until expression$_2$ is passed. In the second case an increment of +1 is implied. In the third case, cycling continues until terminated by some other cause. In the fourth case there is only one cycle. The equivalent for the first case is:

variable = expression$_1$; e$_2$ = expression$_2$; e$_3$ = expression$_3$;
   GO TO S20;
S10: .
     .
     .
LABEL: variable = variable + e$_3$;
S20: IF (e$_3$ >= 0) & (variable <= e$_2$) | (e$_3$ < 0) & (variable >= e$_2$)
     THEN GO TO S10;

As can be seen:

1. Statements within the loop cannot change the starting value, the TO limit, or the increment.

2. Entry by a path other than the DO statement will leave the DO variable and limits undefined.
3. The loop will be skipped if the value of expression₁ is beyond that of expression₂.
4. After completion of the loop, the DO variable will have a value beyond the value of expression₃.

EXERCISES

9. Modify the procedure TABLE3 so that it uses R = 1.3, 1.4, 1.5, 1.6, and 1.7, making it more convenient to use for substances other than the three common glasses.

10. The ratio of relativistic to classical momentum as a function of the fraction of the speed of light is given by the formula $1/\sqrt{1 - f^2}$. Write a procedure that will prepare a table of the function for f = 0.1 to 0.9 in steps of 0.1, 0.91 to 0.99 in steps of 0.01, 0.991 to 0.999 in steps of 0.001, and 0.9991 to 0.9999 in steps of 0.0001.

11. The efficiency in per cent for a power screw, as a function of the thread slope s and the coefficient of friction μ, is given by the formula

$$\frac{100*(1 - \mu s)}{s + \mu}$$

Write a program that produces a table of this efficiency, to the nearest per cent, for s = 0.1, 0.15, 0.2, $\cdots$, 2 together with μ = 0.1, 0.125, 0.15, 0.175, and 0.2.

12. Same as exercise 1 in Case Study 2 except use a DO loop with multiple specifications.

13. Same as exercise 2 in Case Study 2 except use a DO loop with multiple specifications.

14. Same as exercise 4 in Case Study 2 except use a DO loop with multiple specifications.

15. The quantum angles of magnetization are given by

$$\arccos\left[\frac{m}{\sqrt{k(k+1)}}\right]$$

where k = 1, 2, $\cdots$ and m = 1, 2, $\cdots$, k. Write a procedure that will produce a partial table of these angles.

16. The normalized entropy as a function of volume V, temperature T, and the specific heat ratio γ is given by the formula

$$s(V, T, \gamma) = \log V + \frac{\log T}{\gamma - 1}$$

Write a program using a nest of three DO loops to produce a table for all combinations of V varying from 1 to 8 by 0.5, T varying from 100 to 1000 by 25, and γ varying from 1.1 to 1.6 by 0.1.

17. Same as exercise 1 of this case study except include in the table the temperature ratio as given by the formula $r^{-0.285}$.

▼18. (Difficult) The chemical reaction constant is given by the formula

$$\frac{\prod_{i=1}^{m} [A_i]^{k_i}}{\prod_{j=1}^{n} [B_j]^{\ell_j}}$$

where the quantities in brackets are the concentrations for the reaction

$$k_1 A_1 + k_2 A_2 + \cdots k_m A_m \rightarrow \ell_1 B_1 + \ell_2 B_2 + \cdots \ell_n B_n$$

Write a procedure that calculates and puts out the reaction constant for any number of reactions, repetitively getting m followed by m pairs of values $[A]$ and k, then n followed by n pairs of values $[B]$ and ℓ.

▼19. Provided the quantity in parentheses is positive, the plate current of a vacuum triode is given by the formula

$$C\left[E_g + \frac{E_a}{\mu}\right]^{1.5}$$

where C is a constant, E_g is the grid to cathode voltage, E_a is the anode to the cathode voltage, and μ is the amplification factor. For a certain vacuum tube, $C = 40$ and $\mu = 92$. Write a procedure that will prepare a table of the current as a function of E_a varying from 50 to 400 in steps of 10, together with E_g varying from -20 to 0 in steps of 5. However, the tube is outside the proper operating range if $E_g + E_a/\mu$ is negative, if the current exceeds 55, or if the product of the current and E_a exceeds 1600; so omit these cases.

20. The following output indicates one possible approach for when the desired number of columns exceeds the number of print positions. Write a corresponding program assuming six print positions.

```
ANGLE OF REFRACTION IN DEGREES
INDEX:
            1.1         1.2         1.3         1.4         1.5
            1.6         1.7         1.8         1.9
ANG__I = 0.1000;
            0.0909      0.0833      0.0769      0.0714      0.0666
            0.0624      0.0588      0.0555      0.0526
ANG__I = 0.2000;
            0.1818      0.1666      0.1538      0.1428      0.1333
            0.1249      0.1176      0.1111      0.1052
ANG__I = 0.3000;
            0.2727      0.2499      0.2307      0.2142      0.1999
            0.1874      0.1764      0.1666      0.1578
ANG__I = 0.4000;
            0.3636      0.3333      0.3076      0.2857      0.2666
            0.2499      0.2352      0.2222      0.2105
```

Exercises

21. The following output indicates one possible approach for when the desired number of columns exceeds the number of print positions. Write a corresponding program assuming six print positions.

```
ANGLE OF REFRACTION IN DEGREES
   INDEX:           1.1       1.2       1.3       1.4
            1.5     1.6       1.7       1.8       1.9
   ANG__I = 0.1000;  0.0909   0.0833    0.0769    0.0714
            0.0666   0.0624   0.0588    0.0555    0.0526
   ANG__I = 0.2000;  0.1818   0.1666    0.1538    0.1428
            0.1333   0.1249   0.1176    0.1111    0.1052
   ANG__I = 0.3000;  0.2727   0.2499    0.2307    0.2142
            0.1999   0.1874   0.1764    0.1666    0.1578
   ANG__I = 0.4000;  0.3636   0.3333    0.3076    0.2857
            0.2666   0.2499   0.2352    0.2222    0.2105
```

▼22. The following output indicates one possible approach for when the desired number of columns exceeds the number of print positions. Write a corresponding program assuming five print positions.

```
ANGLE OF REFRACTION IN DEGREES

INDEX:              1.10         1.20         1.30         1.40
INCIDENT ANGLE
   0.1000           0.0909       0.0833       0.0769       0.0714
   0.2000           0.1818       0.1666       0.1538       0.1428
   0.3000           0.2727       0.2499       0.2307       0.2142
   0.4000           0.3636       0.3333       0.3076       0.2857
   0.5000           0.4545       0.4166       0.3846       0.3571

INDEX:              1.50         1.60         1.70         1.80
INCIDENT ANGLE
   0.1000           0.0666       0.0624       0.0588       0.0555
   0.2000           0.1333       0.1249       0.1176       0.1111
   0.3000           0.1999       0.1874       0.1764       0.1666
   0.4000           0.2666       0.2499       0.2352       0.2222
   0.5000           0.3333       0.3124       0.2941       0.2777

INDEX:              1.90         2.00
INCIDENT ANGLE
   0.1000           0.0526       0.0499
   0.2000           0.1052       0.0999
   0.3000           0.1578       0.1499
   0.4000           0.2105       0.1999
   0.5000           0.2631       0.2499
```

▼23. (Mathematically Advanced) The binomial coefficients are given by the formula

$$B(N, K) = \frac{N(N-1)(N-2) \cdots (N-K+1)}{K(K-1)(K-2) \cdots 1} = \frac{N!}{K!(N-K)!}$$

Write a corresponding function procedure B(N, K). Use floating-point arithmetic to maximize the useful range of the function.

24. (Additional Language Feature) For the F, D, and SL1 compilers, a label of a DO statement may be listed after an END statement. If so, the END statement ends not only the labeled DO, but also any unended DO statements within. For example,

the nest of DO loops in procedure TABLE3 from this case study could be written:

```
S10: DO THETA__I = .1 TO 10 BY .1, 11 to 89;
    PUT SKIP LIST (THETA__I);
    SINE = SIND (THETA__I);
    DO R = 1.44, 1.52, 1.66;
        U = SINE/R;  THETA__R = ATAND(U/SQRT(1 - U*U));
        PUT LIST (U);
END S10;
```

By internal bookkeeping the computer will know where to go each time the END statement is reached. However, note that it takes a nest of at least three DO loops before this technique saves any typing.

Many programmers list the label of the procedure after the last END statement to clearly mark the procedure boundary. However, DO loops will be prematurely terminated if the procedure END is also used to end them. For this exercise, determine a table that you would like to have, then write a corresponding program using this technique.

▼25. (Additional Language Feature) (Unsupported by SL1 or Student PL) To achieve titling at the beginning of each page, we may include the following statement in the TABLE procedure of this case study:

```
ON ENDPAGE (SYSPRINT) PUT PAGE LIST
    ('REFRACTION TABLE IN DEGREES FOR INDEX OF 1.52:');
```

Alternatively, to suppress the gaps between pages we may include the statement:

```
ON ENDPAGE (SYSPRINT);
```

With the D and Model 20 compilers we must use a shorter name such as SYSPRT together with a declaration such as DECLARE SYSPRT FILE PRINT ENVIRONMENT (MEDIUM(SYSLST, 1403) F(132));

Rewrite TABLE so that it also puts out the subheadings INCIDENT ANGLE and REFRACTED ANGLE at the beginning of each page.

CASE STUDY 9

INTERPOLATION

Singly-Subscripted Variables, the INITIAL and STATIC Attributes

In Case Study 6 we used subscript notation to state the formula for the arithmetic mean:

$$\bar{x} = (x_1 + x_2 + x_3 + \cdots x_n)/n = \left[\sum_{k=1}^{n} x_k\right]/n$$

However, in the program we could use a single unsubscripted variable to represent each of the xs in turn because each value was needed only once. Unfortunately, this technique will not work if we want to evaluate the average deviation as well as the arithmetic mean. The average deviation is a measure of the spread of the values, and it is defined as

$$\frac{1}{n} \sum_{k=1}^{n} |x_k - \bar{x}|$$

The difficulty is that the mean must be completely calculated before the calculation of the average deviation can be started. One technique would be to have duplicate data files—the first one gotten during calculation of the mean and the second during calculation of the average deviation. Fortunately, this duplication can be avoided with subscripted variables.

Subscripted variables enable us to store many related values without having to use a distinct name for each one. One generic name is used for all of the values—for example, X—and individual elements are referred to by subscript—for example,

X(5) or X(K). Subscripts are enclosed in parentheses rather than lowered because of the restriction to one line. By using a counter as the subscript, it is easy to refer successively to each variable without explicitly writing the names for all of them. Variables with one subscript are also called *vectors* or *one-dimensional arrays*.

9.1 STATISTICS EXAMPLE

The following procedure uses a subsubscripted variable to evaluate the arithmetic mean and average deviation:

```
MEANDV:  PROCEDURE OPTIONS (MAIN);
    DECLARE X(10000);
    ON ENDFILE(SYSIN)   GO TO FINISH;
    SUM = 0;   N = 1;
    DO N = 1 BY 1;
        GET LIST (X(N));
        SUM = SUM + X(N);
    END;
    FINISH:   N = N - 1;
    AVG = SUM/N;
    SUM = 0;
    DO K = 1 TO N;
        SUM = SUM + ABS(X(K) - AVG);
    END;
    AVG_DEV = SUM/N;
    PUT PAGE DATA (AVG, AVG_DEV);
END;
```

9.2 STORAGE ALLOCATION

In all of the procedures that we have discussed so far, storage for all of the variables is allocated immediately upon entry to the procedure. However, in our statistics example the number of data values to be stored is unknown until after the input phase of execution is completed. Moreover, N would be different for each execution. Consequently, the approach taken here is to allocate space for 10,000 data values, which is more than enough for most applications, and to use only as much of this space as is necessary for each execution. The important point is that *some* number must be allocated in a DECLARE statement. The declaration also helps the compiler realize that X is not a function. Incidentally, we cannot write "GET LIST(N); DECLARE X(N);" because regardless of where it is located, the DECLARE statement is the first to be executed, and N is not known at that time.

Many computers have only one program in the memory at a time. If so, all of the memory may as well be allocated if there is a chance that it could be used.

Otherwise it is wasted capacity, and too modest a declaration may require subsequent recompilation simply to increase the allocation. The available space varies greatly among implementations, but in this text we will assume there are 10,000 words of storage available for subscripted variables. Actually, it is impractical to predict exactly how much space is available for subscripted variables because the machine language instructions, constants, and unsubscripted variables take space too. However, we shall not make any attempt at an adjustment for these items because they should not require anywhere near 10,000 words for the programs in this text.

An increasing number of large computers are operated on a *multiprogramming* basis, keeping several programs in the memory at once. The computer shifts execution to another program whenever a program is awaiting the completion of input and output. This technique may save a great deal of time because the reading of a punched card or the printing of a line generally takes several thousand times as long as an arithmetic operation. For computers large enough to operate on this basis, there is usually a generous minimum-size portion of memory that may be allocated to a program; so we shall stick with our assumption of 10,000 in this text. A few experiments will establish the corresponding numbers for any specific implementation.

The first storage location for a set of subscripted variables is assumed to correspond to the subscript 1 by default, but we will see later how to declare otherwise. In our example, we have used both N and K as *subscript* variables for the *subscripted* variable X. This is all right because a subscript variable is merely a means of indexing through the elements of a subscripted variable. The name of the index is irrelevant because it does not appear in the final result. In mathematics, variables whose names do not appear in the final results are called *dummy variables*.

Note that there is no data value for the last value of N; so this final value of N is 1 larger than the number of subscripted variables which have valid data. Consequently, N has been reduced by 1 after the input loop.

9.3 SUBSCRIPTED PARAMETERS AND ARGUMENTS

Actually, evaluating the average and average deviation are such common chores that it would be convenient to have separate function procedures for them:

```
AVG:    PROCEDURE (N, X);
    DECLARE X(*);
    SUM = 0;
    DO K = 1 TO N;
        SUM = SUM + X(K);
    END;
    RETURN (SUM/N);
END;
```

```
AVGDEV:  PROCEDURE (N, AVERAGE, X);
   DECLARE X(*);
   SUM = 0;
   DO K = 1 TO N;
      SUM = SUM + ABS(AVERAGE - X(K));
   END;
   RETURN (SUM/N);
END;
```

Note that we denote all of the elements in X as a parameter by listing the name X without any subscripts. Since X is a parameter which is to take on the identity of whatever argument is used, we have used an asterisk in the DECLARE statement for the number of storage locations, signifying that it is to be the number established for the argument. We cannot use this technique in a main procedure because main procedures have no parameters.[1]

As an example of how these procedures may be used, the following program finds the average and average deviation for any number of sets of data:

```
STTSTC:  PROCEDURE OPTIONS (MAIN);
   DECLARE X(10000);
   PUT PAGE LIST ('AVERAGE', 'AVERAGE DEVIATION');
   S10:  GET LIST (N);
      DO K = 1 TO N;
         GET LIST (X(K));
      END;
      AVERAGE = AVG(N, X);   AVDEV = AVGDEV (N, AVERAGE, X);
      PUT SKIP LIST (AVERAGE, AVDEV);
      GO TO S10;
END;
```

9.4 REPETITIVE INPUT AND OUTPUT SPECIFICATIONS[2]

Input and output of subscripted variables are required so frequently that there are some special compact, efficient ways to do them. We may put an *iteration specification* in the list of a GET LIST, PUT LIST, or PUT DATA statement.

---

[1] For the SL1, Model 20 and D compilers, only a constant may be specified for storage allocation, and this constant must not exceed 127 for the SL1 compiler. Supposedly this constant must be the same for each corresponding parameter and argument. However, this rule may be disobeyed for the Model 20, D, and F compilers, by declaring one location for the parameter even though any number is used for the argument. For the F compiler, this technique is more efficient than an asterisk. For Student PL, *only* an asterisk may be used in a DECLARE statement, and the allocation must be performed separately with a statement such as "ALLOCATE X(10000);" or "ALLOCATE X(N);." The ALLOCATE statement is discussed more fully in Case Study 16.

[2] Unsupported by Student PL and SL1. For Student PL we must use the scheme in STTSTC, except with "X(K) = INPUT;" in the DO loop. For SL1 we may not use subscripts in GET or PUT lists, but we may get around this restriction by using "GET LIST(Q); X(K) = Q;" in a DO loop.

For example, the input for **STTSTC** may be entirely taken care of by the single **GET** statement:

S10: GET LIST (N, (X(K) DO K = 1 TO N));

Note that we must get in **N** before using it in a repetitive specification. Note also the differences from a **DO** loop: the data item comes before the keyword "DO," no semicolons are used, and the grouping is accomplished by parentheses rather than with keywords "DO" and "END." The limits of the repetitive specification may depend upon variables previously gotten in the same **GET** statement. This was done with **N** in our example. Conceptually, it helps to consider the entire specification within parentheses to be a single "item" which is treated just like any other item in the list. For example, we could have the statements:

GET LIST ((repetitive specification), variable, (repetitive specification));
PUT LIST ((repetitive specification));

Note that a separate set of parentheses is needed for a repetitive specification even if it is the only item in the list.

To avoid precounting the data for **STTSTC**, we could write:

ON ENDFILE (SYSIN) GO TO DONE;
GET LIST((X(N) DO N = 1 BY 1));
DONE: N = N - 1;

EXERCISES

1. The standard deviation is used more frequently than the average deviation as a measure of spread. It is defined as

$$\sqrt{\sum_{k=1}^{n} (x_k - \bar{x})^2 / n}$$

Write a corresponding function procedure STDEV(N, AVERAGE, X) similar to AVGDEV.

2. Write a procedure similar to **STTSTC** that uses AVG and STDEV from exercise 1 to find the average and standard deviation for any number of sets of data.

3. Manipulation of the formula in exercise 1 gives

$$\sigma = \sqrt{\sum_{k=1}^{n} (x_k)^2 / n - (\bar{x})^2}$$

With this formula, the average and standard deviation may be evaluated simultaneously, avoiding the use of subscripted variables. Write a program similar to MEAN of Case Study 6 that does this.

▼4. Rewrite exercise 5 of Case Study 6 as a subordinate procedure using subscripted variables; then write a main procedure that uses it to find any number of combined centers of gravity.

5. Finding the least squares fit of a straight line is frequently required as part of a larger program; so adapt LINE of Case Study 6, making it a subordinate procedure using subscripted variables.

▼6. Write a procedure that gets in data for two lines, uses LINE of exercise 5 for each line, and then uses the procedure TWOEQN of exercise 3 in Case Study 5 to find the intersection of the two lines, putting out the x and y coordinates of this intersection.

9.5 BOUNDS AND EXTENT OF A SUBSCRIPTED VARIABLE

In Case Study 2 we used one formula to approximate the entire sine function. Alternatively, we may store a table of function values as a subscripted variable, and then use interpolation for angles which fall between the tabulated values. This technique may also be used for empirical functions. The table and graph in Fig. 9-1 display the experimentally determined current-voltage characteristic for an Esaki diode.

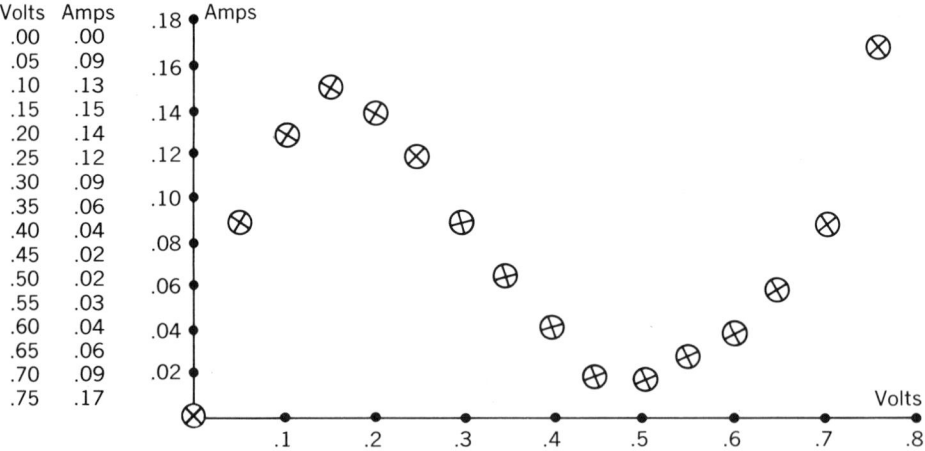

Fig. 9-1 Current-voltage characteristic of an Esaki diode.

If we want to perform a computer analysis of a circuit containing this diode, we may represent the diode's current-voltage characteristic by a function procedure that uses this table, interpolating for voltages which fall between the tabulated values. We may store the values 0.00, 0.09, 0.13, ..., 0.17 as the values of a subscripted variable, AMPS. Since the first value in the table corresponds to a voltage of 0, it is natural to want the first subscript to be 0 rather than the usual 1. We may do this with the declaration:

DECLARE AMPS(0:15);

This declaration means that the first value of the subscript is 0 and the last is 15. These numbers are called the *bounds* of a subscript, and the total number of subscripted values is called the *extent*. The extent is 16 in this case, and in general it is one greater than the difference between the bounds. By default, the lower bound is 1 when not specified.[3]

At first glance it may seem appropriate to have the last subscript be 0.75 with the intermediate ones being 0.05, 0.10, 0.15, ..., 0.70 because these are the corresponding voltage values. However, it must be remembered that a subscripted variable is a list of numbers rather than a function. The subscript tells which item in the list is to be used—the zeroth, the first, the sixth, or the ten thousandth. Fractional subscripts are inappropriate. There is, for example, no item between the second and third. We may use fractional expressions as subscripts, but they will be truncated to a whole value. For example:

AMPS(1.5*3) is equivalent to AMPS(4)

We do not have to store the tabulated voltages as an additional subscripted variable because they are equally spaced. The voltage associated with subscript 0 will be 0.00, the voltage associated with subscript 1 will be 0.05, and the voltage associated with subscript K will be V = 0.05*K. Consequently, the subscript associated with voltage V will be:

K = V/0.05 = 20*V

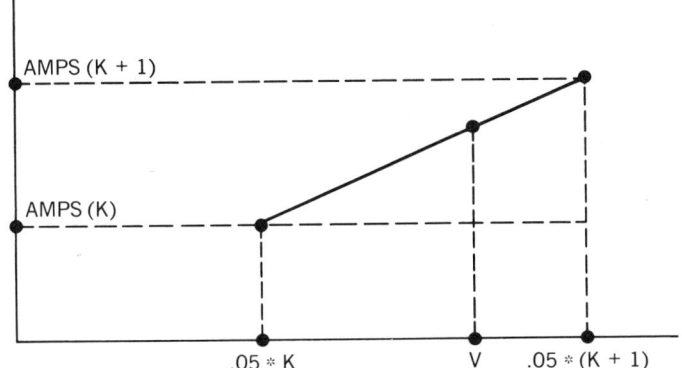

Fig. 9-2 Geometric basis for interpolation.

[3] For the D and Model 20 compilers, the lower bound must be 1, and it may not be explicitly specified. For the SL1 compiler, the lower bound may not be less than −127.

When V lies between two tabulated values, 20*V will lie between the subscripts of the tabulated values to the left and right of X. However, any fractional portion will be dropped when 20*V is assigned to K. Thus, given V, the assignment statement "K = 20*V;" will determine the subscript of the tabulated value to the left. Then, by similar triangles as is shown in Fig. 9-2, the interpolated current will be:

$$\text{AMPS}(K) + \frac{(V - 0.05*K)*[\text{AMPS}(K + 1) - \text{AMPS}(K)]}{0.05}$$

9.6 THE INITIAL AND STATIC ATTRIBUTES

The following function procedure implements these formulas, introducing the INITIAL and STATIC attributes:

```
ESAKI:   PROCEDURE (V);
   DECLARE AMPS(0:15) INITIAL (0, .09, .13, .15, .14, .12, .09, .06,
      .04, .02, .02, .03, .04, .06, .09, .17) STATIC;
   K = 20*V;
   RETURN (AMPS(K) + 20*(V - .05*K)*(AMPS(K+1) - AMPS(K)));
END;
```

AMPS remains the same for each invocation; so there is no need to pass it as a parameter. Therefore, it is established as a variable internal to ESAKI. Rather than establish its values by 16 assignment statements, we use the more compact INITIAL attribute. The successive elements of AMPS beginning with AMPS(0) are initialized to the successive numbers in the list after the keyword INITIAL. Moreover, AMPS has been given the STATIC attribute to avoid having the initialization performed each time the procedure is invoked. By default, variables internal to subordinate procedures are given the AUTOMATIC attribute, meaning that space is allocated for them each time the procedure is entered and freed each time the procedure is left. For example in ESAKI, storage for K is automatically allocated each time ESAKI is entered, and freed each time ESAKI is left. This allows the space to be used for other purposes when ESAKI is not active. The AUTOMATIC attribute saves space, and the STATIC attribute saves time. Consequently, it is best to declare all variables STATIC if there is enough space.

More generally, any number in an initial value list may be preceded by an *iteration factor* enclosed in parentheses. This iteration factor specifies how many successive elements are to be initialized to the value which follows it. For example, to initialize all 50 values of a subscripted variable named VECTOR to zero, we may write:

```
DECLARE VECTOR(50) INITIAL ((50)0);
```

The INITIAL attribute may also be applied to unsubscripted variables.[4]

---

[4] Student PL does not support the INITIAL or STATIC attributes. SL1 does not support iteration factors; only one constant may be listed in the INITIAL attribute; and this constant must have the same precision and scale as the variable. For the Model 20, the INITIAL attribute may not be declared for arrays of the AUTOMATIC class.

9.7 A GENERAL-PURPOSE INTERPOLATION PROCEDURE

The Esaki diode procedure can easily be adapted for any other function, but interpolation is required so frequently that it would be useful to have a general-purpose function procedure. Let us then consider a general function $Y = F(X)$, but with the tabulated values of X unequally spaced. We must then store both X and Y as subscripted variables. We may no longer use a simple formula such as $K = 20*V$ to find the proper interval in which to interpolate. We must now search for the proper interval. Also, we should now pass the subscripted variables X and Y as parameters since we do not know in advance what their values will be. Since the first element of X is not necessarily zero, we will revert to the customary practice of having the lower subscript bounds be one. Accordingly, we may write this general-purpose function procedure as follows:

```
F:  PROCEDURE (XIN, X, Y);
    /* THIS GENERAL PURPOSE FUNCTION PROCEDURE EVALUATES ANY EMPIRICAL
       OR MATHEMATICAL FUNCTION BY LINEAR INTERPOLATION FOR ARBITRARILY
       SPACED VALUES OF THE INDEPENDENT VARIABLE.  */
    DECLARE
       XIN    /* THE INDEPENDENT VARIABLE */,
       X(*)   /* THE TABULATED VALUES OF THE INDEPENDENT VARIABLE */,
       Y(*)   /* THE TABULATED FUNCTION VALUES CORRESPONDING TO THE
                  TABULATED X'S  */;
    DO J = 2 BY 1;
       IF XIN <= X(J)  THEN
           RETURN(Y(J-1)+(Y(J)-Y(J-1))*(XIN-X(J-1))/(X(J)-X(J-1)));
    END;
END;
```

When XIN is within the range of the subscripted variable X, there must be a valid subscript J for which $X(J-1) < XIN <= X(J)$. This value of J is the subscript of the point to the *right* of XIN. To find this subscript, we may start with $J = 2$, increasing it by one until $XIN <= X(J)$. As is shown in Fig. 9-2, the interpolated value of the function then becomes:

$$Y(J-1) + \frac{(Y(J) - Y(J-1))*(XIN - X(J-1))}{X(J) - X(J-1)}$$

In the procedure we have calculated $J - 1$ separately to avoid having the computer evaluate the same subexpression four times.

If XIN is less than X(1), the procedure will give an extrapolated value. If this extrapolation is not valid, the procedure should be modified to warn the user that an invalid value of XIN has been used. If XIN is greater than the last value of X, the search will continue into the invalid adjacent area of memory without warning. If there is a chance of this happening, the procedure should be modified to warn the user that an invalid value of XIN has been used. To do this, we may include a parameter N which specifies the number of valid values of X, then test J against N.

As an example of the use of this procedure, assume that we have experimentally determined unequally spaced values of fish weight as a function of fish length, and we wish to produce a table using interpolation for every inch of length from 18 through 118 inches. The following program does this, using input consisting of the number of fish, followed by the length and weight of the shortest fish, followed by the length and weight of the next shortest fish, and so on:

```
FISHY:  PROCEDURE OPTIONS (MAIN);
   DECLARE (LENGTH(5000), WEIGHT(5000), LENGHIN) FLOAT;
   GET LIST (N, (LENGTH(L), WEIGTH(L) DO L = 1 TO N));
   PUT PAGE LIST ('LENGTH', 'WEIGHT');
   DO LENGTHIN = 18 TO 118;
      PUT SKIP LIST (LENGTHIN, F(LENGTHIN, LENGTH, WEIGHT));
   END;
END;
```

If instead the data were arranged so that all of the weights followed all of the lengths, we could write the input statement:

```
GET LIST (N, (LENGTH(L) DO L = 1 TO N),
             (WEIGHT(L) DO L = 1 TO N));
```

If instead the data were arranged pairwise but in order of *decreasing* length, we could write the input statement:

```
GET LIST (N, (LENGTH(L), WEIGHT(L) DO L = N TO 1 BY -1));
```

SUMMARY

Singly-Subscripted Variables

Singly-subscripted variables consist of a list of variables all referred to by the same name, individual elements being distinguished by a subscript that is the truncated value of an expression in parentheses following the name. Successive integer subscripts denote successive items in the list. The highest valid subscript bound for a subscripted variable must be declared:

DECLARE variable name (upper bound)

The lowest valid bound must also be declared if different from 1:

DECLARE variable name (lower bound: upper bound)

The bounds must be constants if declared in a main procedure. The bounds of a parameter may be indicated by an asterisk. All of a subscripted variable is indicated for an argument or parameter by listing its name without subscript.

Repetitive I/O Specifications

A repetitive specification may be used in place of a single variable in the list of a GET or PUT statement. The general form of a repetitive specification is:

(list of variables DO variable = list of specifications)

where the specifications are like those for the DO loop.

EXERCISES

7. Modify FISHY so that it also uses AVG and AVGDEV to find the average and average deviation of the measured lengths and measured weights.
8. Write a procedure like FISHY which is appropriate to your discipline and show some sample data.
▼9. Change ESAKI so that the smallest value of K used is 0 and the largest value is 14. This causes extrapolation for values of V outside the tabulated range.
10. Change ESAKI into a subroutine ESAKI(V, CURRENT, IER) where CURRENT is the output current and IER is an error indicator given the value 1 if V is outside the tabulated range and 0 otherwise.
▼11. ESAKI unnecessarily evaluates the interpolation formula when V happens to coincide with a tabulated value. Modify ESAKI to directly return AMPS(K) in this instance. Assuming all floating-point values are uniformly distributed between 0 and 0.75 with a mantissa of length six, will this strategy save or lose time on the average?
12. The population of a yeast colony as a function of time is given by the following table. Write a corresponding function procedure similar to ESAKI, then write a main procedure which uses it to repetitively get in values of time and put them out together with the corresponding population.

Time	0	3	6	9	12	15	18	21	24	27	30	33	36	39	42
Population	1	2	5	11	23	48	106	230	390	690	680	630	580	560	500

13. Write a general-purpose function procedure F(N, XIN, XMIN, DX, Y) for interpolation with equally spaced values of the independent variable where N is the number of tabulated values, XMIN is the minimum tabulated value of the independent variable, DX is its uniform increment, and Y, containing the tabulated function values, is a subscripted variable whose lower bound is 0. Use extrapolation for values of XIN outside the tabulated range.
▼14. Same as exercise 13 except write it as a subroutine including an error indicator which is set equal to 1 if XIN is outside the tabulated range and set equal to 0 otherwise.

15. Rewrite the procedure for interpolation with arbitrary spacing so that it extrapolates on both ends.

16. Rewrite the procedure for interpolation with arbitrary spacing as a subroutine. Include an error indicator which is set equal to 1 if XIN is outside the tabulated range and set equal to 0 otherwise.

17. The procedure for interpolation with unequal intervals uses a *linear search* technique to isolate the proper interval, but the *binary search* in the flow chart below is more efficient if N, the number of tabulated points, is greater than about a half dozen.

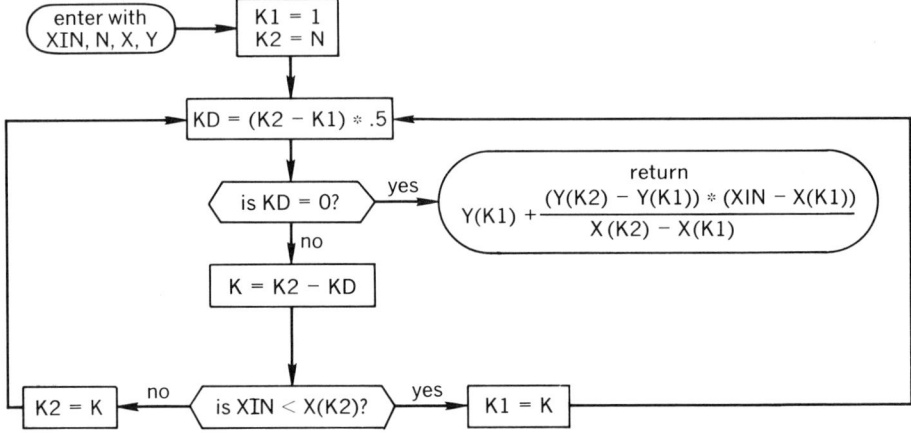

This technique bears a great resemblance to Bolzano's method for refining the root of an equation. Write a corresponding function procedure.

▼18. One of Newton's divided difference formulas giving the interpolated value based upon passing a parabola through three successive points is:

Y(J − 2) + (XIN − X(J − 2))(D1 + (XIN − X(J − 1))(D2 − D1)/(X(J) − X(J − 2)))

where D1 = $\dfrac{Y(J-1) - Y(J-2)}{X(J-1) - X(J-2)}$ and D2 = $\dfrac{Y(J) - Y(J-1)}{X(J) - X(J-1)}$

with   X(J − 2) <= XIN <= X(J)

For smooth functions, this formula is generally more accurate than linear interpolation using the same number of points. Write a corresponding general-purpose interpolation procedure.

19. (Difficult) Same as exercise 18 except derive and use the corresponding formula for equally spaced values of X.

▼20. Write a subroutine SMOOTH (N, X) that modifies successive values of the subscripted variable X from J = 2 through N − 1 according to the formula X(J) = (X(J − 1) + X(J) + X(J + 1))/3. This smoothing reduces the high-frequency noise in the data stored in X.

21. The procedure of exercise 20 is an example of a digital low-pass filter. We may achieve a digital high-pass filter by subtracting from each X(J) the average of all values of X up through X(J). Starting with J = 1 we successively apply the pair of formulas

$$AVG = \frac{(J - 1)*AVG + X(J)}{J}$$

$$X(J) = X(J) - AVG$$

until J = N. Write a corresponding subroutine HIPASS(N, X).

▼**22.** (Mathematically Advanced) Numerical differentiation may be used to find the approximate derivative of tabulated empirical or mathematical functions:

$$DYONDX(J) = \frac{Y(J + 1) - Y(J - 1)}{2*DX}$$

for J = 2 through N - 1 where Y contains the tabulated function values, N is the number of tabulated function values, and DX is the uniform spacing of the independent variable. Write a corresponding function procedure DERIV1(N, DX, Y, DYONDX). (Note that there is a PL/I version of a program called FORMAC that can perform *analytic* differentiation.)

23. (Mathematically Advanced) Same as exercise 22 except for the second derivative:

$$D2YONDX2(J) = \frac{Y(J - 1) - 2*Y(J) + Y(J + 1)}{DX*DX}$$

CASE STUDY 10

SORTING AND MERGING

The BEGIN Block and Array Expressions

Finding the maximum value in a list stored as a subscripted variable is such a frequent requirement that it would be nice to have a function procedure for it. The MAX function will not suffice because it is designed to find the maximum among several arguments rather than the maximum within one subscripted argument. To find the maximum element in a subscripted variable, the following subroutine arbitrarily assumes that it is the first element, then searches through the rest of the elements, revising the assumption whenever a larger value is found:

10.1 MAXIMIZATION EXAMPLE

```
VMAX:   PROCEDURE (N, X);
    DECLARE X(*);
    XMAX = X(1);
    DO K = 2 TO N;
        IF X(K) > XMAX  THEN XMAX = X(K);
    END;
    RETURN (XMAX);
END;
```

However, we may also want to know *which* element is largest. The following subroutine finds both the largest value and its subscript:

```
MAXID:   PROCEDURE (N, X, XMAX, K_OF_XMAX);
   DECLARE X(*);
   K_OF_XMAX = 1;   XMAX = X(K_OF_XMAX);
   DO K = 2 TO N;
      IF X(K) > XMAX   THEN DO;   K_OF_MAX = K;   XMAX = X(K);   END;
   END;
END;
```

More generally, we may write a subroutine which finds the maximum value and its subscript from the Jth position on:

```
MAXJON:  PROCECURE (J, N, X, XMAX, K_OF_XMAX);
   DECLARE X(*);
   K_OF_XMAX = J;   XMAX = X(K_OF_MAX);
   DO K = J+1 TO N;
      IF X(K) > XMAX   THEN DO;   K_OF_XMAX = K;   XMAX = X(K);   END;
   END;
END;
```

10.2 SORTING EXAMPLE

We are all capable of sorting numbers, but let us develop a way that is systematic so that we may write a subroutine for it. Specifically, let us sort the list of numbers in Fig. 10-1 into descending order.

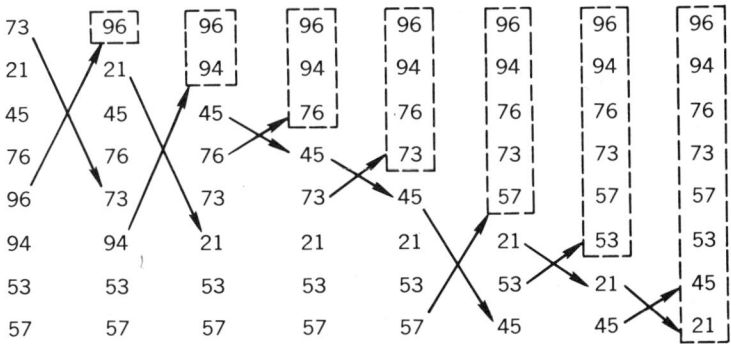

Fig. 10-1 Successive stages of sorting.

First, we may search through the entire list to find the largest value, then interchange it with the first value to bring the largest value to the first position. Next we may search through the remainder of the list to find the largest of the remaining values, then interchange it with the second value to bring the second largest value to the second position. By continuing in this fashion, we may bring the third largest to the third position, and so on, until we have brought the next to the smallest to the next to the last position, leaving the smallest in the last position.

To find the largest value we may use our subroutine **MAXJON** with $J = 1$; then to find the second largest we may use the **MAXJON** with $J = 2$; and so on, up to $J = N - 1$:

```
SRCHST:  PROCEDURE (N, X);
   DECLARE X(*);   /* INPUT & OUTPUT: THE SUBSCRIPTED VARIABLE WHOSE
      ELEMENTS ARE REARRANGED INTO DESCENDING ORDER. */
   DO J = 1 TO N-1;
      CALL MAXJON(J, N, X, XMAX, K_OF_XMAX);
      X(K_OF_XMAX) = X(J);
      X(J) = XMAX;
   END;
END;
```

Actually, **MAXJON** was developed merely to simplify the explanation of sorting. It is unlikely to have much use outside of **SORT**; so it would be more efficient to combine the two. We may do this by replacing the **CALL MAXJON** statement with the procedure itself, less its **PROCEDURE**, **END**, and **DECLARE** statements:

```
SORT:  PROCEDURE (N, X);
   DECLARE X(*);
   DO J = 1 TO N-1;
      K_OF_XMAX = J;  XMAX = X(K_OF_XMAX);
      DO K = J+1 TO N;
         IF X(K) > XMAX THEN DO;  K_OF_XMAX = K;  XMAX = X(K);  END;
      END;
      X(K_OF_XMAX) = X(J);  X(J) = XMAX;
   END;
END;
```

EXERCISES

▼1. Write a subroutine which sorts numbers into *ascending* order by the search-sort technique of Sec. 10.2.

2. The following flow chart illustrates how to sort into descending order by the *insertion* method—an interchange is made every time a larger value is found:

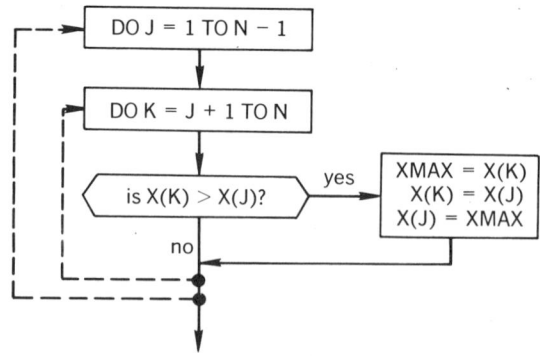

Sec. 10.3   Merging Example   169

Write a corresponding subroutine. Note that it would not do to have "X(K) = X(J); X(J) = X(K);" for the interchange because the first assignment statement would destroy the original value of X(K) before it could be used in the second assignment statement. Interchanges always require an extra storage location.

3. Same as exercise 2 except modify it to sort into *ascending* order.

▼4. (Difficult) The following flow chart illustrates how to sort into descending order by the *exchange* or *bubble* method. Write a corresponding subroutine.

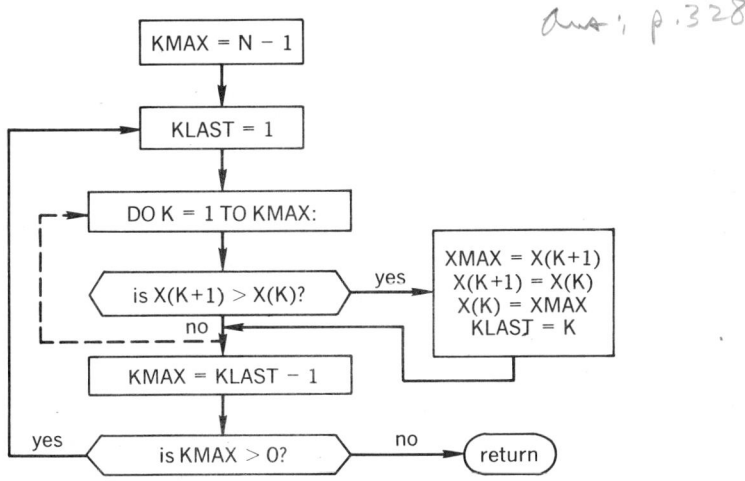

Ans: p.328

5. Fish are not ordinarily caught in order of increasing length; so rewrite the procedure FISHY of the previous case study so that it sorts the weights and lengths in order of increasing lengths before producing the table. Note that the corresponding weights must be interchanged each time two lengths are interchanged or else the correspondence will be lost.

10.3 MERGING EXAMPLE

Frequently we have the task of merging two already sorted lists into a single sorted list. The flow chart in Fig. 10-2 and the following procedure show how to do this for lists arranged in descending order.

```
MERGE:   PROCEDURE (M, N, X, Y, Z);
   DECLARE X(*), Y(*), Z(*);
   X(M+1), Y(N+1) = 1E30;   I, J = 1;
   DO K = 1 TO M+N;
      IF X(I) > Y(J)   THEN DO;   Z(K) = X(I);   I = I + 1;   END;
                       ELSE DO;   Z(K) = Y(J);   J = J + 1;   END;
   END;
END;
```

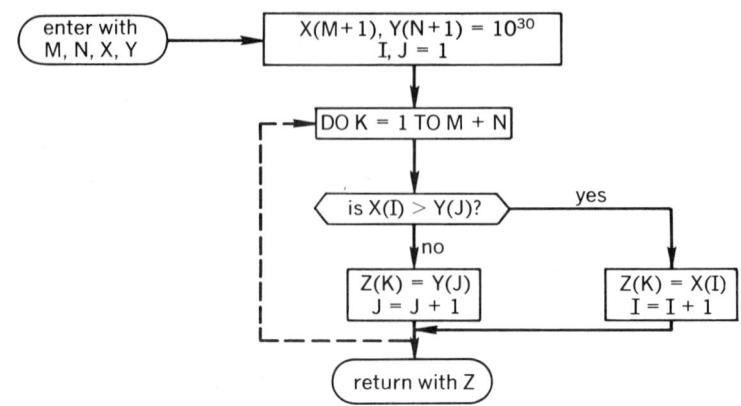

Fig. 10-2 A flow chart for merging.

Note the artifice of establishing very large values of the extra elements $X(M+1)$ and $Y(N+1)$ so that there will always be values of both X and Y for the comparison, even when their real endings $X(M)$ and $Y(N)$ are not reached simultaneously. It is assumed that 10^{30} is larger than any of the actual values of X or Y.

10.4 ADJUSTABLE BOUNDS

In writing a main procedure to use this subroutine, we have the problem of deciding how to partition the 10,000 available locations among the three subscripted variables. If we use

DECLARE X(2500), Y(2500), Z(5000);

we unnecessarily preclude merging a long list of say 4000 elements with a shorter list of up to 1000 elements. On the other hand if we use

DECLARE X(4000), Y(1000), Z(5000);

we unnecessarily preclude merging two medium length lists of say 2000 elements each. One way to get around this dilemma is to have a main procedure which merely gets the bounds of the subscripts, passing them to another procedure which does everything else. For example, the following program uses this technique to get two lists, merge them, and put out the merged list, allocating the subscripted variable storage during execution:

```
BOUNDS:  PROCEDURE OPTIONS (MAIN);
         GET LIST (M, N);
         CALL MAJOR (M, N);
         END;
```

```
MAJOR:  PROCEDURE (M, N);
   DECLARE X(M+1), Y(N+1), Z(M+N);
   GET LIST ((X(K) DO K = 1 TO M), (Y(J) DO J = 1 TO N));
   CALL MERGE (M, N, X, Y, Z);
   PUT PAGE DATA (Z);
END;
```

We may use parameters as bounds for the declaration of internal subscripted variables because storage for internal variables is allocated upon entry into a subordinate procedure.[1] This technique is not possible in a main procedure because it has no parameters and the DECLARE statement is executed before the subscript bounds can be established by a GET statement.

10.5 INPUT AND OUTPUT OF ENTIRE ARRAYS[2]

Note that in the PUT DATA statement of MAJOR we have listed Z without subscripts or a repetitive specification. This will cause all of Z to be put out beginning with Z(1). For example, the output might look somewhat as follows:

```
Z(1)= 8.50000E+02   Z(2)= 7.30000E+02   Z(3)= 2.70000E+02

Z(4)= 7.50000E+01   Z(5)= 9.99999E+29;
```

Since all of the allocated values are put out, we must be careful not to use this form when we are not using all of the allocated values as was the case in DATARED of the previous case study. This form may also be used with input, but it is unsuitable in MAJOR because it would force us to prepare phony data values for X(M + 1) and Y(N + 1).

10.6 THE BEGIN BLOCK[3]

The effect of the allocation at execution time may be accomplished more directly with the BEGIN block:

```
REVISE:  PROCEDURE OPTIONS (MAIN);
   GET LIST (M, N);
   BEGIN;
      DECLARE X(M+1), Y(N+1), Z(M+N);
      GET LIST ((X(K) DO K = 1 TO M), (Y(J) DO J = 1 TO N));
      CALL MERGE (M, N, X, Y, Z);
      PUT PAGE LIST (Z, (X(K) DO K = 1 TO M), (Y(J) DO J = 1 TO M));
   END;
END;
```

[1] Constants must be used for the Model 20, SL1, and D compilers; and asterisks must be used for Student PL.
[2] Unsupported by Student PL.
[3] Unsupported by the Model 20.

The BEGIN block begins with the keyword BEGIN and ends with the keyword END. Like a DO group, it is entered sequentially, and it has access to labels in the procedure which contains it. Like a subordinate procedure, it may declare its own internal variables. Its primary purpose is to regulate the allocation of storage.[4]

Like internal procedures discussed in Case Study 5, BEGIN blocks may be nested. To determine exactly where a name is known in a nest of PROCEDURE blocks and/or BEGIN blocks—the name's *scope*—we must know what is meant by *explicitly declared*. A name is said to be explicitly declared if it appears in a DECLARE statement, in a parameter list, as a statement label, or as a procedure label.

A contained block has access to any names established in the block or blocks which contain it unless the names are explicitly redeclared in the contained procedure. For example, the BEGIN block in REVISE has access to M and N.

A containing block has access to any names in its contained block provided they are not explicitly declared therein. For example, K, but not Z, is known outside the BEGIN block in REVISE. These rules are discussed more thoroughly in Sec. 5.3.

As another example, suppose that we have a non-overlapping need for a 10,000 element vector X and a 10,000 element vector Y with only 10,000 total spaces available. We may use this space for both variables by declaring them in separate BEGIN blocks.

Another thing to note about the procedure REVISE is that we did not use "PUT PAGE LIST (Z, X, Y);" because output of the extra values $X(M + 1) = 10^{30}$ and $Y(N + 1) = 10^{30}$ would cause confusion.

10.7 ARRAY EXPRESSIONS[5]

We have seen how all of the elements of a subscripted variable may be designated for input, output, parameters, or arguments by listing their names without subscripts. The same is true for arithmetic operations. For example, if A, B, C, and D are subscripted variables with equal bounds 1 to N, the following statements are valid with the indicated equivalents:

```
                        ┌DO  J  =  1  TO  N:
C  =  A  +  B;          │    C(J)  =  A(J)  +  B(J);
                        └END;
                        ┌DO  J  =  1  TO  N;
A  =  (A*B/C)**D;       │    A(J)  =  (A(J)*B(J)/C(J))**D(J);
                        └END;
```

---

[4] For the F compiler, another use of the BEGIN block is to include more than one statement in an on-unit. The DO group is not allowed for this purpose.

[5] SL1 and the Model 20 allow only the array assignment "array = array;" or "array = scalar expression;."

```
A, B, = SIN(C);        ⎧DO J = 1 TO N;
                       ⎨   A(J), B(J) = SIN(C(J));
                       ⎩END;
                       ⎧DO J = 1 TO N;
A = MAX(B, C);         ⎨   A(J) = MAX(B(J), C(J));
                       ⎩END;
```

We may use any of the arithmetic operations or built-in functions—they are interpreted on an element-by-element basis. As with input and output of entire arrays, array expressions should not be used when we are using only part of the allocated space. Although all of the arrays must have identical bounds, we may mix arrays and unsubscripted variables or constants. For example, if A and B are subscripted variables with bounds 1 and N, and U is an unsubscripted variable, the following statements are valid with the indicated equivalents:

```
                       ⎧DO J = 1 TO N;
A = B + 3;             ⎨   A(J) = B(J) + 3;
                       ⎩END;
                       ⎧DO J = 1 TO N;
A = 3*B + U;           ⎨   A(J) = 3*B(J) + U;
                       ⎩END;
```

10.8 ARRAY MANIPULATION FUNCTIONS[6]

The standard arithmetic and mathematical built-in functions return an array result for array arguments.[7] There are also several array manipulation functions which return a single value for array arguments. One of these is **SUM(X)** which returns the sum of the elements in the subscripted variable X. For example, *provided we are using all of the allocated storage*, the effect of our function procedures for the average and average deviation written in the previous case study may be accomplished with the two statements:

```
AVG = SUM(X)/N;
AVG_DEV = SUM(ABS(X - AVG))/N;
```

Another array-manipulation function is **PROD(X)** which returns the product of the elements in the subscripted variable X. For example, provided we are using all of the allocated storage, the geometric mean may be calculated by the statement:

```
GMEAN = PROD(X)**-N;
```

---

[6] The Model 20, SL1, and Student PL do not support these functions.
[7] For SL1 and the Model 20, array arguments may not be used for the built-in functions.

This statement is equivalent to the statements:

```
PROD = 1;
DO K = 1 TO N;
    PROD = PROD*X(K);
END;
GMEAN = PROD**-N;
```

10.9 SUBSCRIPTED STATEMENT LABELS

Suppose that we wish to write a function procedure for the current through a junction diode, as given by the expressions in Fig. 10-3. We may not use our interpolation procedure from Case Study 9 because the curve is approximated by straight lines for some segments and by parabolas for others. However, we may use *subscripted statement labels* to assist with the selection of the proper formula:

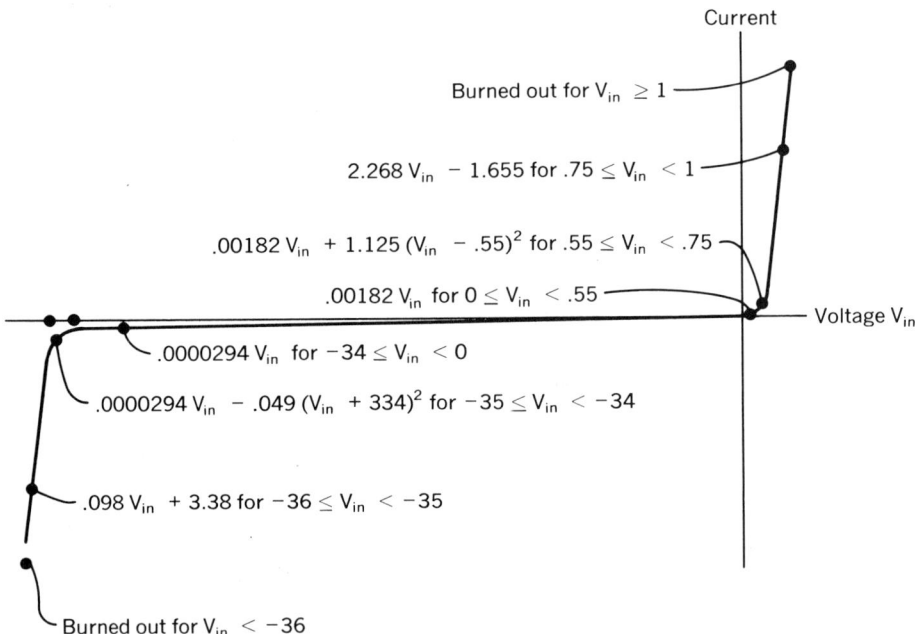

Fig. 10-3 Approximate current-voltage relationship for a junction diode.

```
DIODE:  PROCEDURE (VIN);
    DECLARE STMT(7) LABEL,
        V(7) INITIAL (-36, -35, -34, 0, .55, .75, 1) STATIC;
    DO J = 1 TO 7;
        IF VIN < V(J)  THEN GO TO STMT(J);
    END;
```

Sec. 10.10   *Optional Difficult Application: Indirect Sorting*   175

```
        STMT(1):   PUT LIST ('DIODE BURNED OUT');  STOP;
        STMT(2):   RETURN (.098*VIN + 3.38);
        STMT(3):   RETURN (.0000294*VIN - .049*(VIN+34)**2);
        STMT(4):   RETURN (.0000294*VIN);
        STMT(5):   RETURN (.00182*VIN);
        STMT(6):   RETURN (.00182*VIN + 1.125*(VIN-.55)**2);
        STMT(7):   RETURN (2.268*VIN - 1.655);
    END;
```

STMT is a subscripted label variable; so we have declared its bounds and label attribute.

10.10 OPTIONAL DIFFICULT APPLICATION: INDIRECT SORTING

When several related subscripted variables are to be sorted on the basis of one of them, it is best not to sort at all. Instead it is better to derive an integer subscripted variable which gives, for ascending order, the *subscript* of the smallest value followed by the subscript of the second smallest, and so on. For example, assume that we have a subroutine ASCEND(N, X, LOC) which, for N floating-point values X, derives N fixed-point values LOC such that LOC(1) is the subscript of the smallest X, LOC(2) is the subscript of the second smallest X, and so on. We can then modify the procedure BEARING of Case Study 4 as follows so that it prints out the acceptable bearing specifications in order of increasing cost:

```
BEARING:   PROCEDURE OPTIONS (MAIN);
    DECLARE
        (#(1000)           /* IDENTIFICATION NUMBER       */  FIXED(9),
        (R_LOAD(1000)      /* RADIAL LOAD IN POUNDS       */,
        A_LOAD(1000)       /* AXIAL LOAD IN POUNDS        */,
        SPEED(1000))       /* MAXIMUM RPM                 */  FIXED(5),
        (ID(1000)          /* INSIDE DIAMETER IN INCHES   */,
        OD(1000)           /* OUTSIDE DIAMETER IN INCHES  */,
        T(1000))           /* THICKNESS IN INCHES         */  FIXED(5,3),
        COST(1000)         /* PRICE PER UNIT IN $         */  FIXED(5,2))
                                                              DECIMAL,
        X(1000), LOC(1000);
    PUT PAGE LIST ('BEARING NUMBERS', 'RADIAL LOAD', 'AXIAL LOAD',
        'MAXIMUM RPM', 'INSIDE DIAM.', 'OUTSIDE DIAM.', 'THICKNESS',
        'PRICE/UNIT');
    ON ENDFILE(SYSIN) GO TO S20;
    N = 1;
    NEXT:  GET LIST(#(N), R_LOAD(N), A_LOAD(N), SPEED(N), ID(N), OD(N),
            T(N), COST(N));
        X(N) = COST(N);
        IF (SPEED(N) >= 50000) & (ID(N) = 0.625) & (T(N) >= 0.25)
            THEN N = N +1;
        GO TO NEXT;
    S20:   N = N -1;  CALL ASCEND(N, X, LOC);
        PUT SKIP LIST ((#(LOC(J)), R_LOAD(LOC(J)), A_LOAD(LOC(J)),
            SPEED(LOC(J)), ID(LOC(J)), OD(LOC(J)), T(LOC(J)), COST(LOC(J))
            DO J = 1 TO N));
END;
```

ASCEND may be written as follows, using the search-sort technique:

```
ASECND:  PROCECURE (N, X, LOC);   /* INDIRECT SEARCH-SORT */
   DECLARE X(*), LOC(*);
   DO J = 1 TO N;
      LOC(J) = J;
   END;
   DO J = 1 TO N-1;
      K_OF_XMIN = J;  XMIN = X(LOC(J));
      DO K = J+1 TO N;
         IF X(LOC(J)) < XMIN  THEN DO;
            K_OF_XMIN = K;  XMIN = X(LOC(K));  END;
      END;
      LOC_J = LOC(J);
      LOC(J) = LOC(K_OF_XMIN);
      LOC(K_OF_XMIN) = LOC_J;
   END;
END;
```

LOC is sometimes called an *indirect address*. This concept is carried further with PL/I's list-processing **POINTER** attribute discussed in Case Study 16.

SUMMARY

BEGIN Block

A BEGIN block may be used to achieve allocation of storage during execution. A typical use is:

```
procedure name; PROCEDURE OPTIONS (MAIN);
   GET LIST (subscript bounds);
   BEGIN;
      DECLARE variable 1 (bound 1), variable 2 (bound 2) ...;
      ...
   END;
END;
```

A name is said to be explicitly declared if it appears in a DECLARE statement, in a parameter list, as a statement label, or as a procedure label.

A contained block has access to any names established in the block or blocks which contain it unless the names are explicitly redeclared in the contained procedure.

A containing block has access to any names in its contained block provided they are not explicitly declared therein.

I/O For Entire Array

An entire array may be indicated for input or output by listing its name without subscripts.

Array Expressions

When an array is used without subscripts in an expression, the entire array is designated on an element-by-element basis. All items in an array expression must be either scalars or arrays with identical bounds.

Array Manipulation Functions

SUM(x) returns the sum of all the elements in its subscripted argument.

PROD(x) returns the product of all the elements in its subscripted argument.

EXERCISES

▼6. The dot, inner, or scalar product of two vectors with bounds 1 to N is defined as

$$A(1)*B(1) + A(2)*B(2) + \cdots A(N)*B(N) = \sum_{J=1}^{N} A(J)*B(J)$$

Write a statement to evaluate it assuming A and B have both been allocated exactly N storage locations.

▼7. Same as exercise 6 except write a function procedure to evaluate it assuming more than N storage locations have been allocated.

8. Excluding decay products which are themselves radioactive, the total radiation intensity of a mixture of N radioactive isotopes as a function of time T is

$$R = \sum_{J=1}^{N} A(J) * EXP(-T/TAU(J))$$

where A(J) is the initial radiation intensity of the Jth isotope and TAU(J) is its time constant. Write a statement to evaluate R assuming A and TAU have both been allocated exactly N storage locations.

9. Same as exercise 8 except write a function procedure to evaluate it assuming more than N storage locations have been allocated.

10. The weighted mean of N numbers X(K) is given by the formula

$$WMEAN = \frac{\sum_{K=1}^{N} X(K) * W(K)}{\sum_{K=1}^{N} W(K)}$$

Write a statement to evaluate it assuming exactly N storage locations have been allocated for X and W.

11. Same as exercise 10 except write a function procedure to evaluate it assuming more than N storage locations have been allocated.

12. The Mth norm of a vector with bounds 1 to N is defined as

$$\left[\sum_{K=1}^{N} |X(K)|^M \right]^{1/M}$$

Write a statement to evaluate this norm, assuming exactly N storage locations have been allocated for X.

13. Same as exercise 12 except write a function procedure to evaluate the norm assuming more than N storage locations have been allocated.

▼14. (Mathematically Advanced) The Fourier series is frequently used to approximate periodic functions:

$$Y = C + \sum_{K=1}^{N} \left\{ A(K) * SIN\left[\frac{K*T}{\pi * P}\right] + B(K) * COS\left[\frac{K*T}{\pi * P}\right] \right\}$$

where T is the independent variable and C and P are constants as are the N values of A and B. Write a statement to evaluate Y assuming exactly N storage locations have been allocated for A and for B.

▼15. (Mathematically Advanced) Same as exercise 14 except write a function procedure to evaluate Y assuming more than N storage locations have been allocated. The calculations may be speeded up by taking advantage of the following identities:

$$\cos 2t = 2(\cos t)^2 - 1$$
$$\sin 2t = 2 \sin t \cos t$$
$$\cos (k+1)t = 2 \cos t \cos kt - \cos (k-1)t$$
$$\sin (k+1)t = 2 \cos t \sin kt - \sin (k-1)t$$

16. If the x, y, and z coordinates of point A are stored as A(1), A(2), and A(3), respectively; the x, y, and z coordinates of point B are stored as B(1), B(2), and B(3), respectively; and similarly for point C; then the x, y, and z coordinates of the center of the triangle ABC are given by the vector D = (A + B + C)/3. Declaring A, B, C, and D to have an extent of three, write a program that repetitively gets in the corner coordinates of any number of triangles and puts out their center coordinates.

▼17. Same as exercise 16 except use a BEGIN block so that the program can be used for triangles in both two- and three-dimensional coordinate systems.

▼18. If we have subroutine LINE (N, X, Y, A, B) that determines the least-squares straight-line fit y = Ax + B to the data contained in the subscripted variables X and Y, then the deviations of the data from this line are given by the vector Y − A*X − B, provided exactly N locations are allocated for X and for Y. Write a program that finds the least-squares fit and its average or standard deviation for any number of sets of data.

19. Same as exercise 18 except have the program also sort and put out the data on the basis of increasing X.

▼20. Since N must be passed anyway, why not use the declaration "DECLARE X(N);" in SORT?

21. (Mathematically Advanced) In two dimensions the cross-product of two vectors A and B is defined as the scalar

$$A(1) * B(2) - A(2) * B(1)$$

Write a corresponding function procedure CROSS2(A, B).

22. (Mathematically Advanced) In three dimensions the cross-product of two vectors A and B is defined as the vector

$$C(1) = A(2) * B(3) - A(3) * B(2)$$
$$C(2) = A(3) * B(1) - A(1) * B(3)$$
$$C(3) = A(1) * B(2) - A(2) * B(1)$$

Write a corresponding subroutine CROSS3(A, B, C).

▼23. Write a subroutine PLYADD (M, N, A, B, C) that evaluates the coefficients C of the polynomial which is the sum of the polynomials with coefficients A and B:

$$C(K) * X^K + C(K-1) * X^{K-1} + \cdots C(0) * X^0$$
$$= A(M) * X^M + A(M-1) * X^{M-1} + \cdots A(0) * X^0 +$$
$$B(N) * X^N + B(N-1) * X^{N-1} + \cdots B(0) * X^0$$

Note that the statement "C = A + B;" would only work when M = N.

24. Write a subroutine PLYMLT (M, N, A, B, C) that evaluates the coefficients C of the polynomial which is the product of polynomials with coefficients A and B:

$$C(M+N) * X^{M+N} + \cdots C(0) * X^0 = [A(M) * X^M + \cdots$$
$$A(0) * X^0] * [B(N) * X^N + \cdots B(0) * X^0]$$

The relevant formula is

$$C(K) = \sum_{J=MAX(0, K-N)}^{MIN(K, M)} A(J) * B(K - J)$$

▼25. (Difficult) Write a subroutine PLYDIV (M, N, A, B, Q, R) that evaluates the coefficients Q of the quotient of the polynomial with coefficients A divided by the polynomial with coefficients B. Store the *remainder* as the N − 1 lowest-order coefficients of A.

26. (Difficult) Write a subroutine PLYSUB (M, N, A, B, C) that evaluates the coefficients C of the polynomial resulting from substitution of the polynomial with coefficients B into the polynomial with coefficients A.

▼27. (Mathematically Advanced) Write a subroutine PLYDIF (M, A, B) that evaluates the coefficients B of the polynomial which is the derivative of the polynomial with coefficients A:

$$B(M-1) * X^{M-1} + B(M-2) * X^{M-2} + \cdots B(0) * X^0$$
$$= \frac{d}{dx} [A(M) * X^M + A(M-1) * X^{M-1} + \cdots A(0) * X^0]$$

28. (Mathematically Advanced) Write a subroutine PLYINT (M, A, B) that evaluates the coefficients B of the polynomial which is the indefinite integral of the polynomial with coefficients A:

$$B(M+1) * X^{M+1} + B(M) * X^M + \cdots B(1) * X^1$$
$$= \int^x [A(M) * X^M + A(M-1) * X^{M-1} + \cdots A(0) * X^0] dx$$

29. (Additional Language Feature—F compiler) The built-in array manipulation function POLY(a, x) returns

$$a_m + a_{m+1}x_k + a_{m+2}x_k x_{k+1} + \cdots (a_n x_k x_{k+1} \cdots x_{k+n-m+1})$$

where the bounds of a and x are (m : n) and (k : ℓ) respectively. If x is a scalar, then the function returns

$$a_m + a_{m+1}x^1 + a_{m+2}x^2 + a_{m+3}x^3 + \cdots a_n x^{n-m}$$

Devise an application for this function, and write a corresponding procedure.

30. Invert the equations in Fig. 10-3 to derive an approximate voltage-current relationship, and write a corresponding function procedure.

CASE STUDY 11

MATRIX MULTIPLICATION

Multiply-Subscripted Variables and Array Cross-Sections

Subscripted variables may have more than one subscript, the maximum number being implementation defined.[1] For example, the following table displays some meteorological data for the year 1984, which may be stored as a doubly-subscripted variable RAIN(I, J). The first subscript denotes the row or month, and the second subscript denotes the column or location.

RAINFALL FOR 1984

Station: Month	1	2	3	4	5	6	7
1	9.6	4.7	7.2	8.3	9.6	5.1	3.2
2	7.3	5.7	7.9	9.8	7.7	9.6	8.7
3	6.3	6.8	9.0	4.0	6.2	3.2	7.2
4	5.2	5.1	3.6	7.1	7.0	4.8	5.9
5	4.7	4.0	3.0	2.7	4.7	6.0	7.0
6	3.5	3.6	7.0	4.4	3.9	2.8	7.2
7	6.0	4.0	2.0	3.0	4.1	3.6	5.5
8	3.3	4.8	2.8	3.7	3.2	3.0	4.2
9	2.0	2.2	2.4	2.1	3.3	3.0	4.1
10	5.2	5.1	5.0	3.9	3.1	1.1	1.2
11	1.6	1.7	3.8	3.9	4.2	5.6	6.2
12	9.3	9.4	9.7	7.1	9.2	8.7	9.6

[1] 32 for the F compiler, three for the D and Model 20 compilers, two for SL1, and any number for Student PL.

Note the difference between the number of subscripts and their extents. **RAIN** has two subscripts with extents of 12 and 7. The number of subscripts is sometimes called the number of *dimensions*, but we will avoid this term because it is also ambiguously used for both the extents of the subscripts and the bounds of the subscripts. Multiply-subscripted variables are often called *arrays*, or *matrices*, but the word *matrix* is usually associated with doubly-subscripted variables and the word vector is usually associated with singly-subscripted variables. Mathematicians are more restrictive, using the terms matrix and vector only when certain laws pertain to their arithmetic. Physicists are even more restrictive, using the term vector only when certain transformation laws pertain to a change in coordinate system, and using the more general term *tensor* for any arrays which obey these transformation laws. In this text, all unsubscripted variables will be called scalars; all singly-subscripted variables will be vectors; all doubly-subscripted variables will be called matrices; and in general, subscripted variables will be called arrays.

11.1 DECLARATION OF ARRAYS

The storage for **RAIN** may be allocated by the statement:

DECLARE RAIN(12, 7);

Note that the bounds for the two subscripts are separated by a comma. Alternatively, we may make the program work for any number of locations by the declaration:

DECLARE RAIN(12, N);

provided this declaration is within a **BEGIN** block and **N** has been previously established outside the block.[2]

11.2 ARRAY EXPRESSIONS

Array expressions work for multiply-subscripted variables just as they do for vectors—on an element-by-element basis. Assume, for example, that we have matrices **SNOW, SLEET,** and **HAIL** with similar bounds as **RAIN**. We may add these as follows to get a total precipitation matrix:

PRECIPITATION = RAIN + SNOW + SLEET + HAIL;

[2] Bounds must be constants for the SL1, Model 20, and D compilers.

For bounds of 12 and 7 this statement is equivalent to the nest of DO loops:

```
DO I = 1 TO 12;
   DO J = 1 TO 7;
      PRECIPITATION(I, J) = RAIN(I, J) + SNOW(I, J) + SLEET(I, J) + HAIL(I, J);
   END;
END;
```

Note that the operations are performed in row order with the column subscript varying most rapidly.[3]

11.3 ARRAY MANIPULATION FUNCTIONS[4]

The array manipulation functions work the same way for multiply-subscripted variables as they do for vectors. For example, to find the rainfall for the year averaged over the seven stations we may use the statement:

$$\text{AVERAGE} = \text{SUM(RAIN)}/7;$$

For bounds of 12 and 7 this statement is equivalent to the statements:

```
SUM = 0;
DO I = 1 TO 12;
   DO J = 1 TO 7;
      SUM = SUM + RAIN(I, J);
   END;
END;
AVERAGE = SUM/7;
```

11.4 ARRAY CROSS SECTIONS[5]

We may denote an entire cross section of an array by writing an asterisk in the appropriate subscript position. For example, the rainfall at location three for each of the 12 months is denoted by the column vector RAIN(*, 3), and the rainfall in June for each of the seven stations is denoted by the row vector RAIN(6, *). Replacement of all of the subscripts by asterisks denotes all of the array. For example,

$$\text{RAIN}(*, *) \text{ is equivalent to RAIN}$$

---

[3] SL1 and the Model 20 permit only the array expression "array = array;" or "array = scalar expression;."
[4] These are not supported by Student PL, SL1, or the Model 20.
[5] Cross sections are not supported by the D, Model 20, or SL1 compilers.

As an example of the use of cross sections, the following statements determine the rainfall-rate-per-day matrix if **DAYS** is a vector having the successive values 31, 29, 31, 30, 31, 30, 31, 31, 30, 31, 30, and 31—1984 being a leap year:

```
DO J = 1 TO 7;
   RATE(*, J) = RAIN(*, J)/DAYS;
END;
```

These statements are equivalent to the statements:

```
DO J = 1 TO 7;
   DO I = 1 TO 12;
      RATE(I, J) = RAIN(I, J)/DAYS(I);
   END;
END;
```

Note that we may not write **RATE = RAIN/DAYS** because **RAIN** and **DAYS** do not have identical bounds.

As another example, suppose that we collect similar meteorological data for every year from 1984 through 2000, combining it into a triply-subscripted variable:

```
DECLARE RAINS(1984: 2000, 12, 7);
```

Our original matrix for 1984 is simply the cross section **RAINS(1984, *, *)**. Cross sections such as this are used in the following statements to accumulate in the 12-by-7 matrix **EVEN** the total rainfall in the even years from 1984 through 2000:

```
EVEN = 0;
DO K = 1984 TO 2000 BY 2;
   EVEN = EVEN + RAINS(K, *, *);
END;
```

These statements are equivalent to the statements:

```
DO I = 1 TO 12;
   DO J = 1 TO 7;
      SUM = 0;
      DO K = 1984 TO 2000 BY 2;
         SUM = SUM + RAINS(I, J, K);
      END;
      EVEN(I, J) = SUM;
   END;
END;
```

11.5 INPUT AND OUTPUT OF ARRAYS

Provided all of the allocated space is in use, entire arrays may be denoted for input and output by listing them without subscripts. For example, to establish the values for the matrix **RAIN** which has been declared to have the bounds 12-by-7, we may use the statement:

> GET LIST (RAIN);

This statement is equivalent to the nested repetitive specifications:

> GET LIST (((RAIN(I, J) DO J = 1 TO 7) DO I = 1 TO 12));

This in turn is equivalent to the nest of **DO** loops:

```
DO I = 1 TO 12;
   DO J = 1 TO 7;
      GET LIST (RAIN(I, J));
   END;
END;
```

Note that the rightmost subscript varies most rapidly; so we must prepare the data in that order to prevent assignment to incorrect locations. This *natural order* is the order in which the elements are stored in the memory. For matrices this corresponds to row order if we associate the rightmost subscript with the column and the leftmost with the row. We should not use the form GET LIST (RAIN) if the data are already prepared in another order or if we do not intend to prepare data for all of the allocated locations. For example, if the data are prepared in column order we may use the statement:

> GET LIST (((RAIN (I, J) DO I = 1 TO 12) DO J = 1 TO 7));

If we wished to get in data only for the even numbered months in row order, we may use the statement:

> GET LIST (((RAIN(J, I) DO J = 1 TO 7) DO I = 2 TO 12 BY 2));

If a number of elements are likely to be the same number, for example, zero for **SLEET**, some keypunching time may be saved by initializing the entire matrix to zero, then using the null field for those elements which are in fact zero:

```
SLEET = 0;
GET LIST (SLEET);
   or
DECLARE SLEET (12, 7) INITIAL ((12*7)0);
GET LIST (SLEET);
```

The latter form would be suitable only when **SLEET** is gotten once.[6]

Output of arrays is similar to input. We may use "PUT LIST (RAIN); or "PUT DATA (RAIN);" if we want the values of every allocated element in the natural

---

[6] Student PL does not support the INITIAL attribute, and only a constant is allowed for an iteration factor for the D, Model 20, and SL1 compilers. The Model 20 does not permit the INITIAL attribute for AUTOMATIC arrays.

order. However, the output of matrices is such a common chore that it would be nice to have a subroutine which puts them out neatly aligned and identified as follows:

```
COLUMN:              1                2                3                4
ROW
      1        3.67879E+05      3.45671E+00      1.98566E-06      3.45621E+01
      2        3.00000E+00      2.21019E+09      5.00000E+00      4.00000E+00
      3        3.55233E+01      1.42499E+00      3.44230E+05      2.55302E+08

COLUMN:              5                6                7                8
ROW
      1        9.44351E-07      1.23455E-06      9.87653E+04      5.55753E+02
      2        2.32322E-03      2.22219E+02      6.95320E+07      2.53234E+00
      3        3.54524E-03      2.55301E+04      2.66234E-07      2.55425E+05

COLUMN:              9               10               11
ROW
      1        2.00123E+05      7.88561E+03      2.88650E+06
      2        2.53213E+03      7.56229E-05      5.33200E+04
      3        2.25300E+03      2.66199E+00      2.23559E+02
```

The following subroutine accomplishes this for any number of rows and columns:

```
MTXOUT:   PROCEDURE (M, N, A);
    /* THIS SUBROUTINE PUTS OUT AN M BY N MATRIX IN ROW ORDER
       PARTITIONED INTO 4 COLUMN SEGMENTS ASSUMING THAT THERE ARE AT
       LEAST 5 TAB STOPS.  THE OUTPUT IS IN LIST DIRECTED FLOATING-
       POINT FORM WITH ROW AND COLUMN HEADINGS.  */
    DECLARE A(*,*);
    NLEFT = 1;
S10:  NRIGHT = MIN(NLEFT+3, N);
    PUT SKIP(2) LIST ('COLUMN:', (K  DO K = NLEFT TO NRIGHT));
    PUT SKIP LIST ('ROW');
    DO L = 1 TO M;
        PUT SKIP LIST (L, (A(L,K) DO K = NLEFT TO NRIGHT));
    END;
    NLEFT = NLEFT + 4;
    IF NRIGHT ¬= N   THEN GO TO S10;
END;
```

[7] Remember that bounds must be constants for the D, model 20, and SL1 compilers. Moreover, the subordinate procedure needs all but the rightmost bound for address computation; so we cannot use the trick of simply using bounds of one as we did for singly-subscripted parameters. Consequently, the best policy is to use standard bounds for all multiply-subscripted parameters and arguments—say ten. We may then use only part of the space for smaller arrays, and we must recompile for larger arrays. An alternative is to store all arrays as singly-subscripted variables and to do our own address computation.

EXERCISES

▼1. Write a program that gets in all of RAIN in natural order and puts out the maximum rainfall among the locations for each month as calculated by the VMAX function of Case Study 10.

2. Write a program that gets in all of RAIN in natural order and puts out the maximum rainfall among the months for each location as calculated by the VMAX function of Case Study 10.

▼3. Same as exercise 1 except use the MAX function.

4. If RAIN, SNOW, SLEET, and HAIL are in inches, they should be multiplied by their specific gravities—1, 0.1, 0.9, and 0.7, respectively—before being added together. Write a program that gets in RAIN, SNOW, SLEET, and HAIL in their natural orders; calculates PRECIPITATION; and puts out PRECIPITATION using MTXOUT.

▼5. Same as exercise 4 except assume that the input order is RAIN(1, 1), SNOW(1, 1), SLEET(1, 1), HAIL(1, 1), RAIN(1, 2), SNOW(1, 2) ...—row order, but all together.

6. Actually, the specific gravity of snow varies considerably; so repeat exercise 4, getting in also a 12-by-7 matrix giving the average snow specific gravity for each month and location.

▼7. Write a program that gets in RAINS with the leftmost subscripts varying most rapidly, and puts it out with the rightmost subscripts varying most rapidly using MTXOUT.

8. Same as exercise 7 except get in and put out only the data for leap years.

11.6 MATRIX-VECTOR MULTIPLICATION

The following table gives the quantity of certain nutrients per gram of certain foods. A more realistic table would have many more rows and columns, but we wish to keep things simple so that numerical results may be easily verified.

	Yeast	Liver	Milk	Candy	Gin
Niacin	4	9	1	0	-2
Calcium	3	1	9	1	0
Iron	5	8	1	0	0
Leucine	5	7	2	0	0

How many units of niacin would a meal of 0 grams of yeast, 30 grams of liver, 20 grams of milk, 10 grams of candy, and 30 grams of gin provide? The answer is:

$$4*0 \;+\; 9*30 \;+\; 1*20 \;+\; 0*10 \;+\; -2*30 \;=\; 230$$

More generally, if we consider the food quantities in the meal to be represented by a vector X with X(1) being the quantity of yeast, X(2) being the quantity of liver, and so on; and if we store the table as a subscripted variable A(I, J), where I denotes the row or nutrient and J denotes the column or food; then the quantity of niacin is given by the expression:

$$A(1,1)*X(1) \;+\; A(1,2)*X(2) \;+\; A(1,3)*X(3) \;+\; A(1,4)*X(4) \;+\; A(1,5)*X(5)$$

This expression is called the *dot, scalar, or inner product* of vector A(1, *) with vector X. We may call this quantity B(1), standing for the quantity of the first nutrient. Similarly, we may associate B(2) with the amount of calcium, and so on. In general, if the matrix A has M rows and N columns, we may evaluate the nutrient vector B from the food vector X by the following formulas:

$$A(1, 1)*X(1) + A(1, 2)*X(2) + \cdots A(1, N)*X(N) = B(1)$$
$$A(2, 1)*X(1) + A(2, 2)*X(2) + \cdots A(2, N)*X(N) = B(2)$$
$$\vdots$$
$$A(M, 1)*X(1) + A(M, 2)*X(2) + \cdots A(M, N)*X(N) = B(M)$$

Writing all of this out is rather tedious; so let us develop a more compact notation. First, note that "X(1) +" applies to every term in the first column; so we may reduce the amount of writing by writing X(1) at the top of the column with "*" and "+" understood. Doing the same for X(2) through X(N) gives:

$$[X(1) \quad X(2) \quad \cdots \quad X(N)]$$

$$\begin{bmatrix} A(1, 1) & A(1, 2) & \cdots & A(1, N) \\ A(2, 1) & A(2, 2) & \cdots & A(2, N) \\ \vdots & \vdots & & \vdots \\ A(M, 1) & A(M, 2) & \cdots & A(M, N) \end{bmatrix} = \begin{bmatrix} B(1) \\ B(2) \\ \vdots \\ B(M) \end{bmatrix}$$

We have also let one equal sign stand for all of them, and we have used brackets to make the organization clearer. Now, we may swing the row of Xs through 90°, making an X column above the B column.

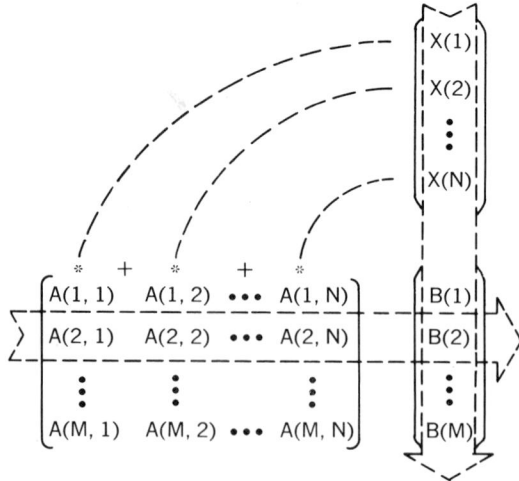

For our specific numerical example, we have

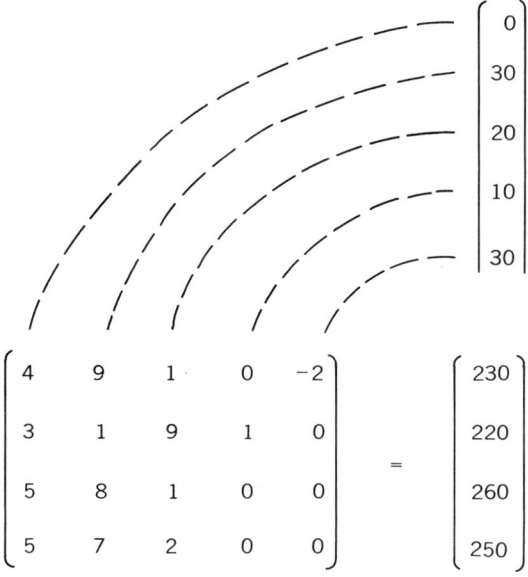

$$\begin{bmatrix} 4 & 9 & 1 & 0 & -2 \\ 3 & 1 & 9 & 1 & 0 \\ 5 & 8 & 1 & 0 & 0 \\ 5 & 7 & 2 & 0 & 0 \end{bmatrix} \begin{bmatrix} 0 \\ 30 \\ 20 \\ 10 \\ 30 \end{bmatrix} = \begin{bmatrix} 230 \\ 220 \\ 260 \\ 250 \end{bmatrix}$$

To save space, the X column is usually written next to the A matrix:

$$\begin{bmatrix} A(1,1) & A(1,2) & \cdots & A(1,N) \\ A(2,1) & A(2,2) & \cdots & A(2,N) \\ \vdots & \vdots & & \vdots \\ A(M,1) & A(M,2) & \cdots & A(M,N) \end{bmatrix} \begin{bmatrix} X(1) \\ X(2) \\ \vdots \\ X(N) \end{bmatrix} = \begin{bmatrix} B(1) \\ B(2) \\ \vdots \\ B(M) \end{bmatrix}$$

Each element of B is equal to the inner product of the corresponding row vector of A with the column vector X; so the vector B is said to be the inner product of the matrix A with the vector X. With inner products—

$$\text{vector times vector} = \text{scalar}$$
$$\text{matrix times vector} = \text{vector}$$

The inner product is not a product in the ordinary sense because it involves addition as well as multiplication. Nevertheless, the terminology is appropriate because it turns out that this product obeys most of the same rules as the ordinary product. Symbolically, we may denote this product as

$$A \cdot X = B$$

This shorthand notation is quite economical.

In a specific problem, we must worry about all of the individual coefficients, but we may do a lot of preliminary manipulation using the shorthand notation. For example, it turns out that matrix-vector multiplication is distributive:

$$A \cdot (X + Y) = A \cdot X + A \cdot Y$$

Algebra using this shorthand notation is called *matrix algebra*.

Let us now develop a subroutine to perform this matrix-vector multiplication. If all of the allocated space is in use we may write:

```
B(I)  =  SUM(A(I, *)*X);
```

If all of the allocated space is not in use we may write the equivalent DO loop:

```
SUM  =  0;
DO  J  =  1  TO  N;
     SUM  =  SUM  +  A(I, J)*X(J);
END;
B(I)  =  SUM;
```

We may eliminate the final statement by using B(I) everywhere else in place of SUM; but unless the compiler is rather sophisticated, this will slow down execution by causing over 2*N references to the same subscripted variable. For most compilers, references to subscripted variables—especially multiply-subscripted variables—are slower than references to unsubscripted variables. Regardless of which way we use to calculate B(I), we may enclose the calculation in a DO loop on I to calculate all of the B values:

```
DO  I  =  1  TO  M;
     B(I)  =  SUM(A(I, *)*X);
END;
```

or

```
DO  I  =  1  TO  M;
   SUM  =  0;
   DO  J  =  1  TO  N;
        SUM  =  SUM  +  A(I, J)*X(J);
   END;
   B(I)  =  SUM:
END;
```

Mathematically, matrix-vector multiplication may be defined as

$$B(I) = \sum_{J=1}^{N} [A(I,J) * X(J)] \quad \text{for } I = 1, 2, \ldots M$$

This may be remembered by noting that the adjacent subscripts (J) are the same and summed upon, the remaining subscript being left over to appear in the answer. As a final refinement, we may use the MULTIPLY function to accumulate a double-precision inner product. This substantially reduces the cumulative chopoff error for large matrices without appreciably increasing the storage requirements:

```
MVPROD:  PROCEDURE (M, N, A, X, B);
    DECLARE A(*,*) /* M BY N INPUT MATRIX              */,
            X(*)   /* INPUT VECTOR WITH N ELEMENTS     */,
            B(*)   /* OUTPUT VECTOR WITH M ELEMENTS    */,
            SUM FLOAT(16)   /* PRECISION HERE AND IN THE MULTIPLY
                            FUNCTION ARE IMPLEMENTATION DEPENDENT */;
    DO I = 1 TO M;
        SUM = 0;
        DO J = 1 TO N;
            SUM = SUM + MULTIPLY(A(I,J), X(J), 16);
        END;
        B(I) = SUM;
    END;
END;
```

11.7 MATRIX-MATRIX MULTIPLICATION

Suppose now that we also wish to find the nutrient vector B(*, 2) corresponding to a second food vector X(*, 2) where we have introduced a second subscript to distinguish them from the first vectors which we will now call B(*, 1) and X(*, 1). In fact, imagine that we wish to find the nutrient vectors for L such food vectors. We may portray this with a display similar to matrix-vector multiplication, but with L columns of X and B:

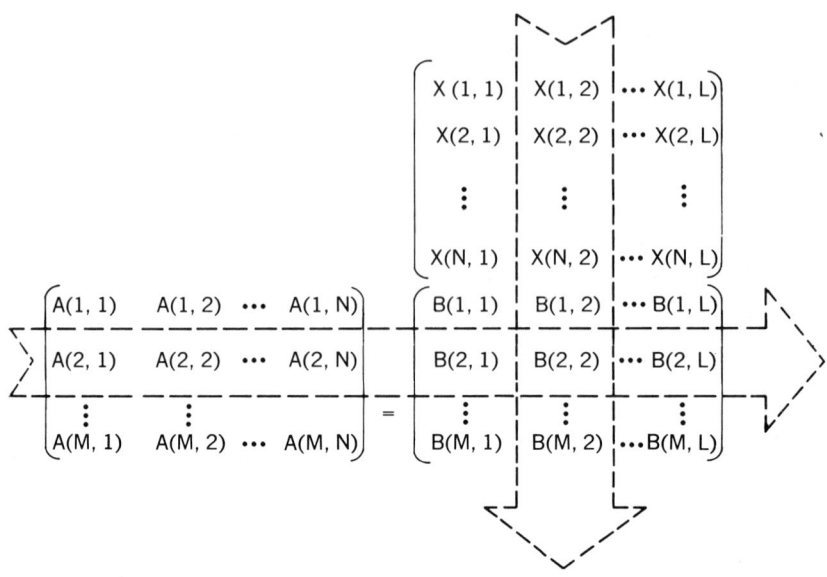

Each element of B will be given by the inner product of the corresponding row of A and column of X. This is called matrix-matrix multiplication. Symbolically, this product may be denoted

$$A \cdot X = B$$

As an illustrative example, it is easily verified that

$$\begin{bmatrix} 3 & 2 & 3 & 1 \\ 2 & 2 & 1 & 2 \\ 3 & 1 & 1 & 2 \end{bmatrix} \begin{bmatrix} 3 & 1 & 1 \\ 1 & 2 & 1 \\ 1 & 3 & 3 \\ 2 & 2 & 1 \end{bmatrix} = \begin{bmatrix} 16 & 18 & 15 \\ 13 & 11 & 9 \\ 15 & 13 & 9 \end{bmatrix}$$

To write a subroutine for matrix-matrix multiplication, all we need to do is add another DO loop to MVPROD for the additional subscript running from 1 to L, giving the matrix-vector product for all L columns:

```
MMPROD:  PROCEDURE (M, N, L, A, X, B);
         DECLARE  A(*,*)     /* M BY N GIVEN MATRIX        */,
                  X(*,*)     /* N BY L GIVEN MATRIX        */,
                  B(*,*)     /* M BY L RESULT MATRIX       */,
                  SUM   FLOAT(16)  /* IMPLEMENTATION DEPENDENT DOUBLE
                                      PRECISION SUM       */;
```

```
      DO I = 1 TO M;
         DO K = 1 TO L;
            SUM = 0;
            DO J = 1 TO N;
               SUM = SUM + MULTIPLY(A(I,J), X(J,K), 16);
            END;
            B(I,K) = SUM;
         END;
      END;
   END;
```

Mathematically, matrix multiplication may be defined as

$$B(I, K) = \sum_{J=1}^{N} A(I, J) * X(J, K) \qquad \begin{array}{l} I = 1, 2, \cdots M \\ K = 1, 2, \cdots L \end{array}$$

An easy way to remember this is that the inner two subscripts (J) are common and summed upon, the outer two (I, K) being left over to appear in the answer. Note that this matrix multiplication is quite different from the PL/I array assignment statement B = A*X which is not even valid unless A, B, and X have identical bounds.

11.8 THE GET DATA STATEMENT[7]

For input of matrices which are sparse, consisting mostly of zeros, there may be less data preparation required if the elements are initialized to zero; then only the nonzero values are gotten with a GET DATA statement. For example, the following program gets in any number of matrices A and X, putting them out together with their product:

```
PRODCT:  PROCEDURE OPTIONS (MAIN);
   NEXT:   GET LIST (M, N, L);
   BEGIN;
      DECLARE A(M,N) INITIAL((M*N)0), B(M,L) INITIAL((M*L)0), X(N,L);
      GET DATA (A, X);
      CALL MTXOUT(M, N, A);  CALL MTXOUT(N, L, X);
      CALL MMPROD(M, N, L, A, X, B);
      CALL MTXOUT(M, L, B);
   END;
   GO TO NEXT;
END;
```

The data for the GET DATA statement consist of a series of assignments separated by a *comma* and/or one or more blanks. The final item for each execution of the GET DATA statement must be a semicolon. For example, we could have the following data for the procedure PRODCT:

2, 3, 3, A(1, 1) = 3 A(1, 3) = 6   X(2, 2) = 17.3   A(2, 2) = 1E−4
X(3, 1) = 4   X(2, 1) = −6.7;

[7] Not available with Student PL, SL1, the Model 20, or the D compiler.

Note that the numbers may be scrambled, and not all of the variables in the list of the **GET DATA** statement need to be specified. However, no variables that are not in the list may be specified. Alternatively, we may simply write:

GET DATA;

and input a value for any variable in the procedure. This saves some keypunching, but it obscures the purpose of the statement to the human reader, and it is generally less efficient since it forces the procedure to remember the name (not just the address) of every variable in the program. Even with the list, **GET DATA** is less efficient than **GET LIST**; but the **GET DATA** statement is occasionally quite convenient.

SUMMARY

Multiply-Subscripted Variables

A typical declaration for a doubly-subscripted variable is:

DECLARE variable name (bound1, bound2);

Cross Sections

Cross sections may be designated by using an asterisk in place of one or more of the subscripts.

Natural Order

The natural order according to which multiply-subscripted variables are stored is with the rightmost subscripts varying most rapidly. This is the order of execution when entire arrays are designated for input, output, or expressions.

GET DATA Statement

The data for a GET DATA statement consist of a series of assignments separated by commas and/or blanks. Data need not be gotten for all of the items in the list, and the last data item for each execution of the statement is followed by a semicolon. Any variable in the procedure may be gotten if there is no list with the GET DATA statement.

EXERCISES

▼9. As an application of matrix-vector multiplication, assume that each rain gauge location for the RAIN matrix discussed earlier has an associated surrounding land area given by a vector AREA. Then, since volume is equal to height times area, the total volume of the rain for each month will be given by the matrix-vector product of RAIN with AREA. Write a program using MVPROD that gets in RAIN and AREA, then calculates and puts out the total volume vector.

10. (a) Evaluate the two products

$$\begin{bmatrix} 3 & 5 & 2 \\ 1 & 2 & 4 \end{bmatrix} \left\{ \begin{bmatrix} 3 & 1 & 2 \\ 2 & 2 & 3 \\ 1 & 2 & 1 \end{bmatrix} \begin{bmatrix} 1 & 2 \\ 2 & 1 \\ 0 & 2 \end{bmatrix} \right\} \quad \text{and} \quad \left\{ \begin{bmatrix} 3 & 5 & 2 \\ 1 & 2 & 4 \end{bmatrix} \begin{bmatrix} 3 & 1 & 2 \\ 2 & 2 & 3 \\ 1 & 2 & 1 \end{bmatrix} \right\} \begin{bmatrix} 1 & 2 \\ 2 & 1 \\ 0 & 2 \end{bmatrix}$$

to show that they are equal. (Matrix multiplication is *associative*.)

(b) Evaluate the two products

$$\begin{bmatrix} 4 & 3 & 1 \\ 2 & 0 & 2 \\ 1 & 2 & 1 \end{bmatrix} \begin{bmatrix} 5 & 2 & 1 \\ 1 & 1 & 2 \\ 1 & 2 & 2 \end{bmatrix} \quad \text{and} \quad \begin{bmatrix} 5 & 2 & 1 \\ 1 & 1 & 2 \\ 1 & 2 & 2 \end{bmatrix} \begin{bmatrix} 4 & 3 & 1 \\ 2 & 0 & 2 \\ 1 & 2 & 1 \end{bmatrix}$$

to show that they are not equal. (Matrix multiplication is *not commutative*.)

11. Assume that AMOUNT(I, J) gives the amount of component I per barrel of crude oil from source J where there are M components and N sources. The components might include gasoline, kerosene, and asphalt; the sources might include Kuwait, Venezuela, and Texas. Write a program that gets in as a vector CRUDE the quantity of each crude and puts out as vector COMPONENT the corresponding quantity of each component, for any number of crude vectors in succession.

▼**12.** Same as exercise 11 except assume that all of the crude vectors are gotten in at once as the columns of a crude matrix, and all of the component vectors are put out as the columns of a component matrix.

13. As another application of matrix-vector multiplication assume that A(I, J) is the cosine of the angle from axis I of an unrotated Cartesian coordinate system to axis J of a rotated system where I, J = 1, 2, 3. Then, the coordinates of a point in the rotated system are given by the product A·X if the coordinates in the unrotated system are given by X. Write a program that gets in A, then gets in, converts, and puts out the converted coordinates for any number of points.

14. (Mathematically Advanced) As another application of matrix-vector multiplication, if the inertia tensor of a rigid body is given by the 3-by-3 matrix T and the angular velocity is given by the 3-element vector W, then the angular momentum is given by the 3-element vector T·W. Write a program that gets in T, then gets in any number of W vectors, evaluating and putting out the corresponding angular momentum.

▼**15.** (Mathematically Advanced) If coordinate rotation is given by the matrix A defined in exercise 11, then the inertia tensor defined in exercise 12 becomes the matrix-matrix-matrix product:

$$A \cdot T \cdot A$$

Write a program that gets in any number of As and Ts in succession, putting out the rotated inertia tensors.

▼16. Write a function procedure that evaluates the *diagonal product* of N-by-N matrix:

$$\text{DIAG} = \prod_{K=1}^{N} A(K, K)$$

17. Write a function procedure that evaluates the *trace* of an N-by-N matrix:

$$\text{TRACE} = \sum_{K=1}^{N} A(K, K)$$

▼18. Write a function procedure that evaluates the *infinity-norm* of an N-by-N matrix:

$$\|A\|_{\infty} = \max_{I=1}^{N} \left[\sum_{J=1}^{N} |A(I, J)| \right] = \text{maximum row absolute sum}$$

19. Write a function procedure that evaluates the *one-norm* of an N-by-N matrix:

$$\|A\|_{1} = \max_{J=1}^{N} \left[\sum_{I=1}^{N} |A(I, J)| \right] = \text{maximum column absolute sum}$$

▼20. (Mathematically Advanced) In game theory there is a *saddlepoint* if for the payoff matrix the minimum row maximum equals the maximum column minimum. For an M-by-N matrix A, these are defined as

$$\text{min row max} = \min_{I=1}^{M} \left\{ \max_{J=1}^{N} [A(I, J)] \right\}$$

$$\text{max col min} = \max_{J=1}^{N} \left\{ \min_{I=1}^{M} [A(I, J)] \right\}$$

If they are equal, they represent the *value* of the game. Write a program which repetitively gets in M, N, and A, uses MTXOUT to put out A, and puts out the value or the message "NO SADDLEPOINT."

▼21. (Mathematically Advanced) One of the contractions of the Riemann tensor of general relativity is given by the formula

$$Q(J, L) = \sum_{K=1}^{4} R(K, J, K, L) \quad \text{for } J, L = 1, 2, 3, 4$$

Write a subroutine CNTRCT(R, Q) that calculates Q.

22. (Mathematically Advanced) The stress tensor T as a function of the strain tensor S and the elasticity tensor E is given by the formula

$$T(I, J) = \sum_{K=1}^{3} \left\{ \sum_{L=1}^{3} [E(I, J, K, L) * S(K, L)] \right\} \quad \text{for } I, J = 1, 2, 3$$

Write a subroutine STRSTN(E, S, T) that calculates T.

23. The GET DATA statement is convenient when we wish to change the values of

only a few elements. For example, if we expect to revise the matrix A quite frequently, we may write:

   GET LIST (((A(I, J) DO J = 1 TO M) DO J = 1 TO N)); GET DATA (A);

We may then add the revisions onto the end of the data set without having to tamper with the original data. Modify the procedure PRODCT of this case study accordingly.

24. Most structural analysis is done for steel with a Young's modulus of 30E6. We may avoid forcing the user to specify this, yet give him the freedom to specify otherwise by writing DECLARE E INITIAL (30E6);.... GET DATA(E);. Devise a similar example relevant to your discipline.

25. (Mathematically Advanced) If a set of basis vectors is stored as the rows of an M-by-N matrix A, the basis may be orthonormalized by the Gram-Schmidt technique:

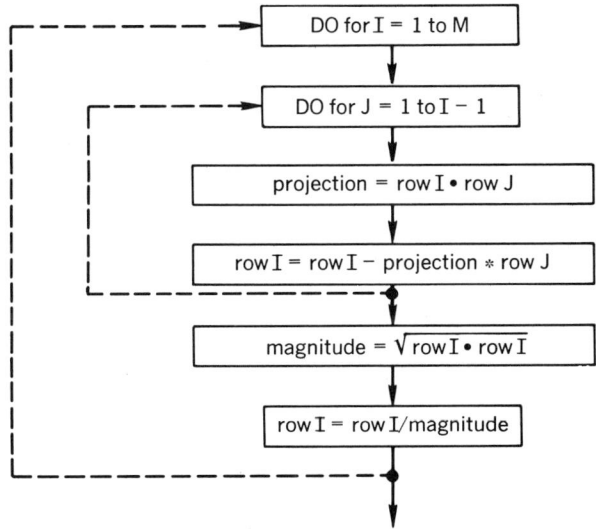

where the dot indicates the inner product. Write a subroutine ORTHNM (M, N, A) that performs this process.

▼26. The merging procedure of Case Study 10 can be generalized to merge lists stored as L rows of an L-by-N matrix. The number of elements used from each list may be stored as a vector. Write a corresponding subroutine.

▼27. (Mathematically Advanced) Just as $e^{at}x_0$ is a solution of the linear differential equation $a\,dx/dt = x$ where x_0 is the initial condition, $e^{At}X_0$ is a solution of the N simultaneous linear differential equations $A \cdot dX/dt = X$. X is a column vector with initial values X_0, A is an N-by-N matrix, and e^{At}, the transition matrix, is defined by the matrix analog of the exponential Taylor series:

$$B = e^{At} = I + At + A \cdot At^2/2! + A \cdot A \cdot At^3/3! + \cdots$$

where I is the identity matrix which is all zeros except for ones along the diagonal. The series may be continued until the largest absolute relative change in two successive values of any element in B is less than a specified tolerance. For moderate values of t, this series converges quite rapidly, and once this series is evaluated for a certain $t = \Delta t$, the solution at subsequent multiples of that time increment may be evaluated by simple repetitive multiplication: $X(K*\Delta t) = B \cdot X((K-1)*\Delta t)$. Write a subroutine MTXEXP(N, A, DT, RELTOL, B) which evaluates the transition matrix B for time DT. Have the subroutine halve DT until convergence is achieved within ten terms.

▼28. Assume that we have a function of two variables, $z(x, y)$ represented by a table Z(I, J) for equally spaced values of x and y;

x = XMIN, XMIN + DX, XMIN + 2DX, ⋯ XMIN + (M−1)DX
y = YMIN, YMIN + DY, YMIN + 2DY, ⋯ YMIN + (N−1)DY

Write a subroutine EQINTP(M, N, XIN, XMIN, DX, YIN, YMIN, DY, Z, ZOUT, IER) which calculates ZOUT by interpolation, setting IER = 1 if the input point XIN, YIN is outside the proper range and setting it equal to zero otherwise.

29. Same as exercise 28 except assume that x and y are unequally spaced, as given by the subscripted variables X and Y.

CASE STUDY 12

LINEAR EQUATIONS

Further Practice with Subscripted Variables

Perhaps the most frequent engineering and scientific computer application is the solution of simultaneous linear algebraic equations. For example, we may wish to solve the following equations for w, x, y and z:

$$9.0w + 6.0x + 8.3y + 1.5z = 3.7$$
$$1.5w + 8.5x + 9.0y + 6.7z = 8.3$$
$$7.0w + 6.5x + 1.3y + 9.0z = 4.0$$
$$6.5w + 1.1x + 9.7y + 8.7z = 6.0$$

However, it is not worth writing a program to solve a specific set of linear equations. It is better to write a program which will solve any four linear equations, getting the coefficients on the left and the constants on the right as data. It is better still to write a program which will solve any number of linear equations. To do this, we must switch from using separate names for each variable to using one subscripted name for all of them. For our example, we may associate X(1) with w, X(2) with x, X(3) with y, and X(4) with z. We may also denote all of the constants on the right-hand side by a subscripted variable B. Thus, we may associate B(1) with 3.7, B(2) with 8.3, B(3) with 4.0, and B(4) with 6.0. Finally, let us denote the left-hand coefficients by a doubly-subscripted variable A where the first subscript denotes the equation or row and the second subscript denotes the unknown or column. For our example, A(1, 1) = 9.0, A(1, 2) = 6.0, and so on. With this scheme, we may represent our equations as

$$A(1,1)*X(1) + A(1,2)*X(2) + A(1,3)*X(3) + A(1,4)*X(4) = B(1)$$
$$A(2,1)*X(1) + A(2,2)*X(2) + A(2,3)*X(3) + A(2,4)*X(4) = B(2)$$
$$A(3,1)*X(1) + A(3,2)*X(2) + A(3,3)*X(3) + A(3,4)*X(4) = B(3)$$
$$A(4,1)*X(1) + A(4,2)*X(2) + A(4,3)*X(3) + A(4,4)*X(4) = B(4)$$

We can see that this is nothing but the matrix-vector product $A \cdot X = B$, except now the Xs are the unknowns rather than the Bs.

12.1 GAUSSIAN ELIMINATION

There are a great many ways to solve simultaneous linear equations. Theoretically-oriented courses in matrix or linear algebra often emphasize Cramer's rule, but it is hopelessly slow. Most general-purpose library subroutines use one of the class of methods typified by *Gaussian elimination*. We will demonstrate this technique using both the general notation for N equations in N unknowns and our numerical example. The numerical example will be performed in chopped-off two-digit decimal floating-point arithmetic so that the results may be easily verified. Accordingly, we begin with the following situation:

$$\begin{bmatrix} A(1,1) & A(1,2) & A(1,3) & \cdots & A(1,N) \\ A(2,1) & A(2,2) & A(2,3) & \cdots & A(2,N) \\ A(3,1) & A(3,2) & A(3,3) & \cdots & A(3,N) \\ \vdots & \vdots & \vdots & & \vdots \\ A(N,1) & A(N,2) & A(N,3) & \cdots & A(N,N) \end{bmatrix} \begin{bmatrix} X(1) \\ X(2) \\ X(3) \\ \vdots \\ X(N) \end{bmatrix} = \begin{bmatrix} B(1) \\ B(2) \\ B(3) \\ \vdots \\ B(N) \end{bmatrix}$$

$$\begin{bmatrix} 9.0 & 6.0 & 8.3 & 1.5 \\ 1.5 & 8.5 & 9.0 & 6.7 \\ 7.0 & 6.5 & 1.3 & 9.0 \\ 6.5 & 1.1 & 9.7 & 8.7 \end{bmatrix} \begin{bmatrix} X(1) \\ X(2) \\ X(3) \\ X(4) \end{bmatrix} = \begin{bmatrix} 3.7 \\ 8.3 \\ 4.0 \\ 6.0 \end{bmatrix}$$

First we may subtract a multiple of the first equation from the second equation, choosing the multiple to make the first coefficient of the second equation zero:

$$\begin{bmatrix} A(1,1) & A(1,2) & A(1,3) & \cdots & A(1,N) \\ 0 & A(2,2) & A(2,3) & \cdots & A(2,N) \\ A(3,1) & A(3,2) & A(3,3) & \cdots & A(3,N) \\ \vdots & \vdots & \vdots & & \vdots \\ A(N,1) & A(N,2) & A(N,3) & \cdots & A(N,N) \end{bmatrix} \begin{bmatrix} X(1) \\ X(2) \\ X(3) \\ \vdots \\ X(N) \end{bmatrix} = \begin{bmatrix} B(1) \\ B(2) \\ B(3) \\ \vdots \\ B(N) \end{bmatrix}$$

Gaussian Elimination

$$\begin{bmatrix} 9.0 & 6.0 & 8.3 & 1.5 \\ 0 & 7.6 & 7.7 & 6.5 \\ 7.0 & 6.5 & 1.3 & 9.0 \\ 6.5 & 1.1 & 9.7 & 8.7 \end{bmatrix} \begin{bmatrix} X(1) \\ X(2) \\ X(3) \\ X(4) \end{bmatrix} = \begin{bmatrix} 3.7 \\ 7.8 \\ 4.0 \\ 6.0 \end{bmatrix}$$

The second row of A and B are modified by this elimination. The proper multiple of the first equation is $1.5/9.0 = A(2, 1)/A(1, 1)$. The divisor $A(1, 1)$ is called the *pivot*. We may proceed in a similar fashion to eliminate the first terms in the third through Nth equations:

$$\begin{bmatrix} A(1,1) & A(1,2) & A(1,3) & \cdots & A(1,N) \\ 0 & A(2,2) & A(2,3) & \cdots & A(2,N) \\ 0 & A(3,2) & A(3,3) & \cdots & A(3,N) \\ \cdot & \cdot & \cdot & & \cdot \\ \cdot & \cdot & \cdot & & \cdot \\ \cdot & \cdot & \cdot & & \cdot \\ 0 & A(N,2) & A(N,3) & \cdots & A(N,N) \end{bmatrix} \begin{bmatrix} X(1) \\ X(2) \\ X(3) \\ \cdot \\ \cdot \\ \cdot \\ X(N) \end{bmatrix} = \begin{bmatrix} B(1) \\ B(2) \\ B(3) \\ \cdot \\ \cdot \\ \cdot \\ B(N) \end{bmatrix}$$

$$\begin{bmatrix} 9.0 & 6.0 & 8.3 & 1.5 \\ 0 & 7.6 & 7.7 & 6.5 \\ 0 & 1.9 & -5.0 & 7.9 \\ 0 & -3.2 & 3.8 & 7.7 \end{bmatrix} \begin{bmatrix} X(1) \\ X(2) \\ X(3) \\ X(4) \end{bmatrix} = \begin{bmatrix} 3.7 \\ 7.8 \\ 1.2 \\ 3.4 \end{bmatrix}$$

Next, a multiple of the second equation may be used to eliminate the second term in the third equation. The proper multiple of $1.9/7.6 = A(3, 2)/A(2, 2)$. This elimination will not disturb the first zero in the third equation because any multiple of the second equation will have a zero as its first coefficient.

$$\begin{bmatrix} A(1,1) & A(1,2) & A(1,3) & \cdots & A(1,N) \\ 0 & A(2,2) & A(2,3) & \cdots & A(2,N) \\ 0 & 0 & A(3,3) & \cdots & A(3,N) \\ \cdot & \cdot & \cdot & & \cdot \\ \cdot & \cdot & \cdot & & \cdot \\ \cdot & \cdot & \cdot & & \cdot \\ 0 & A(N,2) & A(N,3) & \cdots & A(N,N) \end{bmatrix} \begin{bmatrix} X(1) \\ X(2) \\ X(3) \\ \cdot \\ \cdot \\ \cdot \\ X(N) \end{bmatrix} = \begin{bmatrix} B(1) \\ B(2) \\ B(3) \\ \cdot \\ \cdot \\ \cdot \\ B(N) \end{bmatrix}$$

$$\begin{bmatrix} 9.0 & 6.0 & 8.3 & 1.5 \\ 0 & 7.6 & 7.7 & 6.5 \\ 0 & 0 & -6.9 & 6.3 \\ 0 & -3.2 & 3.8 & 7.7 \end{bmatrix} \begin{bmatrix} X(1) \\ X(2) \\ X(3) \\ X(4) \end{bmatrix} = \begin{bmatrix} 3.7 \\ 7.8 \\ -0.7 \\ 3.4 \end{bmatrix}$$

Similarly, appropriate multiples of the second equation may be used to eliminate the second terms in the fourth through Nth equations.

$$\begin{bmatrix} A(1,1) & A(1,2) & A(1,3) & \cdots & A(1,N) \\ A(2,1) & A(2,2) & A(2,3) & \cdots & A(2,N) \\ 0 & 0 & A(3,3) & \cdots & A(3,N) \\ \vdots & \vdots & \vdots & & \vdots \\ 0 & 0 & A(N,3) & \cdots & A(N,N) \end{bmatrix} \begin{bmatrix} X(1) \\ X(2) \\ X(3) \\ \vdots \\ X(N) \end{bmatrix} = \begin{bmatrix} B(1) \\ B(2) \\ B(3) \\ \vdots \\ B(N) \end{bmatrix}$$

$$\begin{bmatrix} 9.0 & 6.0 & 8.3 & 1.5 \\ 0 & 7.6 & 7.7 & 6.5 \\ 0 & 0 & -6.9 & 6.3 \\ 0 & 0 & 7.0 & 10 \end{bmatrix} \begin{bmatrix} X(1) \\ X(2) \\ X(3) \\ X(4) \end{bmatrix} = \begin{bmatrix} 3.7 \\ 7.8 \\ -0.7 \\ 6.6 \end{bmatrix}$$

Then, multiples of the third equation may be used to eliminate the third terms in the fourth through Nth equations. This pattern may be continued until we have eliminated the N − 1th term in the Nth equation, leaving an *upper triangular* set of equations:

$$\begin{bmatrix} A(1,1) & A(1,2) & A(1,3) & \cdots & A(1,N) \\ 0 & A(2,2) & A(2,3) & \cdots & A(2,N) \\ 0 & 0 & A(3,3) & \cdots & A(3,N) \\ \vdots & \vdots & \vdots & & \vdots \\ 0 & 0 & 0 & \cdots & A(N,N) \end{bmatrix} \begin{bmatrix} X(1) \\ X(2) \\ X(3) \\ \vdots \\ X(N) \end{bmatrix} = \begin{bmatrix} B(1) \\ B(2) \\ B(3) \\ \vdots \\ B(N) \end{bmatrix}$$

$$\begin{bmatrix} 9.0 & 6.0 & 8.3 & 1.5 \\ 0 & 7.6 & 7.7 & 6.5 \\ 0 & 0 & -6.9 & 6.3 \\ 0 & 0 & 0 & 16 \end{bmatrix} \begin{bmatrix} X(1) \\ X(2) \\ X(3) \\ X(4) \end{bmatrix} = \begin{bmatrix} 3.7 \\ 7.8 \\ -0.7 \\ 5.9 \end{bmatrix}$$

Now, since the last equation involves only one unknown, and since in reverse order, successive equations involve only one additional unknown each, the equations may be solved in one reverse-order sweep.

X(N) = B(N)/A(N, N)
X(N−1) = [B(N−1) − A(N−1, N) ∗ X(N)]/A(N−1, N−1)
X(N−2) = [B(N−2) − A(N−2, N−1) ∗ X(N−1) − A(N−2, N) ∗ X(N)]/A(N−2, N−2)
.
.
.
X(1) = [B(1) − A(1, 2) ∗ X(2) − A(1, 3) ∗ X(3) ⋯ − A(1, N) ∗ X(N)]/A(1, 1)

Sec. 12.1   Gaussian Elimination

$$X(4) = 5.9/16 = .36$$
$$X(3) = (-0.7 - 6.3 * .36)/-6.9 = .42$$
$$X(2) = (7.8 - 7.7 * .42 - 6.3 * .36)/7.6 = 3.0$$
$$X(1) = (3.7 - 6.0 * .30 - 8.3 * .42 - 1.5 * .36)/9.0 = -.22$$

The reduction to triangular form is called the *elimination*; the simultaneous modification of the Bs is called the *forward substitution*; and the final phase is called the *back substitution*. In general, for the elimination of coefficient A(I, J), we have the following situation:

$$\begin{bmatrix} A(1,1) & \cdots & A(1,J) & A(1,J+1) & \cdots & A(1,N) \\ \cdot & & \cdot & \cdot & & \cdot \\ \cdot & & \cdot & \cdot & & \cdot \\ 0 & \cdots 0 & A(J,J) & A(J,J+1) & \cdots & A(J,N) \\ \cdot & & \cdot & \cdot & & \cdot \\ \cdot & & \cdot & \cdot & & \cdot \\ 0 & \cdots 0 & A(I,J) & A(I,J+1) & \cdots & A(I,N) \\ \cdot & & \cdot & \cdot & & \cdot \\ \cdot & & \cdot & \cdot & & \cdot \\ 0 & \cdots 0 & A(N,J) & A(N,J+1) & \cdots & A(N,N) \end{bmatrix} \begin{bmatrix} X(1) \\ \cdot \\ \cdot \\ X(J) \\ \cdot \\ \cdot \\ X(I) \\ \cdot \\ \cdot \\ X(N) \end{bmatrix} = \begin{bmatrix} B(1) \\ \cdot \\ \cdot \\ B(J) \\ \cdot \\ \cdot \\ B(I) \\ \cdot \\ \cdot \\ B(N) \end{bmatrix}$$

We can see that the general step in the elimination and forward substitution is the subtraction of A(I, J)/A(J, J) times row J from row I.

Actually, we need only subtract for coefficients to the right of column J because A(I, J) becomes zero and the elements to the left of it remain zero. We may accomplish this with the statements:

```
F = A(I, J)/A(J, J);
DO K = J + 1 TO N;
    A(I, K) = A(I, K) - F*A(J, K);
END;
B(I) = B(I) - F*B(J);
```

We wish to do this for every row from I = J + 1 down to N:

```
DO I = J + 1 TO N;
    F = A(I, J)/A(J, J);
    DO K = J + 1 TO N;
        A(I, K) = A(I, K) - F*A(J, K);
    END;
    B(I) = B(I) - F*B(J);
END;
```

Finally, we wish to do this for every column from J = 1 through N − 1:

```
DO J = 1 TO N − 1;
  DO I = J + 1; TO N;
    F = A(I, J)/A(J, J);
    DO K = J + 1 TO N;
      A(I, K) = A(I, K) − F*A(J, K);
    END;
    B(I) = B(I) − F*B(J);
  END;
END;
```

This leaves us with the following upper triangular matrix:

$$\begin{bmatrix} A(1,1) & \cdots & A(1,I) & A(1,I+1) & \cdots & A(1,N) \\ \vdots & & \vdots & \vdots & & \vdots \\ 0 & \cdots & A(I,I) & A(I,I+1) & \cdots & A(I,N) \\ \vdots & & & & & \vdots \\ 0 & \cdots & 0 & 0 & \cdots & A(N,N) \end{bmatrix} \begin{bmatrix} X(1) \\ \vdots \\ X(I) \\ \vdots \\ X(N) \end{bmatrix} \begin{bmatrix} B(1) \\ \vdots \\ B(I) \\ \vdots \\ B(N) \end{bmatrix}$$

In general, we wish to subtract from B(I) all terms to the right of the diagonal, then divide by the diagonal coefficient A(I, I). We may do this with the statements:

```
SUM = B(I);
DO J = I + 1 TO N;
  SUM = SUM − A(I, J)*X(J);
END;
X(I) = SUM/A(I, I);
```

We wish to do this for every row I in reverse order from N to 1:

```
DO I = N TO 1 BY −1;
  SUM = B(I);
  DO J = I + 1 TO N;
    SUM = SUM − A(I, J)*X(J);
  END;
  X(I) = SUM/A(I, I);
END;
```

We may now combine these steps into the following subroutine:

Gaussian Elimination

```
GAUSS:  PROCEDURE (N, A, B, X);  /* THIS SUBROUTINE SOLVES N LINEAR
        EQUATIONS IN N UNKNOWNS USING GAUSSIAN ELIMINATION.  */
   DECLARE A(*,*) /* GIVEN COEFFICIENTS (ALTERED BY SOLUTION)  */,
           B(*)   /* GIVEN CONSTANTS (ALTERED BY SOLUTION)  */,
           X(*)   /* SOLUTIONS */;
   /* ELIMINATION AND FORWARD SUBSTITUTION:  */
   DO J = 1 TO N-1;    /*  SET COLUMN  */
      JP1 = J+1;  PIVOT = A(J,J);
      DO I = JP1 TO N;  /* SET ROW */
         F = A(I,J)/PIVOT;  /* CALCULATE MULTIPLE  */
         DO K = JP1 TO N;  /* SUBTRACT MULTIPLE OF ROW J FROM ROW I  */
            A(I,K) = A(I,K) - F*A(J,K);
         END;
         B(I) = B(I) - F*B(J);
      END;
   END;
   /*  BACK SUBSTITUTION:  */
   DO I = N TO 1 BY -1;  /* SET ROW  */
      SUM = B(I);
      DO J = I+1 TO N;  /* SUBTRACT TERM  */
         SUM = SUM - A(I,J)*X(J);
      END;
      X(I) = SUM/A(I,I);
   END;
END;
```

The exercises show several ways to improve this procedure. Note that we have introduced the auxiliary variables **JP1** and **PIVOT** to avoid redundant calculations and redundant subscript reference. As an example of the use of this subroutine, the following program uses it to solve any number of sets of simultaneous linear equations:

```
LNEQNS:  PROCEDURE OPTIONS (MAIN);
         /* THIS PROGRAM SOLVES ANY NUMBER OF SETS UP TO 99 SIMULTANEOUS
            LINEAR ALGEBRAIC EQUATIONS.  FOR EACH SET, THE INPUT SHOULD
            CONSIST OF THE NUMBER OF EQUATIONS, THEN THE FIRST ROW OF
            COEFFICIENTS FOLLOWED BY THE FIRST CONSTANT, THEN THE SECOND
            ROW OF COEFFICIENTS FOLLOWED BY THE SECOND CONSTANT, AND SO ON
            IN LIST FORM.  THIS PROGRAM REQUIRES SUBROUTINES MTXOUT AND
            GAUSS.  */
   DECLARE A(99,99), B(99), X(99);
S10:  GET LIST (N, ((A(I,J) DO J = 1 TO N), B(I) DO I = 1 TO N));
      PUT PAGE LIST ('COEFFICIENTS:');
      CALL MTXOUT (N, N, A);
      PUT SKIP(2) LIST ('CONSTANTS:');
      PUT SKIP(2) DATA ((B(I) DO I = 1 TO N));
      CALL GAUSS (N, A, B, X);
      PUT SKIP(2) LIST ('SOLUTIONS:');
      PUT SKIP(2) DATA ((X(I) DO I = 1 TO N));
      GO TO S10;
END;
```

The input illustrates the flexibility of the repetitive specification. The output illustrates the good practice of putting out the input for the record. In cases like this where there may be a great deal of input, the likelihood of a data error is especially high, and it will be easier to spot in nicely arranged output than on the input medium. Also, there is no point using a **BEGIN** block since the space partitions in a set pattern between **A**, **B**, and **X**.

12.2 ILL-CONDITIONED AND SINGULAR MATRICES

Let us now solve the following equations by hand, using two-digit floating-point arithmetic with chopoff:

$$\begin{bmatrix} 6.0 & -3.6 & 4.2 \\ 5.6 & 1.6 & -2.8 \\ 8.9 & -0.3 & -1.1 \end{bmatrix} \begin{bmatrix} X(1) \\ X(2) \\ X(3) \end{bmatrix} = \begin{bmatrix} 9.5 \\ 5.3 \\ 1.4 \end{bmatrix}$$

Successive eliminations give:

$$\begin{bmatrix} 6.0 & -3.6 & 4.2 \\ 0 & 4.9 & -6.7 \\ 0 & 4.7 & -6.9 \end{bmatrix} \begin{bmatrix} X(1) \\ X(2) \\ X(3) \end{bmatrix} = \begin{bmatrix} 9.5 \\ -3.5 \\ -12 \end{bmatrix}$$

then

$$\begin{bmatrix} 6.0 & -3.6 & 4.2 \\ 0 & 4.9 & -6.7 \\ 0 & 0 & -0.6 \end{bmatrix} \begin{bmatrix} X(1) \\ X(2) \\ X(3) \end{bmatrix} = \begin{bmatrix} 9.5 \\ -3.5 \\ -9.0 \end{bmatrix}$$

Back substitution gives $X(3) = 15$, $X(2) = 20$, and $X(1) = 3.0$. If we now check these solutions by substituting back into the original equations without chopoff, we get:

$$A \cdot X = \begin{bmatrix} 9.0 \\ 6.8 \\ 4.2 \end{bmatrix} \text{ whereas } B = \begin{bmatrix} 9.5 \\ 5.3 \\ 1.4 \end{bmatrix}$$

The discrepancy between the calculated and true solutions is also large. The true solution vector to two significant digits is

$$X = \begin{bmatrix} 3.2 \\ 24 \\ 18 \end{bmatrix}, \text{ whereas we calculated } \begin{bmatrix} 3 \\ 20 \\ 15 \end{bmatrix}$$

Of course computers generally work to more than two decimal places, but no matter how many places there are, it is possible to contrive examples which will yield large discrepancies. Fortunately, magnification of chopoff error is not always such a serious problem. The trouble with our example is that the third coefficient row is nearly equal to one half the first coefficient row plus the second coefficient row. (The Bs do not matter.) When any coefficient row is nearly such a *linear combination* of any others, it causes arithmetic subtraction of two nearly equal coefficient rows at some stage of the elimination. This causes a large relative error in a value which must later be used as a divisor—thus propagating the large relative error to all subsequent calculations. In our case the inaccurate divisor is -0.6.

A matrix is said to be *ill-conditioned* when any row is nearly, but not exactly, equal to a linear combination of other rows. A matrix is said to be *singular*, meaning peculiar, when any row is exactly equal to a linear combination of any other rows. Theoretically, singular matrices result in a coefficient row of all zeros at some stage of the elimination. The solutions are undefined since the subsequent divisions by zero are undefined. Computationally, chopoff usually alters singular matrices into highly ill-conditioned matrices.

We may also consider singularity and ill-conditioning from a geometric point of view. The solution of two linear equations may be viewed as finding the intersec-

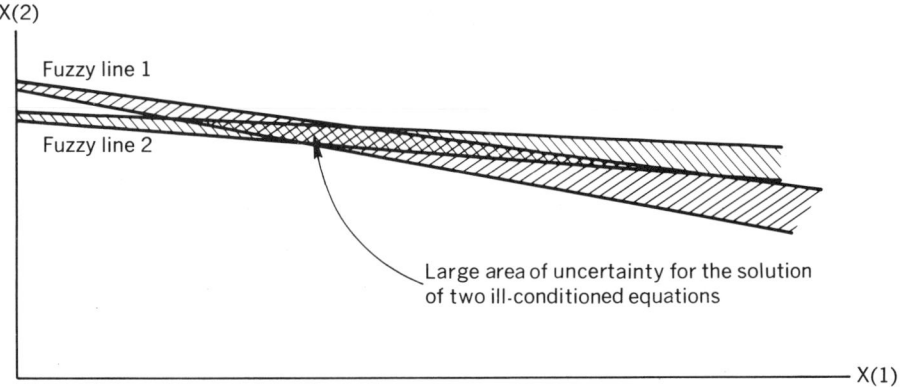

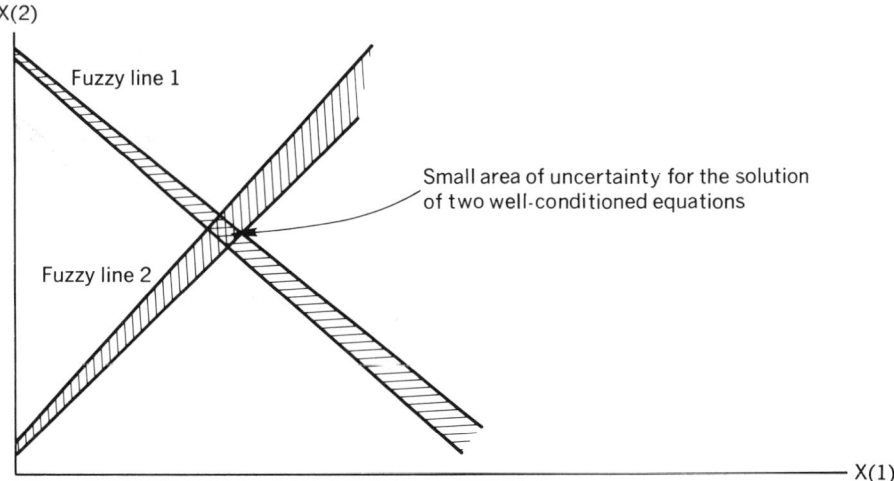

Figure 12-1

tion of two lines in a plane. The two coefficient rows will be proportional if one is a linear combination of the other. This means the lines are parallel; so they do not intersect! When the lines are nearly parallel, small experimental errors in the As or Bs or small chopoff errors in combining them cause large errors in the Xs or point of intersection as is shown in Fig. 12-1.

The solution of three linear equations may be viewed as finding the intersection of three planes in space. There is no unique point of intersection when one of the rows is a linear combination of the others, because one of the planes is then parallel to the intersection of the others as is shown in Fig. 12-2.

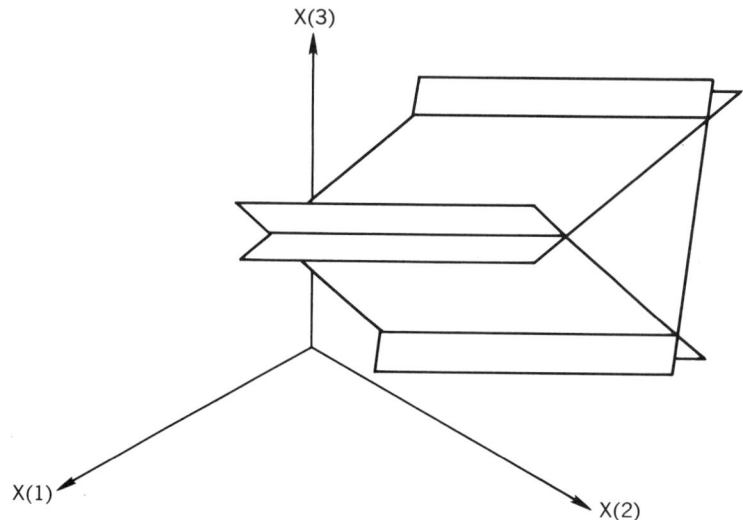

Fig. 12-2 Graphical portrayal of a singular set of three simultaneous linear equations.

The same geometric concept holds for any number of simultaneous equations. Their solution may be viewed as finding the intersection of hyperplanes in hyperspace, and the coefficient matrix is singular when any of the hyperplanes is parallel to another or to the intersection of others. The coefficient matrix is ill-conditioned when this is almost the case.

Double precision will not reduce magnified experimental errors unless the coefficients are known to more than single precision. However, roundoff errors occurring during the solution are also greatly magnified; so double precision is worthwhile if the magnified roundoff errors predominate over the magnified experimental errors. With Gaussian elimination, this may be the case if the relative experimental errors are less than N times default precision, for example default precision is 10^{-6} to 10^{-7} for six-digit arithmetic.

Actually, we do not need to choose the diagonal element as a pivot. We may choose any element in the same column whose row has not already been used for pivoting. We may minimize the error magnification by choosing from these the element which has the largest magnitude relative to the remaining elements in its row. This strategy is called *scaled partial pivoting*, and it is a must for a general-purpose subroutine because otherwise a zero diagonal element may cause termination for a nonsingular matrix. However, this strategy complicates the programming; so it is deferred until Case Study 18.

EXERCISES

▼1. The subroutine GAUSS may be improved to take advantage of zero multipliers, skipping to the end of the I loop if the numerator of F is zero. Rewrite GAUSS accordingly. (It is not worth also testing for zero A(I, K) within the inner loop unless the portion of zeros is known to be quite large.)

2. In back substitution, B(I) is often a different order of magnitude than the terms which are subtracted from it. Consequently, it is more accurate to form the sum of the terms, and *then* subtract from B(I). Rewrite GAUSS accordingly.

3. In the back substitution, chopoff may be substantially reduced with a negligible increase in storage requirements by accumulating the inner product SUM in double precision. Rewrite GAUSS accordingly.

▼4. Since B is altered anyway, space may be saved by using the same vector for B and X. Rewrite GAUSS accordingly.

5. Write a version of GAUSS which does not alter A or B.

▼6. The coding of GAUSS may be simplified by storing B and X as the same extra column of A. Rewrite GAUSS accordingly.

7. Write a version of GAUSS which eliminates the *upper* triangular portion of A, leaving a lower triangular matrix:

$$\begin{bmatrix} A(1,1) & 0 & 0 & \cdots & 0 \\ A(2,1) & A(2,2) & 0 & & \vdots \\ A(3,1) & A(3,2) & A(3,3) & \cdots & 0 \\ \vdots & \vdots & \vdots & & \vdots \\ A(N,1) & A(N,2) & A(N,3) & \cdots & A(N,N) \end{bmatrix}$$

▼8. When a physical system described by linear equations is given a different input or loading vector, only the right-hand constants change. If we wish to find the solutions for several different B vectors, a great deal of time may be saved by collecting the corresponding B and X vectors into the columns of B and X matrices, performing the forward and backward substitution on an entire row of B or X wherever we deal with the one-element "row" in GAUSS. Rewrite GAUSS accordingly.

9. Gauss-Jordan elimination is similar to Gaussian, but the I loop is run from 1 to J − 1 and from J + 1 to N, thus eliminating all but the diagonal coefficients. The back substitution may then be replaced by simple division by the diagonal coefficients. Write a corresponding subroutine GSSJDN. (Although instructive to write, this subroutine will take about 50 per cent longer to execute than GAUSS does.)

12.3 ITERATIVE SOLUTION OF LINEAR EQUATIONS

Imagine that we have a uniform brick-shaped block of material; we impose a distribution of temperature over its surface; and we wish to know the distribution of temperature inside the brick. We may represent the temperature distribution by the temperatures at regularly spaced points within the brick. In general, we may number the points with three subscripts from 0 through L + 1 in the x direction, 0 through M + 1 in the y direction and 0 through N + 1 in the z direction as is shown in Fig. 12-3. We may then represent the temperature by a subscripted variable T(I, J, K).

To find the temperatures of the internal points, we may take advantage of the fact that the temperature at each internal point is approximately equal to the

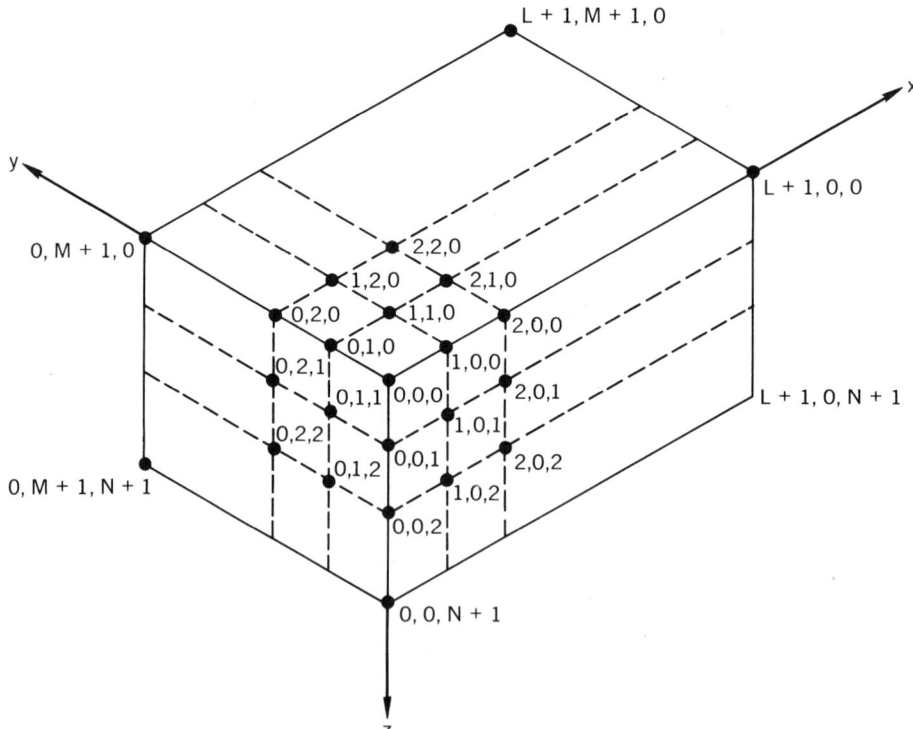

Fig. 12-3 Node numbering scheme for a rectangular parallelopiped

average of the temperatures at the adjacent points. Taking as adjacent points those directly in front, in back, to the right, to the left, above, and below, we may say that:

T(I, J, K) = [T(I, J, K + 1) + T(I, J, K − 1) + T(I, J + 1, K) + T(I, J − 1, K) + T(I + 1, J, K) + T(I − 1, J, K)]/6

If we write equations like this for all the interior points we would have a set of L * M * N simultaneous equations. However, for L, M, and N of only ten each,

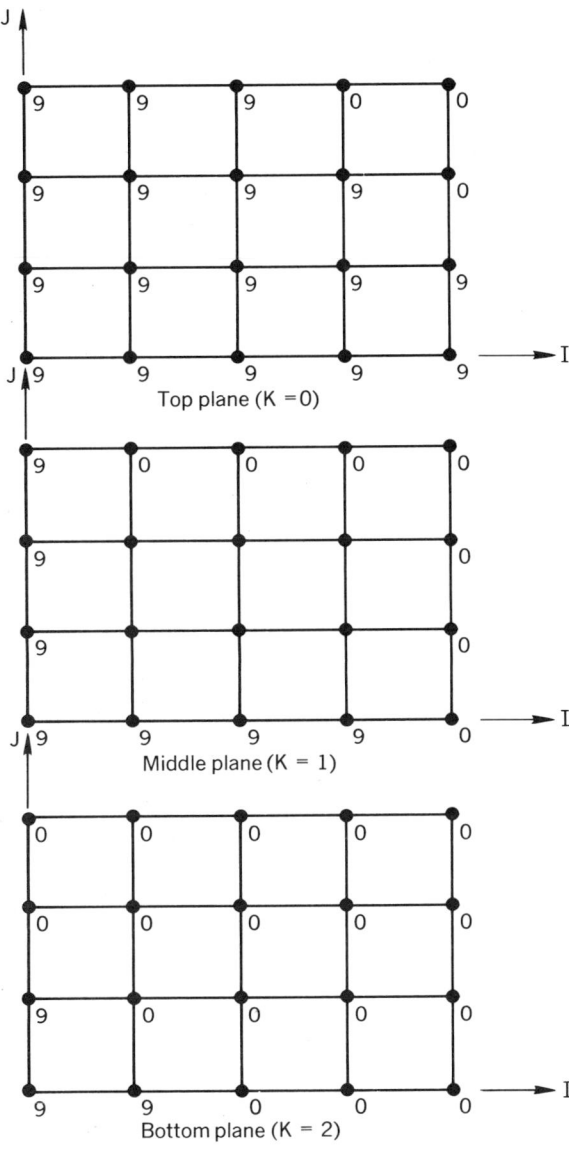

Fig. 12-4 Example surface temperatures.

this would be 1000 equations in 1000 unknowns, requiring one million words to store the coefficient matrix. Very few computers have one million words of main storage. Actually, most of the coefficients would be zero, then the remaining ones would be either one or six. With such sparseness and regularity, it is better not to store the coefficients at all. A simpler approach is to start with a guess at the solution, and iteratively apply our averaging equation to each internal point in succession. As we continue to do this, sweeping through all of the internal points many times, the internal temperatures will converge toward the solution. We

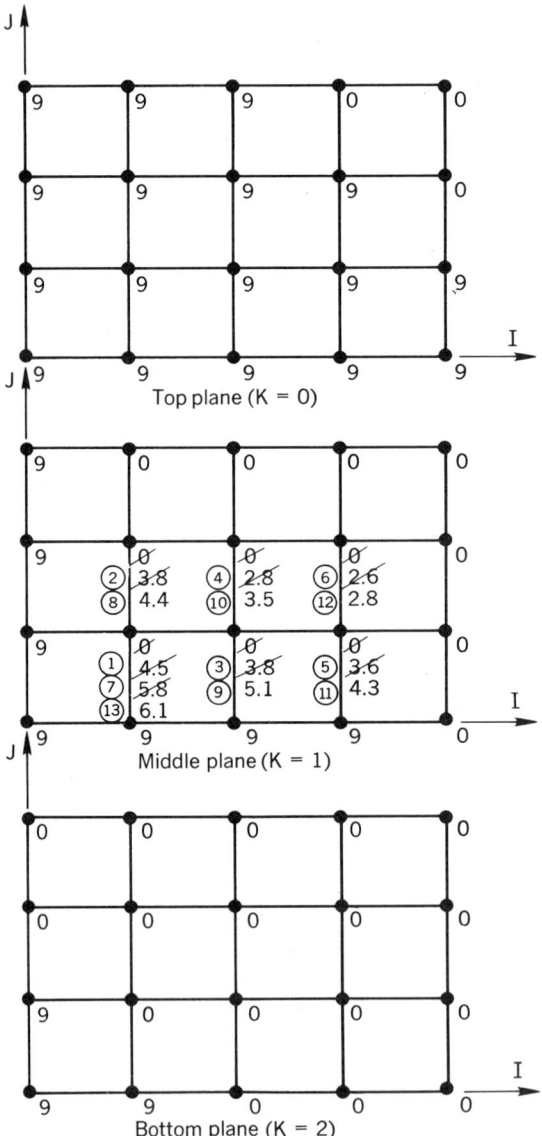

Fig. 12-5 Iterative derivation of internal temperatures.

may terminate the process when successive changes are less than some specified tolerance. This process bears a great resemblance to the resubstitution technique of exercise 9 in Case Study 7. Convergence is certain with this temperature problem regardless of the initial temperature guess, but convergence may take longer than we can afford. Consequently, it is a good idea to place a limit on the number of iterations.

As an example of this technique, Fig. 12-4 shows the given temperature distribution for a brick with one portion held at a temperature of nine units and the other portion held at a temperature of zero units, with L = 3, M = 2, and N = 1. Fig. 12-5 shows successive estimates of the internal temperatures beginning with guesses of zero, calculated in the order indicated by the encircled numbers. It is instructive to continue these calculations using a slide rule until no temperature changes by more than 0.1.

The following subroutine incorporates these ideas:

```
TEMP:   PROCEDURE (L, M, N, ITMAX, ABTOL, T);
   DECLARE
      T(*,*,*)  /* TEMPERATURE:  GIVEN EXACTLY FOR SURFACE POINTS WHERE
                   ANY SUBSCRIPT EQUALS 0, L+1, M+1, OR N+1.  GIVEN
                   APPROXIMATELY FOR OTHER POINTS WHICH ARE INTERNAL.
                   RESULT FOR INTERNAL POINTS.  */,
      L, M, N   /* GIVEN:  L+1, M+1, AND N+1 ARE THE HIGHEST USED
                   VALUES OF THE 1ST, 2ND, AND 3RD SUBSCRIPTS
                   RESPECTIVELY.  (LOWEST USED VALUES ARE ZERO.)  */,
      ITMAX     /* GIVEN:  MAXIMUM ALLOWABLE NUMBER OF ITERATIONS.  SET
                   LESS THAN T/(L*M*N*DT) WHERE DT IS THE TIME TO
                   EXECUTE 1 CYCLE OF THE INNER LOOP; AND WHERE T IS
                   THE MAXIMUM ALLOWABLE EXECUTION TIME.  */,
      ABTOL     /* GIVEN & RESULT:  DESIRED ABSOLUTE ACCURACY; IF IT
                   IS NOT ACHIEVED WITHIN ITMAX ITERATIONS, THEN ABTOL
                   IS RESET TO THE ACHIEVED ACCURACY.  */;
   DO IT = 1 TO ITMAX;
      BIGGEST = 0;
      DO I = 1 TO L;
         DO J = 1 TO M;
            DO K = 1 TO N;
               TNEW = (T(I,J,K+1)+T(I,J,K-1)+T(I,J+1,K)+T(I,J-1,K)
                  +T(I+1,J,K)+T(I-1,J,K))*.166667;
               BIGGEST = MAX(BIGGEST, ABS(TNEW-T(I, J, K)));
               T(I,J,K) = TNEW;
            END;
         END;
      END;
      IF BIGGEST < RELTOL   THEN RETURN;
   END;
   ABTOL = BIGGEST;
END;
```

This method of representing a physical problem is an example of *finite-difference* methods, and this method of solving the equations is called the *method of successive displacements*, the *Gauss-Seidel method*, or *relaxation*. The same averaging principle applies to many problems in gravitation, electromagnetic theory, fluid mechanics, solid mechanics, and other fields. For a thorough but mathematically advanced treatment of these techniques, see the books by Varga or by Forsythe and Wasow.[1]

---

[1] R. Varga, *Matrix Iterative Analysis* (Englewood Cliffs, N. J.: Prentice-Hall, Inc., 1962); G. Forsythe and W. Wasow, *Finite-Difference Methods for Partial Differential Equations* (New York: John Wiley & Sons, Inc., 1960).

12.4 SYMMETRIC MATRICES

In many applications, we deal with *symmetric* matrices where $A(I, J) = A(J, I)$ for all I and J, such as in the following example:

$$\begin{bmatrix} 3.1 & 0.1 & 0.3 & 5.9 \\ 0.1 & 4.0 & 1.9 & 7.2 \\ 0.3 & 1.9 & 5.8 & 6.0 \\ 5.9 & 7.2 & 6.0 & 8.9 \end{bmatrix}$$

When this is so, we may cut the storage requirement almost in half by storing only the lower triangular portion of the matrix in *compacted* form as a singly-subscripted variable in the following order:

$$\begin{bmatrix} A(1) & & & \\ A(2) & A(3) & & \\ A(4) & A(5) & A(6) & \\ \vdots & \vdots & \vdots & \vdots \end{bmatrix}$$

Since there is one element in the first row, two in the second, up to $I - 1$ in the $I - 1$th row, the subscript of the element in the Ith row and Jth column will be

$$K = 1 + 2 + 3 + \cdots (I - 1) + J$$

Using the formula for the sum of an arithmetic series, we may express this as

$$K = 0.5 * I * (I - 1) + J$$

We may then use the same DO loops we would with a doubly-subscripted variable, using this expression wherever we ordinarily use the subscripts I and J. For example, the *diagonal product* of any N-by-N matrix A is defined as

$$\prod_{I=1}^{N} A(I, I)$$

Consequently, if A is symmetric and stored in compacted form as a singly-subscripted variable B, the following function procedure evaluates the diagonal product:

```
SYMPROD:  PROCEDURE (N, B);
    DECLARE B(*);
    PRD = 1;
    DO I = 1 TO N;
        PRD = PRD*B(.5*I*(I-1) + I);
    END;
    RETURN(PRD);
END;
```

Actually, the subscript expression simplifies to B(.5 * I * (I + 1)), but we can save 2 * N multiplications with the following technique:

```
SYMPRD:  PROCEDURE (N, B);
    DECLARE B(*);
    PRD = 1;  M = 0;
    DO I = 1 TO N;
        M = M + I;
        PRD = PRD*B(M);
    END;
    RETURN(PRD);
END;
```

As another example, the *infinity-norm* of an N-by-N matrix is defined as:

$$||A||_\infty = \max_{I=1}^{N} \left[\sum_{J=1}^{N} |A(I, J)| \right] = \text{maximum row absolute sum}$$

The following function procedure evaluates the one-norm for a symmetric matrix stored in compacted form as a singly-subscripted variable B:

```
FINNRM:  PROCEDURE (N, B);
    DECLARE B(*);
    BIGGEST = 0;
    DO I = 1 TO N;
        SUM = 0;   L = .5*I*(I-1);
        DO J = 1 TO I;
            SUM = SUM + ABS(B(L+J));
        END;
        DO J = I+1 TO N;
            SUM = SUM + ABS(B(.5*J*(J-1)+I));
        END;
        BIGGEST = MAX(SUM, BIGGEST);
    END;
    RETURN (BIGGEST);
END;
```

Note how we must reverse I and J whenever they correspond to an upper triangular element.

EXERCISES

▼10. Suppose the brick for our temperature problem has holes and corners cut out of it. We may still use the same scheme if we have an integer "type" variable ID(I, J, K) which is a one if the point (I, J, K) is internal and zero otherwise. We may then test each point in the inner loop, evaluating a new T(I, J, K) only if ID(I, J, K) = 1. Rewrite TEMP accordingly.

11. Convergence may be greatly speeded for TEMP by magnifying the changes in T, thus getting the temperature to wherever it is going faster. However, if we magnify the change too much, we may overshoot to such an extent that the process becomes

unstable, diverging instead of converging. The optimum magnification factor is not generally known in advance, but it increases with the number of points, and it is generally between one (no magnification) and two (doubling the change). Rewrite TEMP using a magnification factor of $2 - 1/(L + M + N)$.

12. (Difficult) Same as exercise 11 except have the procedure keep changing the magnification factor so as to minimize the ratio of BIGGEST at the end of successive sweeps.

▼13. The closer the initial guesses for the internal temperatures, the sooner TEMP will converge. The simplest guess to program is zero for the internal temperatures, but a better guess is the average of all the surface temperatures. Write a subroutine INITL which initializes all of the internal temperatures to this average.

14. Same as exercise 13 except assume that the brick has holes and corners cut out as described in exercise 10.

15. Same as exercise 13, except take the average only of the six surface points above, below, to the right, to the left, in front, and in back of the point—weighted according to their inverse distance from the internal point:

$$T(I, J, K) = \frac{\dfrac{T(0, J, K)}{I} + \dfrac{T(L + 1, J, K)}{L + 1 - I} + \dfrac{T(I, 0, K)}{J} + \dfrac{T(I, M + 1, K)}{M + 1 - J} + \dfrac{T(I, J, 0)}{K} + \dfrac{T(I, J, N + 1)}{N + 1 - K}}{\dfrac{1}{I} + \dfrac{1}{L + 1 - I} + \dfrac{1}{J} + \dfrac{1}{M + 1 - J} + \dfrac{1}{K} + \dfrac{1}{N + 1 - K}}$$

▼16. Write an input routine to get in L, M, N and the surface temperatures with a minimum of data preparation.

▼17. Write a main program which uses TEMP, an initialization subroutine such as in exercise 13, an input routine such as in exercise 16, and puts out the temperature at particular points specified in the input. (There are too many points to put the temperature for all of them. We must guard against swamping ourselves with so much output that we cannot comprehend it. This is a good application for graphical output such as is described in Case Study 14.)

18. If the thermal conductivity of the brick varies with position, given by a subscripted variable C(I, J, K), then we must multiply the six temperatures on the right-hand side of our averaging equation by the following six factors, respectively:

$$1 + \alpha,\ 1 - \alpha,\ 1 + \beta,\ 1 - \beta,\ 1 + \gamma,\ 1 - \gamma$$

where $\alpha = 0.25[C(I, J, K + 1) - C(I, J, K - 1]/C(I, J, K)$

$\beta = 0.25[C(I, J + 1, K) - C(I, J - 1, K)]/C(I, J, K)$

$\gamma = 0.25[C(I + 1, J, K) - C(I - 1, J, K)]/C(I, J, K)$

Rewrite TEMP accordingly.

▼19. The Gauss-Siedel method is rarely used on a computer when the whole coefficient matrix is stored, but it is instructive to write such a subroutine. The applicable iteration equation may be derived by transposing all terms except the diagonal one, then dividing by the diagonal coefficient:

$$X(I) = \frac{B(I) - A(I,1)X(1) - A(I,2)X(2) \cdots -A(I,I-1)X(I-1) - A(I,I+1)X(I+1) \cdots A(I,N)X(N)}{A(I,I)}$$

Write a corresponding subroutine GSSSDL(N, A, B, ITMAX, RELTOL, X);.

20. The *trace* of an N-by-N matrix A is defined as

$$\sum_{I=1}^{N} A(I, I)$$

Write a function procedure for it assuming A is symmetric, stored in compacted form as a singly-subscripted variable B.

21. The *one-norm* of an N-by-N matrix is defined as

$$\|A\|_1 = \max_{J=1}^{N} \left[\sum_{I=1}^{N} |A(I, J)| \right] = \text{maximum column absolute sum}$$

Write a function procedure for it assuming A is symmetric, stored as a singly-subscripted variable B.

22. (Difficult) Write a subroutine which multiplies two symmetric matrices stored in compacted form. (Note that in general, the product is not symmetric.)

▼23. (Difficult) If a matrix A(I, J) is predominately zeros, space may be saved by storing only the nonzero elements in row order as the successive element of a vector B(K), and also storing integer vectors IB(K) and JB(K) which give the true subscripts for each B element. Write a function procedure which finds the infinity-norm of a matrix stored in this fashion.

24. (Difficult) Same as exercise 23 except assume that instead of IB and JB, there is an integer vector INTERVAL(K) which gives the number of steps in A rowwise, between successive elements of B.

▼25. With many physical systems, we know in advance that the only nonzero coefficients will be on the diagonal or immediately adjacent to it. In such cases, the matrix is called *tridiagonal*, and a great deal of time and space may be saved by storing the coefficients as three vectors:

$$\begin{bmatrix} E(1) & F(1) & & & & & \\ D(2) & E(2) & F(2) & & & & \\ & D(3) & E(3) & F(3) & & & \\ & & \cdot & \cdot & \cdot & & \\ & & & \cdot & \cdot & \cdot & \\ & & & & \cdot & \cdot & F(N-1) \\ & & & & & D(N) & E(N) \end{bmatrix}$$

The elimination process does not introduce nonzero elements outside these diagonals; so the loops need not run beyond them. Write a corresponding subroutine TRIGSS.

▼26. (Difficult) With many physical systems, we know in advance that the only nonzero coefficients will be within a band of width 2M + 1 centered on the diagonal. If M is considerably less than N/2, time and space may be saved by storing the matrix as a matrix C with bounds (N, −M : M) where C(I, J) = A(I, I + J). The elimination process does not introduce nonzero elements outside the band, so the loops need not run beyond them. Write a corresponding subroutine BNDGSS.

CASE STUDY 13

BOOLEAN ALGEBRA

Bit-Strings[1]

13.1 THE NOT OPERATOR

A string of zeros and ones such as 1011100 may be considered a binary number for numerical applications, but it may also be considered a *bit-string* for non-numerical applications. For example, the value "false" may be associated with the digit "0" and the value "true" with the digit "1." George Boole was the first to do this, and he also associated the symbol "+" with the "or" operation, the symbol "·" (multiplication) with the "and" operation, and the symbol "−" with the "not" operation (not-true is false and not-false is true). The brevity of this notation enabled Boole to build up a systematic "algebra" which facilitates the manipulation of logic propositions. Boolean algebra has since been found to have other applications in engineering and science. PL/I has these Boolean capabilities except that it uses the symbols &, |, and ⌐ which are less confusing.[2]

---

[1] Unsupported by the Model 20.
[2] These are written AND, OR and NOT in the 48-character set. Operators written in alphabetic characters are reserved identifiers.

13.2 BIT-STRING CONSTANTS

Bit-string constants within a program or as data are written as a string of zeros and ones enclosed in apostrophes (single quotes) and immediately followed by the letter "B"—for example '1'B, '0'B, or '1011100'B. The constant '10101010'B may also be written (4)'10'B where the 4 enclosed in parentheses is a *repetition factor*.[3] Repetition factors must be unsigned decimal constants.

13.3 BIT-STRING VARIABLES

Any variable may be declared a bit-string variable by using the keyword BIT— for example: DECLARE ANSWER BIT(1);. The "1" in parentheses establishes the length as one bit. The example in the first sentence of this case study has a length of seven.[4] Bit-strings of length longer than one may be associated with all kinds of things, but the original Boolean algebra was concerned only with bit-strings of length one, and we will limit ourselves to that at first.

The result of any comparison is a bit-string of length one. For example, the comparison ABS(DX) <= ABS(RELTOL*X) has either the value '1'B (true) or '0'B (false). Used in an IF statement, it is this one-bit string which determines whether or not the THEN clause will be executed. However, we may also assign the result of a comparison to a bit-string variable and use the bit-string variable in an IF statement:

```
CONVERGED = ABS(DX) <= ABS(RELTOL * X);
IF CONVERGED THEN ...;
```

This technique saves time and memory space when the same comparison must be made more than once. For example, if we are given a longitude with absolute value <= 180° and a latitude with absolute value <= 90°, then the point is in the western hemisphere if

$$\text{longitude} >= 0$$

and the point is in the northern hemisphere if

$$\text{latitude} >= 0$$

---

[3] Unsupported by SL1.

[4] The F compiler permits bit-string constants of from zero to at least 8056 bits and variables of from zero through 32,767 bits. The D compiler permits constants and variables from one through 64 bits. The SL1 compiler permits constants and variables from 1 through 127 bits. The Student PL interpreter permits only one-bit strings declared simply BIT or written TRUE or FALSE as constants.

Accordingly, the following function procedure returns a 0, 1, 2, or 3 according to whether a point is in the northwestern, southwestern, northeastern, or southeastern quarter-sphere:

```
LOC:    PROCEDURE (DEG_LONG, DEG_LAT);
        DECLARE (EAST, SOUTH, WEST) BIT(1);
        EAST = DEG_LONG < 0;
        SOUTH = DEG_LAT < 0;
        IF EAST & SOUTH     THEN RETURN (3);
        WEST = ¬EAST;
        IF WEST & SOUTH     THEN RETURN (1);
        IF WEST & ¬SOUTH    THEN RETURN (0);
        RETURN (2);
        END;
```

Without the bit-string variables we must use either redundant comparisons or extra GO TO statements—either of which are slower.

Note that ¬ is a prefix operator whereas & and | are infix operators. In fact, ¬ shares priority with exponentiation, prefix +, and prefix −. (If ¬ was applied to a numeric value, the value would first be converted to a bit-string).[5] Consequently, the parentheses are meaningful in the following IF statement:

IF ¬(X > Y) THEN...;

This IF statement is equivalent to, but slower than,

IF X ¬> Y THEN...;

Also note that it is natural for & to have higher priority than | because & is distributive with respect to | just as multiplication is distributive with respect to addition:

A & (B | C) = A&B | A&C

This is why Boole associated multiplication with the & operator and addition with the | operator.

The &, |, and ¬ operators give the results we would expect, associating 0 with false and 1 with true as is summarized in the following table:

A	B	¬A	A&B	A|B
1	1	0	1	1
1	0	0	0	1
0	1	1	0	1
0	0	1	0	0

[5] For Student PL, the ¬ operator has a priority between the relational operators and the & operator, conversion from numeric to bit being illegal.

13.4 CONCATENATION

The *concatenation operator* consisting of two adjacent *or* symbols, ||, may be used to join bit-strings together into longer strings.[6] For example, we may write:

$$(DEG\_LONG < 0) \;||\; (DEG\_LAT < 0)$$

The parentheses are necessary because concatenation has a priority between that of addition-subtraction and the relational operators. Our expression will have the value '00'B if neither the longitude nor latitude is negative, '01'B if only the latitude is negative, '10'B if only the longitude is negative, and '11'B if both are negative.

These two-bit strings correspond to the binary numbers 0, 1, 2, and 3 respectively, which are the numbers that we previously established for the corresponding quarter-spheres. Consequently, we may rewrite LOC more efficiently as follows:

```
LOC:   PROCEDURE (DEG_LONG, DEG_LAT);
       RETURN ((DEG_LONG < 0) || (DEG_LAT < 0));
END;
```

The two-bit string will be converted to the binary integer implied by the name LOC.[7]

13.5 MIXTURES OF BIT-STRINGS AND NUMBERS

Bit-strings are converted to arithmetic values when used in arithmetic expressions. For example, suppose we have a function Y defined as

$$Y = SIN(X) + X \quad \text{if } X > 0$$
$$\text{and } Y = SIN(X) \quad \text{if } X <= 0$$

We may write a corresponding function procedure as follows:

```
Y:    PROCEDURE (X);
      RETURN (SIN(X) + (X>0)*X);
END;
```

13.6 WATER DISTRIBUTION EXAMPLE

As another use of one-bit strings, consider the simplified city water system shown in Fig. 13-1. The check valves permit flow only in the indicated directions,

---

[6] This is written CAT with the 48-character set. Operators written in alphabetic characters are reserved identifiers.

[7] Conversion between bit and arithmetic is not allowed in SL1, and only one-bit strings are permitted in Student PL.

Sec. 13.6   Water Distribution Example

and often one or more of the gate valves must be closed during repair or expansion of the system. By making V1 through V8, HOSPITAL, SCHOOL, and FACTORY be one-bit variables, and by getting data for the valves that are closed, the following program determines whether or not the hospital, the school, and the factory receive water:

```
WATER:  PROCEDURE OPTIONS (MAIN);
    DECLARE (V1, V2, V3, V4, V5, V6, V7, V8, HOSPITAL, SCHOOL, FACTORY)
    BIT(1);
    V1, V2, V3, V4, V5, V6, V7, V8 = '1'B;
/* DATA CONSISTS ONLY OF '0'B FOR THOSE VALUES THAT ARE CLOSED */
    GET DATA;
    HOSPITAL = V6 & V4 & (V1 | ((V2 | V5) & V3));
    SCHOOL   = V7 & (V3 | ((V4 | V5 | V2) & V1));
    FACTORY  = V8 & (V3 | ((V4 | V5 | V2) & V1));
    PUT DATA (HOSPITAL, SCHOOL, FACTORY);
END;
```

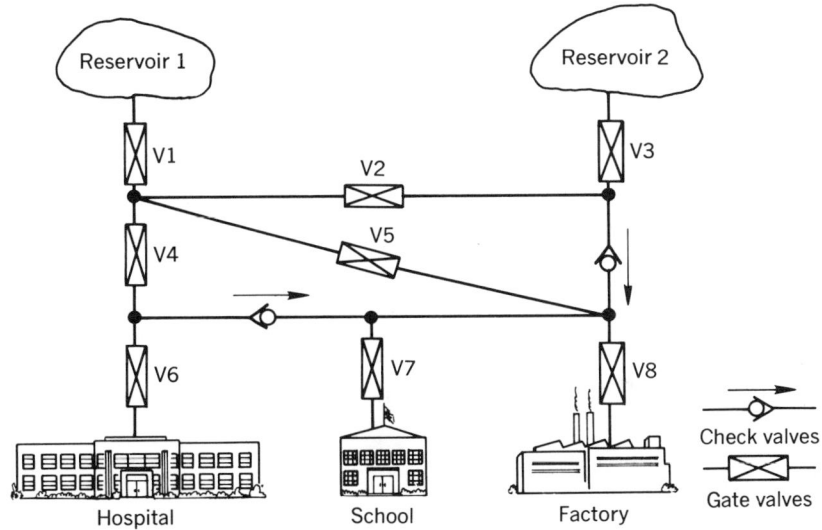

Fig. 13-1 Diagram of a water distribution system.

In the INITIAL attribute we listed the repetition factor even though it is merely one, to avoid ambiguity with the iteration factor of eight.

The Boolean expressions can be simplified somewhat by using intermediate variables to eliminate redundancy and by taking advantage of the priority of "&" over "|":

```
V2ORV5   = V2 | V5;
HOSPITAL = V6 & V4 & (V1 | V2ORV5 & V3);
A        = V3 | (V4 | V2ORV5) & V1;
SCHOOL   = V7 & A;
FACTORY  = V8 & A;
```

Networks such as this where elements have only two states—on or off—are called switching networks. Other examples are relay systems such as a telephone exchange and the circuits within a digital computer. Note, however, that a "closed" electrical switch is analogous to an "open" valve.

EXERCISES

▼1. Rewrite exercise 23 in Case Study 4 using IF statements and one or more one-bit variables.

2. Rewrite LOC using IF statements and one-bit variables, except renumber the Northeast, Northwest, Southwest, and Southeast quarter-spheres 1, 2, 3, and 4, respectively.

▼3. Rewrite exercise 23 in Case Study 4, using the concatenation operator. Note that the standard quadrant-numbering scheme does not correspond to that used for LOC in this case study.

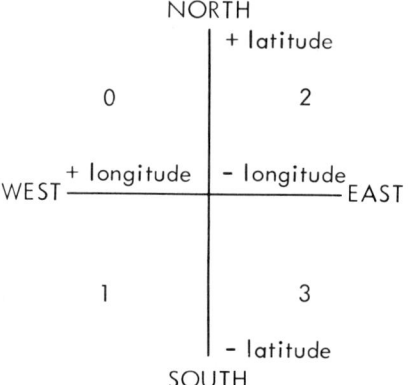

4. Rewrite exercise 24 in Case Study 4 without using NQUAD, but using the concatenation operator.

▼5. Write a function procedure LOC(X, Y) which returns a 1, 2, 3, or 4 according to which region is occupied by the point (X, Y) in the adjacent figure. (Consider points on a line to be equivalent to points below a line.)

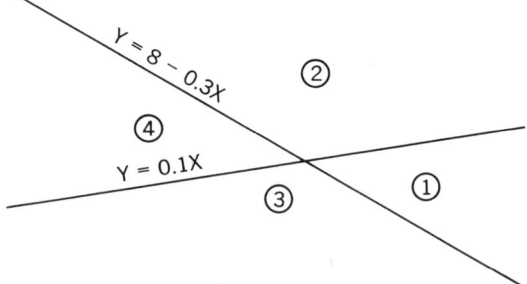

▼6. Rewrite exercise 30 in Case Study 4 so that the function returns the bit-string '1'B if the point is in the triangle and returns the bit-string '0'B otherwise.

7. Rewrite exercise 31 in Case Study 4 so that the function returns the bit-string '1'B if the point is in the shaded area and returns the bit-string '0'B otherwise.

▼8. The figure below shows a road map. Because of a natural disaster, many of the

Sec. 13.7   *Subscripted Bit-String Variables*   225

roads are not open. The Red Cross has received radio reports as to which roads are closed, and it would like to know if emergency supplies can be trucked from point A to point B. Fortunately, the electricity is still available; so they can get their answer by computer, Write a corresponding program.

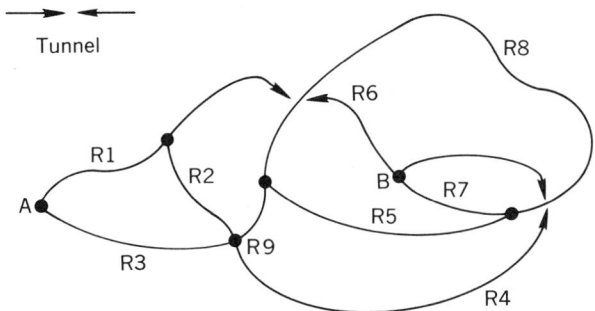

9. The figure below shows an electrical switching network. Write a program which gets in the state of the switches ('1'B for closed and '0'B for open), and puts out whether or not current could flow from A to B provided a voltage difference and a return path were provided.

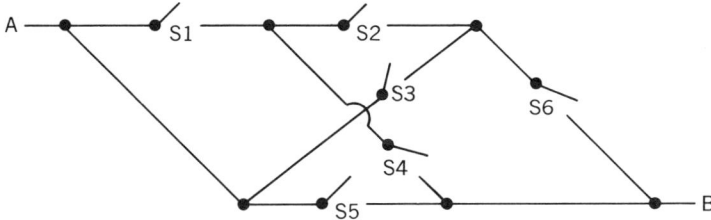

13.7 SUBSCRIPTED BIT-STRING VARIABLES

While instructive, our water distribution program is inflexible and too clumsy for a full-scale water system. For a worthwhile program, *subscripted* one-bit variables enable a more systematic approach. In developing this approach we will use the same network as an example, but we will consider the more general problem of determining for each junction whether or not there is an open flow path to each of the other junctions. In general, we may number the junctions or nodes in the piping network one through N. We may then fill in a one-bit *incidence* matrix FLOW(I, J) with a one wherever there is an open one-branch (or less) flow path from node I to node J. For our example, if V1, V2, and V7 are closed, whereas V3, V4, V5, V6, and V8 are open for the system of Fig. 13-1, the numbering scheme of Fig. 13-2 gives the indicated incidence matrix.

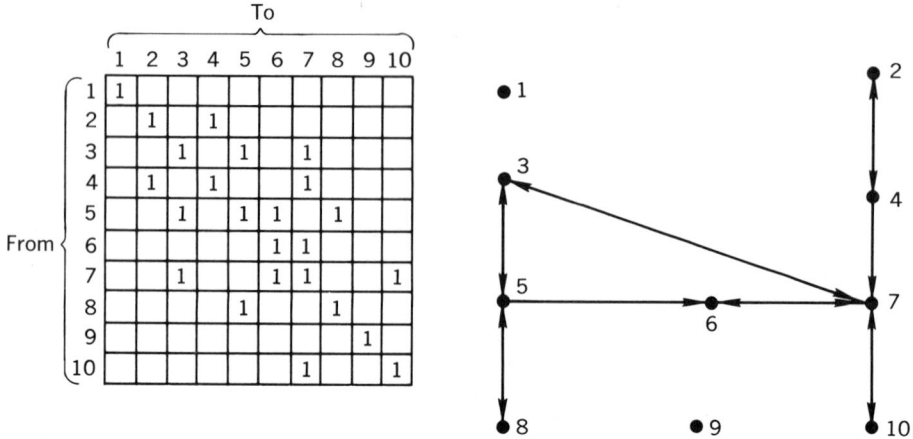

Fig. 13-2 Incidence matrix and graph for a particular state of the water distribution system.

A branch is a direct connection between nodes, and in Fig. 13-2 we have indicated the open branches by a *linear graph* of arrows—double-headed when the absence of a check valve permits flow in both directions. Note that there is always a one-branch (or less) path from a node to itself; so the diagonal of the incidence matrix consists of ones. Note also that the matrix is symmetrical with respect to the diagonal except where there is a check valve.

We may fill in the remainder of the matrix with zeros, meaning there is not a corresponding open one-branch (or less) path. We may now say that the truth or falsehood of an open two-branch (or less) path from any node I to any node J is given by the Boolean expression:

FLOW(I,1)&FLOW(1,J) | FLOW(I,2)&FLOW(2,J) | ... FLOW(I,10)&FLOW(10,J)

More generally, if there are N nodes, we may evaluate the corresponding expression and assign its value to the one-bit variable PATH by the following DO loop:

```
PATH = FLOW(I,1)&FLOW(1,J);
DO K = 2 TO N;
    PATH = PATH | FLOW(I,K)&FLOW(K,J);
END;
```

Next, we may assign the value of PATH to FLOW(I, J) meaning that we have modified FLOW(I, J) so as to take into account all possible paths of two or less branches. We may enclose these statements in a nest of two DO loops on I and J to do this for all I and J:

```
DO I = 1 TO N;
    DO J = 1 TO N;
        PATH = FLOW(I,1)&FLOW(1,J);
```

```
            DO K = 2 TO N;
               PATH = PATH | FLOW(I,K)&FLOW(K,J);
            END;
            FLOW(I,J) = PATH;
         END;
      END;
```

This nest of **DO** loops may be repeated until there is no change in any elements of **FLOW**, meaning the opportunity to include more branches does not connect any two previously unconnected nodes. Thereafter, there will be no change in **FLOW** because the nest of **DO** loops will have the same values to work with each time. This terminal value of **FLOW** called the *dispersion* matrix is the one-bit matrix expressing the truth or falsehood of an open path of any length between pairs of nodes. This is the answer we seek, and the following subroutine finds it:

```
WATER1: PROCEDURE (N, FLOW);
        DECLARE (FLOW(*,*), PATH, CHANGE) BIT(1);
   S10: CHANGE = '0'B;
        DO I = 1 TO N;
           DO J = 1 TO N;
              PATH = FLOW(I,1) & FLOW(1,J);
              DO K = 2 TO N;
                 PATH = PATH | FLOW(I,K) & FLOW(K,J);
              END;
              IF PATH ¬= FLOW(I,J)  THEN DO;
                 FLOW(I,J), CHANGE = '1'B;  END;
           END;
        END;
        IF CHANGE  THEN GO TO S10;
END;
```

13.8 THE ALIGNED ATTRIBUTE

As an example of how WATER1 may be used, the following program gets in the nodes between which there are one-branch paths, calls WATER1, and then puts out the dispersion matrix:

```
HYD: PROCEDURE OPTIONS (MAIN);
     DECLARE FLOW(200,200) BIT(1);
     FLOW = '0'B;
     ON ENDFILE(SYSIN)  GO TO S20;
     GET LIST (N);
     DO K = 1 TO N;
        FLOW(K,K) = '1'B;
     END;
 S10: GET LIST (I, J);
     FLOW(I,J) = '1'B;
     GO TO S10;
 S20: CALL WATER1(N, FLOW);
     PUT PAGE DATA(((FLOW(I,J) DO J = 1 TO N) DO I = 1 TO N));
END;
```

The D and F compilers store one-bit strings 32 to a word (eight per byte); so we have allocated 320,000 variables rather than our customary 10,000. At some expense in storage capacity, execution speed may be increased by forcing alignment of each element on a byte boundary:[8]

DECLARE FLOW(200, 200) BIT(1) ALIGNED;

The other notable feature of this program is the care which has been taken to minimize the amount of data preparation. The program initializes the matrix to zeros, then initializes the diagonal to ones; so only the nodes of the open one-branch paths need be gotten.

13.9 THE ALL FUNCTION AND THE ANY FUNCTION[9]

Provided all of **FLOW** is in use, another way of evaluating **PATH** in the subroutine **WATER1** is:

PATH = ANY(FLOW(I, *) & FLOW(*, J));

We may have bit-string array cross sections and expressions just as we may have numerical array cross sections and expressions. Moreover, PL/I has the built-in bit-string array manipulation function **ANY**, which in our case returns a one if *any* of the elements of its array argument is one, returning a zero otherwise. More generally, it returns a bit-string of length equal to that of the elements and each bit of this returned string is a one or zero according to whether or not any of the corresponding bits of the elements is a one.

There is a similar built-in function **ALL** which returns a bit-string with ones whenever *all* of the corresponding bits of the elements are one.

13.10 STRING AND SUBSTR BUILT-IN FUNCTIONS AND PSEUDO-VARIABLES

With most compilers, the **FLOW** matrix may be more efficiently handled if each of the rows is combined into an N-bit string, resulting in a singly-subscripted variable which we shall call **FROW**. Then, provided all of the allocated space is in use, we may rewrite **WATER1** as:

---

[8] Unsupported by SL1 and Student PL. The D compiler does not permit aligned *bit-strings*.
[9] Unsupported by SL1 and Student PL.

```
WATER2:  PROCEDURE OPTIONS (MAIN);
   GET LIST (N);
   BEGIN;
   DECLARE (CHANGE, PATH) BIT(1), (ZEROS, FROW(N)) BIT(N) INITIAL
      ((N)(1)'0'B, (N)(1)'0'B); 10
   S10:  CHANGE = '0'B;
      DO I = 1 TO N;
         DO J = 1 TO N;
            PATH = FROW(I) &        (SUBSTR(FROW, J, 1)) ¬= ZEROS;
            IF SUBSTR(FROW(I), J, 1) ¬= PATH  THEN DO;
               SUBSTR(FROW(I), J, 1), CHANGE = '1'B;   END;
         END;
      END;
      IF CHANGE  THEN GO TO S10;
   END;
END;
```

Among other things, this procedure introduces the substring function SUBSTR.[11] This function extracts a substring from its first argument. The substring begins at the position given by its second argument with a length given by its third argument. In our case the first argument is the array FROW; so SUBSTR returns an array of values equal to the Jth bits of the elements of FROW. This array is simply column J of our equivalent matrix FLOW. The STRING function then combines or concatenates this array of bit-strings into a single bit-string which is combined with ROW(I) by the & operator.[12] The &, |, and ¬ operators apply on a bit-by-bit basis to strings of length greater than one. Consequently, the result of this & operation is a string of length N each of whose bits is a one if and only if the corresponding bits of FROW(I) and "column" J are both one. This resultant string is compared to ZEROS which has been initialized to N zeros. This resultant string will not equal ZEROS if any of its bits are one, meaning there is a corresponding path. Consequently, the result of this comparison is assigned to PATH. Then PATH is compared with the former value of the Jth bit in FROW(I). If there is a change, it must be from a zero to a one; so this bit and the variable CHANGE are assigned a one. SUBSTR used on the left side of an assignment statement such as this is called a *pseudo-variable* because the value is assigned to a portion of the first argument of SUBSTR rather than to SUBSTR itself.

13.11 THE VARYING ATTRIBUTE AND THE LENGTH FUNCTION[13]

The main procedure using WATER2 would presumably have a BEGIN block so that all of the allocated space is in use. As an alternative, we could declare a constant maximum size together with the VARYING attribute:

[10] The D and SL1 compilers do not permit a variable or an asterisk for the length or extent of a variable.
[11] SL1 does not support SUBSTR.
[12] Student PL and SL1 do not support STRING.
[13] Unsupported by the SL1, Student PL, Model 20 and D compilers.

```
DECLARE (CHANGE, PATH) BIT(1), (ZEROS, FROW(560)) BIT(560) VARYING;
ZEROS = (N)'0'B;
```

The maximum length is allocated, but the computer automatically keeps track of how much is actually in use. Consequently, strings with the **VARYING** attribute may be concatenated, compared, and put out even though not all of the allocated space is in use.

For the D and F compilers, the bytes of storage for a bit-string of length L is CEIL(L/8); so multiples of length eight are most efficient.

We may use the function **LENGTH(a)** when we wish to know the current length of a varying length string.

13.12 THE BOOL FUNCTION

Let us consider a diagnosis procedure. We will consider simplified medical diagnosis, but other applications would be chemical diagnosis, species diagnosis, or malfunction diagnosis of a large system. We will consider the patient's symptoms to be denoted by a bit-string **SYMPTOMS** where the first bit denotes the presence or absence of nausea, the second denotes the presence or absence of coma, and so on. We may devote more than one bit to symptoms which require a scale of severity. For example, we may devote three bits to temperature with 000 denoting low, 001 denoting normal, 011 denoting moderately high, and 111 denoting severely high. Each disease will then be described by an analogous bit-string **DISEASE(K)** giving its nominal symptoms. The computer may search this list of diseases, looking for the closest match.[14] Corresponding bits match if both are one or neither are one. Neither the &, |, or ⌐ operator alone will properly indicate this match, but we could use the combination:

(SYMPTOMS & DISEASE(K)) | (⌐SYMPTOMS & ⌐DISEASE(K))

However, this may be indicated more efficiently by use of the **BOOL** function as:

BOOL(SYMPTOMS, DISEASE(K), '1001'B)[15]

The **BOOL** function enables us to specify any of 16 elementary Boolean operations. In general, BOOL(A, C, 'ijkm'B) is defined by the table:

A	C	BOOL(A,C, 'ijkm'B)
0	0	i
0	1	j
1	0	k
1	1	m

[14] This approach to diagnosis is rather elementary. For a more thorough approach see Robert S. Ledley, *Use of Computers in Biology and Medicine* (New York: McGraw-Hill, 1965).
[15] Student PL and SL1 do not support the BOOL function.

As special cases we can see that:

$$
\begin{array}{rcl}
A \,\&\, C &=& BOOL(A, C, \text{'0001'B}) \\
A \mid C &=& BOOL(A, C, \text{'0111'B}) \\
\neg A &=& BOOL(A, C, \text{'1100'B}) \\
\neg C &=& BOOL(A, C, \text{'1010'B})
\end{array}
$$

SYMPTOMS and DISEASE(K) have more than one bit; so BOOL will be applied to them on a bit-by-bit basis, returning a bit-string of equal length. The number of ones in this returned string is an indication of how well the symptoms and disease match. The SUBSTR function then permits us to count the number of ones, determining the best match. The following program gets in the disease strings, then any number of symptom strings, putting out the most likely disease for each:

```
DIGNOS:   PROCEDURE OPTIONS (MAIN);
   GET LIST (M, N);
   BEGIN;
        DECLARE (DISEASE(N), SYMPTOMS, COMPARISON) BIT(M);
        GET LIST (DISEASE);
   S10:  GET LIST (SYMPTOMS);
        LARGEST = 0;
        DO K = 1 TO N;
           COMPARISON = BOOL(SYMPTOMS, DISEASE(K), '1001'B);
           MATCH = 0;
           DO J = 1 TO M;
              MATCH = MATCH + SUBSTR(COMPARISON, J, 1);
           END;
           IF MATCH > LARGEST  THEN DO;
              LARGEST = MATCH;  ID = K;   END;
        END;
        PUT LIST (ID, SYMPTONS, DISEASE(ID));
        GO TO S10;
   END;
END;
```

Although COMPARISON and BOOL (SYMPTOMS, DISEASE(K), '1001'B) are the same length here, more generally bit-strings are truncated or padded with zeros on the right when assigned to variables of different length. Case Study 14 presents some additional string manipulation techniques which may be used with bit-strings.

13.13 BINARY ARITHMETIC CONSTANTS

Many instruments produce digital output which may be fed directly into a computer or onto a tape or another medium for subsequent direct input to a computer. However, the digital code is often not the same as that of the computer. For example, a *gray code* is often used because only one bit changes between successive numbers. If the original signal is an analog quantity, the ordinary binary code may cause ambiguity because it is impossible to align the encoder so that two or more bits change exactly simultaneously.

The following table presents a four-bit gray code:

	ENCODING TABLE		DECODING TABLE	
Decimal	Gray Code	Binary Code	Gray Code (in binary order)	Decimal
0	0000	0000	0000	0
1	0001	0001	0001	1
2	0011	0010	0010	3
3	0010	0011	0011	2
4	0110	0100	0100	7
5	0111	0101	0101	6
6	0101	0110	0110	4
7	0100	0111	0111	5
8	1100	1000	1000	15
9	1101	1001	1001	14
10	1111	1010	1010	12
11	1011	1011	1011	13
12	1010	1100	1100	8
13	1011	1101	1101	9
14	1001	1110	1110	11
15	1000	1111	1111	10

The following program segment indicates how we may get in and decode a series of four-bit gray code numbers:

```
DECLARE  NUMB(0:15)  INITIAL  (0, 1, 3, 2, 7, 6, 4, 5, 15, 14, 12, 13, 8, 9, 11, 10);
S10:  GET  LIST  (N);
N  =  NUMB(N);
      .
      .
      .
GO TO S10;
```

The data would consist of a stream of binary fixed-point integers separated by blanks and/or commas. Binary fixed-point integers are written as a series of zeros and ones followed immediately by the letter B—for example, **1011B**. More generally, a noninteger binary fixed-point constant may include a binary point such as **1.1101B**.[16] A binary floating-point constant includes a decimal exponent such as **1.1E−13B** meaning $(1\ 1/2) \times 2^{-13}$. Actually, most instruments would not be kind enough to insert the Bs in the data stream; so it would usually be necessary to use the B bit-string format with the **GET EDIT** statement which is not discussed in this text, relying upon conversion from a bit-string to a binary arithmetic number. Alternatively, it may be possible to use the more efficient **READ** statement which is not discussed in this text.

[16] Noninteger binary fixed-point constants are unsupported by the D compiler. All binary constants are unsupported by Student PL, the Model 20, and SL1. For the latter, decoding could be managed by using bit-strings.

SUMMARY

Priority of Operations

$$**, \text{ prefix } -, \text{ prefix } +, \neg$$
$$*, /$$
$$+, -$$
$$||$$
$$<, =, >, <=, >=, \neg<, \neg=, \neg>$$
$$\&$$
$$|$$

Bit-Strings

Bit-strings consist of a string of ones and zeros enclosed in single quotes and immediately followed by the letter B.

The &, |, ¬, =, and ¬= operators apply on a bit-by-bit basis with the shorter string being padded on the right with zeros.

Upon assignment to a bit-string variable, extra bits are truncated or zeros are added on the right if necessary.

Bit-strings are converted to arithmetic values when used in arithmetic expressions.

Bit-String Functions

The function BOOL(a, b, 'ijkl'B) may be used to achieve any of the 16 basic Boolean operations.

The function ANY(a) returns a 1 if any of the bits in its argument is a 1, returning a 0 otherwise.

The function ALL(a) returns a 1 if all of the bits in its argument are 1s, returning a 0 otherwise.

The function SUBSTR(a, i, j) returns a string of length given by its third argument beginning at the position given by its second argument in the string given by its first argument. The substring extends to the end of the string if the third argument is omitted. When used on the left of an assignment statement as a pseudo-variable, SUBSTR(a, i, j) specifies the substring of its first argument which is to receive the assignment.

The function LENGTH(a) returns the length of its argument.

The array manipulation function STRING(a) returns a string which is the concatenation of all of the elements in its array argument.

The VARYING and ALIGNED Attributes

The VARYING attribute permits strings to be used in expressions even though all of their allocated space is not in use.

The ALIGNED attribute permits faster execution at the expense of storage capacity.

Binary Arithmetic Constants

Binary arithmetic constants are written as a string of zeros and ones, optionally including a binary point and/or a decimal exponent, immediately followed by the letter B (e.g., 1.1E−13B = $(1^1/_2) \times 2^{-13}$).

EXERCISES

▼10. Write a version of HYD for systems without check valves, that takes advantage of the symmetry to reduce the data preparation.

11. WATER1 usually may be speeded up by checking to see if there is already known to be a path before performing the inner loop, and by exiting from the inner loop as soon as a path is found. Modify the procedure accordingly.

12. Write a procedure similar to WATER1 with terminology oriented towards switching circuits. Account for the possibility of diodes, which are analogous to check valves.

▼13. Write a version of WATER2 for systems without check valves, that takes advantage of the symmetry to reduce the time.

14. When a dispersion matrix FLOW has already been found, the modification of it due to a few new paths given by a matrix ADDITIONS may be most easily found by calling WATER1 after executing the following statements for all I together with all J:

```
PATH  =  FLOW(I,1)  &  ADDITIONS(1,J);
DO K = 2 TO N;
    PATH  =  PATH | FLOW(I,K)&ADDITIONS(K,J);
END;
FLOW(I,J)  =  PATH;
```

Write an appropriate program to do this.

15. (Difficult) Our water distribution analysis has the disadvantage that it does not record what the path is when it finds that there is one. Devise a method that does.

16. DIGNOS may be written without the BOOL function by simply checking for equality between corresponding bits of SYMPTOMS and DISEASE(K). Rewrite the procedure accordingly.

▼17. Why did we not allocate only two bits to temperature for our diagnosis example, with 00 denoting low, 01 denoting normal, 10 denoting moderately high, and 11 denoting severely high?

18. Devise a computer date-making system using bit-strings. Show sample instructions, a sample questionnaire (preferably machine readable), and the program.

19. Devise a diagnosis system for your discipline.

Exercises

▼20. The operation ¬(A | B) is called the NOR operation from the contraction of NOT and OR. Show the corresponding BOOL function.

▼21. (Difficult) Show how all of the other 15 elementary Boolean operations may be expressed using only the NOR operator.

22. The operation ¬(A & B) is the *Sheffer stroke* or NAND operation from the contraction NOT and AND. Show the corresponding BOOL function.

23. (Difficult) Show how all the other 15 elementary Boolean operations may be expressed using only the NAND operator.

24. The operation (A & ¬B) | (B & ¬A) is called the *exclusive or* operation. Show the corresponding BOOL function.

▼25. Computers may be used to simulate other computers, at a much reduced execution speed, thus assisting in their design. As an example of the type of logic circuits involved, the following table shows the two outputs for a binary *half adder* circuit which adds two one-digit numbers A and B:

A	B	carry digit	sum digit
0	0	0	0
0	1	0	1
1	0	0	1
1	1	1	0

We can see that the carry digit is merely the result of the & operation and that the sum digit is merely the result of the "exclusive or" operation discussed in exercise 24. The binary half adder is often indicated by the following diagram:

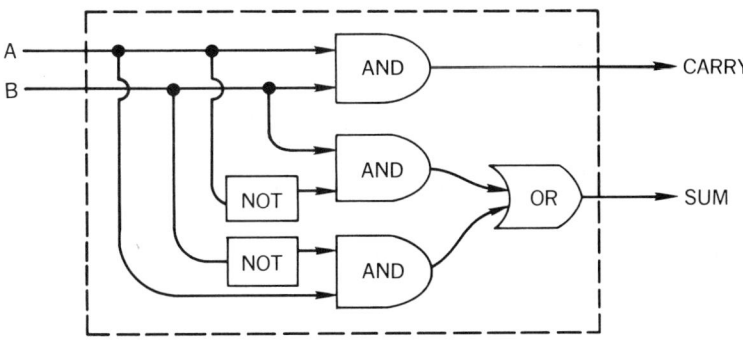

Write a subroutine HAFADD(A, B, CARRY, SUM).

26. When we add other than the lowest digits of two multi-digit numbers, we must include the carry digit from the addition of the previous digits. We may do this by combining two half adders and an OR operation into a *full adder*. In general, the full adder adds the Kth digits of A and B together with the K − 1 carry digit to produce the Kth sum and carry digits.

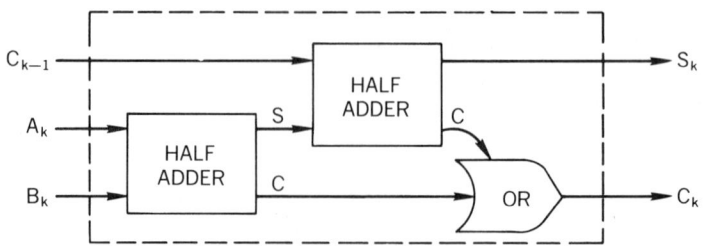

Write a subroutine FULADD(CKM1, AK, BK, SK, CK) that uses the subroutine HAFADD from exercise 25.

▼27. Write a function procedure PLUS(A, B) that uses HAFADD and FULADD from the previous two exercises to completely add two 40-bit strings, raising FIXEDOVERFLOW if the 40th carry digit is one.

CASE STUDY **14**

PATTERN RECOGNITION AND GRAPHICAL OUTPUT

Character-Strings

14.1 CHARACTER-STRING VARIABLES

We have already used character-string constants for output headings and messages. We may also use *character-string variables* and perform certain manipulations with them. For example, we may establish a 255-character variable by the declaration:

DECLARE ARTICLE CHARACTER (255);[1]

To establish a value for ARTICLE—for example a journal reference—we may assign it one within a program:

ARTICLE = 'MOTZKIN, "EVALUATION OF POLYNOMIALS AND EVALUATION OF RATIONAL FUNCTIONS", BULL. AMER, MATH. SOC., VOL 61, 1955, PP. 163FF.';

Note that every blank within the constant is counted as a character; so if the constant will not fit on one line of the input medium, we should extend it clear to the last allowable column of each line and continue it at the first allowable column of each line to avoid introducing spurious blanks. Note also that two adjacent single quotes (apostrophes) are used where we wish one single quote to appear.

[1] Character-string variables may have lengths from 0 to 32,767 for the F compiler, 1 to 255 for the D and Model 20 compilers, and 1 to 127 for the SL1 compiler. The length is not explicitly declared for Student PL.

(We need not use this dodge if the hardware recognizes a double quote.) However, for determining the length of the string, only one of these adjacent quotes is counted, and the first and last single quotes are not counted. The length of this particular constant is, then, 123 characters. The value of ARTICLE will be padded on the right with 132 blanks to make a total of 255. The excess characters on the right would be chopped off if the constant had been longer than ARTICLE.

Alternatively, we may get in this value for ARTICLE as data:

```
            GET LIST (ARTICLE);
```

The data would consist of the same constant. It is well to remember, however, that most implementations have different column limitations for program and data.

14.2 THE INDEX FUNCTION[2]

Let us assume now that we have a whole data file of journal references, and we would like to perform a literature search, finding all of the articles with titles mentioning the chemical phenolphthalein. We could do this with the following program:

```
SEARCH:   PROCEDURE OPTIONS (MAIN);
          DECLARE ARTICLE CHARACTER (255);
          PUT PAGE;
    S10:  GET LIST (ARTICLE);
          IF INDEX(ARTICLE, 'PHENOLPHTHALEIN') ¬= 0   THEN
              PUT SKIP LIST (ARTICLE);
          GO TO S10;
END;
```

The INDEX function returns a zero if the second string is not contained in the first. Otherwise it returns the position where the second string begins.

We have to be more careful if we are searching for a short word which may form part of larger words. For example, INDEX(ARTICLE, 'ION') could give us references dealing with unions, Zion National Park, and even plays written by Ionesco. We may partially resolve this problem by using INDEX(ARTICLE, ' ION ') with blanks before and after ION. However, we will then miss any occurrences immediately preceded or followed by punctuation marks. Consequently, it would be sensible to leave spaces around all punctuation when preparing the data for this application. However, the data are often taken directly from linotype tapes or optical scanners (TV). If so, we could use a statement such as:

```
IF INDEX(ARTICLE, 'ION') ¬= 0 | INDEX(ARTICLE, '"ION') ¬= 0 |
   INDEX(ARTICLE, 'ION"') ¬= 0 THEN PUT SKIP LIST (ARTICLE);
```

In titles, the surrounding quotes are the only punctuation we are likely to have.

[2] Unsupported by SL1 and the Model 20.

14.3 CONCATENATION AND THE SUBSTR FUNCTION

Suppose the data for our search did originate in a form with the single quotes adjacent to the title, and we wanted to preprocess the data, inserting blanks inside these quotes to simplify all subsequent searches. Assume that the first single quote occurs at position K + 1 and the second single quote occurs at position L as is shown below:

```
       position 1     position K                          position L
           ↓              ↓                                  ↓
         MOTZKIN   ...  'EVALUATION  ...  FUNCTIONS'  ...  163FF.
```

We may then insert blanks inside the quotes with the following statements which introduce the *concatenation* operator and the **SUBSTR** function:

```
K = INDEX(ARTICLE, ' ''');  L = INDEX(ARTICLE, ''' ');
ARTICLE = SUBSTR(ARTICLE, 1, K + 1) || ' ' ||
          SUBSTR(ARTICLE, K + 2, L - K - 2) || ' ' || SUBSTR(ARTICLE, L);
```

The **SUBSTR** function extracts a substring from its first argument. This substring begins at the position given by the second argument with a length given by the third argument. When the third argument is not given, the substring extends to the end of the string.[3] The concatenation operator, formed from two adjacent | symbols, joins or concatenates two strings.[4] Concatenation has a priority between that of addition–subtraction and the relational operators. (If two numerical expressions are concatenated, they will first be converted to strings.) Having added two blanks to the middle of **ARTICLE**, two characters will be dropped from the right side of the original value of **ARTICLE**. Consequently, care should be taken that the content of the original data is no more than 253 characters.

14.4 THE TRANSLATE FUNCTION[5]

An alternative to adding blanks for facilitating searches is to translate the punctuation to blanks using the **TRANSLATE** function:

```
ARTICLE = TRANSLATE(ARTICLE, ' ', '''');
```

Every occurrence in the first argument of the character in the third argument

---

[3] The D compiler, the Model 20, and Student PL do not support the two-argument version, and their third arguments must be constants. SL1 does not support the SUBSTR function.

[4] This operator is written CAT in the 48-character set. Operators written in alphabetic characters are reserved identifiers.

[5] Unavailable with Student PL, SL1, the D and Model 20 compilers.

(single quote) will be replaced with the character in the second argument (blank). More generally, the second and third arguments may be multiple-character strings, in which case each of the characters in the third argument is translated to the corresponding character in the second argument.

14.5 PATTERN RECOGNITION EXAMPLE

We have mentioned the possibility of using an optical scanner to enter typewritten material directly into a computer. Accordingly, let us consider how we may write a function procedure which returns the character corresponding to a given pattern produced by an optical scanner. We will presume that the optical scanner scans one line at a time using five vertical slices per space. Each vertical slice produces seven bits or characters, consisting of ones where the corresponding area of the image is mostly black, and zeros where the area is mostly white. For example, a scan of the image in Fig. 14-1 would give the indicated string.

1111111000100000101000100010100001

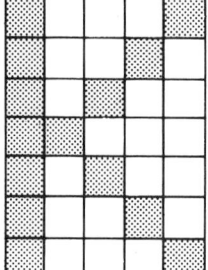

Fig. 14-1 Bit-string representation of a scanned letter K.

We may store all of the standard patterns that we wish to consider, as the **N** elements of a subscripted bit- or character-string variable **PATTERNS**. In addition, we may store the corresponding characters as successive characters in a string **ALPHABET**. We may compare an unknown pattern with each of the standards to see which it matches; then we may use the **SUBSTR** function to isolate the corresponding character in **ALPHABET**:

```
RECOG:  PROCEDURE (N, UNKNOWN, PATTERNS, ALPHABET) CHARACTER(1);
    DECLARE (UNKNOWN, PATTERNS(*)) BIT(*), ALPHABET CHARACTER(*);
    DO K = 1 TO N;
        IF UNKNOWN = PATTERNS(K)  THEN RETURN(SUBSTR(ALPHABET, K, 1));
    END;
    RETURN('?');
END;
```

14.6 THE SUBSTR PSEUDO-VARIABLE[6]

For numerical applications it is easy to swamp ourselves with so much output that we can draw no conclusions. For example, we may ask for the stresses in every member of a large building when we are really interested in knowing only which members are overstressed and which are understressed. As another example, we may ask for 10,000 function values when we are really interested in knowing only the last one, the largest one, or the average. Humans are capable of digesting only a very few numbers together. Massive printout of numbers helps no one but the paper manufacturers. We may save a great deal of time and money by asking ourselves exactly what we intend to do with each and every item of output. This introspection often reveals that we have asked the wrong questions.

Often we are interested in knowing only the *character* of a function, rather than knowing thousands of individual values. A graph is the ideal way to display this character—we may see at a glance how the function zigs and zags. A picture is worth *ten* thousand numbers.

Many computer installations have *plotters* and subroutines for drawing graphs with them. However, it is often more economical and simpler to make graphs directly on the standard output medium.

Let us consider how we may write a general-purpose output subroutine to plot a function Y(X) for uniformly spaced values of X. It is best to have X running down the page and Y running across, because then we may plot each Y value as it is calculated. With Y running up the page, to determine the proper printing order we would have to calculate and save all of the Y values before printing any. (We cannot skip backward by using the SKIP option with a negative number.) Also, with X running down the page, we do not have to decide where to terminate it in advance. For example, we may wish to continue until Y reaches a certain value.

Each time we calculate a new Y value, we may call a subroutine which puts out an asterisk in the appropriate column of the next line. To do this we must establish a scale for Y. Ideally, the extremes of Y will take it to the extremes of the page width, but not beyond, because this will give us the maximum resolution without losing any points off the page. We do not generally know the exact extremes of Y in advance; so we must make a conservative guess based on physical considerations. However, if we are too conservative, our graph will be rather compressed in the Y direction, without much resolution. As a precaution we may "wrap around" any points which run off the page, as is shown in Fig. 14-2.

With our approach, we cannot generally count on the origin lying within the page; so rather than print axes, we have printed a grid and its coordinates. This

---

[6] Unsupported by SL1.

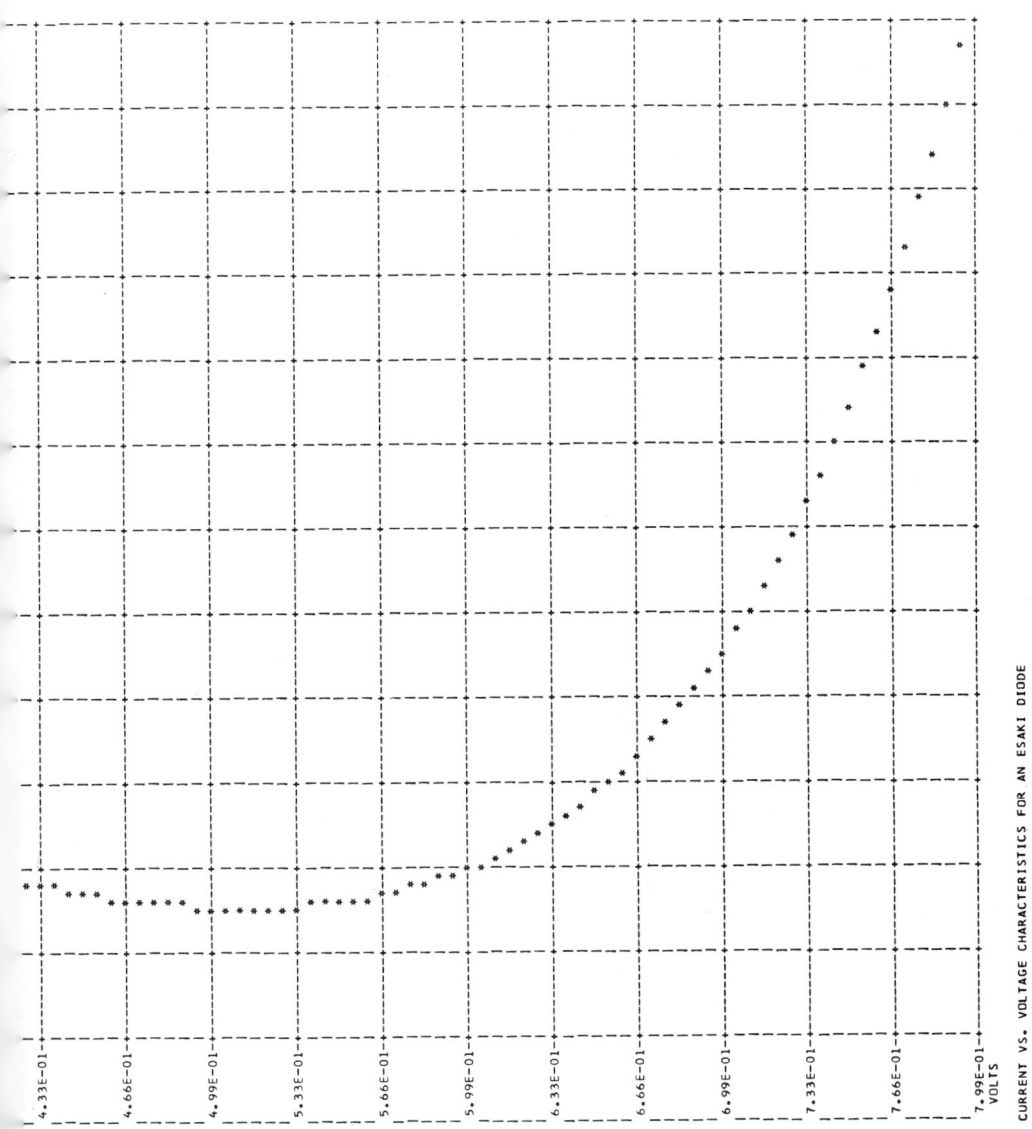

Fig. 14-2 A graph produced on a printer.

graph was produced using the following subroutine:

```
GRAPH:   PROCEDURE (Y, YMIN, YMAX, XMIN, XMAX, XINC, TITLE, YUNITS,
                XUNITS);
   /* FOR INTEGER GRID COORDINATES MAKE (YMAX-YMIN) A MULTIPLE OF 13
      AND XINC A MULTIPLE OF 1/6.  */
   DECLARE   /* PARAMETERS   */
      Y  ENTRY                 /*  GIVEN:  FUNCTION WHOSE VALUES ARE TO BE
                                            PUT OUT  */,
      YMIN, YMAX               /*  GIVEN:  ESTIMATES FOR THE EXTREMES
                                            OF Y  */,
      XMIN, XMAX, XINC         /*  GIVEN:  EXTREMES AND INCREMENT FOR THE
                                            INDEPENDENT VARIABLE  */,
      TITLE CHARACTER(*)       /*  TITLE FOR THE GRAPH  */,
      XUNITS CHARACTER(*)      /*  UNITS OF INDEPENDENT VARIABLE  */,
      YUNITS CHARACTER(*)      /* UNITS OF DEPENDENT VARIABLE  */;
   DECLARE   /*  INTERNAL VARIABLES  */
      HGRID  CHARACTER(131) INITIAL((14)'+---------'),
      VGRID  CHARACTER(131) INITIAL((14)'|             '),
      LINE CHARACTER(131), COORD FLOAT(3);
   ON ENDPAGE (SYSPRINT);

   /* PUT OUT YUNITS AND ORDINATES:  */
   LINE = ' ';  SUBSTR(LINE,100) = YUNITS;  PUT LIST (LINE);
   LINE = ' ';  DIFFY = YMAX - YMIN;  YINC = DIFFY/13;
   DO K = 1 TO 12;
      COORD = YMIN + K*YINC;  SUBSTR(LINE,10*K-4,9) = COORD;
   END;
   PUT LIST (LINE);

   /* PUT OUT PLOT:  */
   DO LINE# = 0 BY 1;
      X = XMIN + LINE# * XINC;
      IF X > XMAX  THEN GO TO S10;
      IF MOD(LINE#, 6) = 0   THEN LINE = HGRID;
                             ELSE LINE = VGRID;
      K = MOD(130*(Y(X)-YMIN)/DIFFY, 130) + 1;
      SUBSTR(LINE, K, 1) = '*';
      IF MOD(LINE#, 6) = 0  THEN DO;
         COORD = X;  SUBSTR(LINE,1,9) = COORD;  END;
      PUT SKIP LIST(LINE);
   END;

   /* ADD XUNITS AND TITLE:  */
S10:  PUT SKIP LIST (XUNITS);
      PUT SKIP(2) LIST (TITLE);
      PUT PAGE;
END;
```

This procedure introduces the SUBSTR *pseudo-variable*. When used on the left side of an assignment statement, SUBSTR is called a pseudo-variable because the assignment is to a portion of the first argument rather than to the identifier SUBSTR.

We have used the MOD function to facilitate printing the horizontal grid every sixth line and to "wrap around" the graph in case it should run off the page. Also, we have used the null ON ENDPAGE statement to avoid having gaps in the middle of the graph. Another feature illustrated by the last PUT statement is that we may concatenate arithmetic values, in which case they are converted to decimal character strings.

Sec. 14.8 Comparison of Character-Strings

```
            PUT SKIP LIST (ROW);
            DO J = 1 TO 131;
               IF SUBSTR(ROW, J, 1) = '*'  THEN SUBSTR(ROW, J, 1) = '$';
                                           ELSE SUBSTR(ROW, J, 1) = ' ';
            END;
            PUT SKIP(0) LIST (ROW);
         END;
         DO I = 154 TO 160;
            GET LIST (ROW);  PUT SKIP LIST (ROW);
         END;
      END;
   APHSRT: PROCEDURE (M, N, WORD);
      DECLARE WORD(*) CHARACTER(*)   /* THE SUBSCRIPTED CHARACTER STRING
         VARIABLE WHOSE ELEMENTS ARE TO BE SORTED INTO ALPHABETIC ORDER*/,
         M  /* THE LENGTH OF THE STRINGS */,
         N  /* THE NUMBER OF STRINGS */,
         FIRST CHARACTER(M) /* TEMPORARY STORAGE TO FACILITATE INTERCHANGE
                                                                       */;
      DO J = 1 TO N-1;
         K_OF_FIRST = J;  FIRST = WORD(J);
         DO K = J+1 TO N;
            IF WORD(K) < FIRST  THEN DO;  K_OF_FIRST = K;
               FIRST = WORD(K);  END;
         END;
         WORD(K_OF_FIRST) = WORD(J);  WORD(J) = FIRST;
      END;
   END;
```

The statement:

IF SUBSTR(ROW), J, L) = '*' THEN GO TO S10;

causes the immediately following statements to be skipped if the character in the Ith row and Jth column is an asterisk. We use the expression SUBSTR(ALPHA, (L—960)/3, 1) to isolate the proper letter from the alphabet. We then assign this letter to the Jth position of the string ROW. Later in the program we again use SUBSTR to increase the border contrast by overprinting '*' with '$.'

14.8 COMPARISON OF CHARACTER-STRINGS

Character-strings may be compared just like numbers, the letter A being considered less than the letter B and so on. This enables us to perform alphabetic sorting by the same techniques used for numbers. For example, the following subroutine adapts the search-sort technique of Case Study 10 to sort character-strings into alphabetic order:

```
APHSRT: PROCEDURE (M, N, WORD);
   DECLARE WORD(*) CHARACTER(*)    /* THE SUBSCRIPTED CHARACTER STRING
      VARIABLE WHOSE ELEMENTS ARE TO BE SORTED INTO ALPHABETIC ORDER*/,
      M  /* THE LENGTH OF THE STRINGS */,
      N  /* THE NUMBER OF STRINGS */,
```

[7] Declared lengths and extents must be constants for the SL1, Model 20, and D compilers.

```
        FIRST CHARACTER(M) /* TEMPORARY STORAGE TO FACILITATE INTERCHANGE
                                                                   */;
   DO J = 1 TO N-1;
      K_OF_FIRST = J;  FIRST = WORD(J);
      DO K = J+1 TO N;
         IF WORD(K) < FIRST  THEN DO;  K_OF_FIRST = K;
            FIRST = WORD(K);  END;
      END;
      WORD(K_OF_FIRST) = WORD(J);  WORD(J) = FIRST;
   END;
END;
```

Blanks are generally considered less than A; so SMITH properly is considered less than SMITHSON. In general, comparisons are judged according to an implementation-defined collating sequence. The collating sequences for the IBM 360 PL/I BCD and EBCDIC codes are indicated in Fig. I-1 in the Introduction.

Case Study 13 contains some additional string-manipulation techniques.

SUMMARY

Character-Strings

A character-string consists of a string of any characters recognized by the hardware, including blanks, enclosed in single quotes. A single quote within the string may be denoted by two adjacent single quotes. The length includes blanks but not the surrounding quotes or any extra internal single quotes.

Character-string variables are declared using the attribute CHARACTER(length).

Upon assignment to a character-string, characters are truncated or padded on the right with blanks if necessary.

The result of character-string comparisons is determined according to an implementation-defined collating sequence.

Built-in Functions

The function INDEX(a, b) searches its first argument for the substring given by its second argument, returning the position at which it first occurs or a zero if it does not occur.

The function SUBSTR(a, i, j) returns a substring of length given by its third argument beginning at the position given by its second argument in the string given by its first argument. The length is the remainder of the string when the third argument is omitted. As a pseudovariable on the left side of an assignment or in an input list, SUBSTR denotes the portion of its first argument which is to receive the value.

The function TRANSLATE(a, b, c) returns its first argument with every character in its second argument translated into the corresponding character in its third argument.

EXERCISES

1. Adapt the subroutine MERGE of Case Study 10 for alphameric merging.

2. Assuming that book references include a Library of Congress call number before the author's name, write a program to find any references to "game theory" or "theory of games" between call numbers QA260 and QA270.

▼3. PL/I includes a built-in function DATE, without arguments, that returns a character-string in the form *yymmdd* where *yy* is the year, *mm* is the month, and *dd* is the day. For example, June 3, 1972 would appear as "720603." Write a function procedure which returns a character-string giving the date in spelled-out form such as "JUNE 3, 1972."

4. PL/I includes a built-in function TIME, without arguments, that returns a character-string giving the time in the form *hhmmssttt* where *hh* is the hour (0 through 23), *mm* is the minutes, *ss* is the seconds, and *ttt* is the milliseconds. Write a function procedure that returns a character-string giving the time in the form *hh:mm:ss.ttt* AM (or PM).

▼5. A psychologist has decided to automate his free-association tests. He has a tape giving a list of words, and his secretary punches the response words. (Alternatively, the psychologist may use a machine for interpreting spoken words.) Successive responses are gotten in and compared with each element in the corresponding row of a matrix SIGNIFICANT which contains the corresponding responses that the psychologist regards as significant. If there is a match, the response and corresponding stimulus are put out. For example, significant responses to the stimulus "mother" might be "love," "hate," "shame," "father," "brother," or "apple pie." Write a corresponding program.

6. One way for a linguist to compile a dictionary for an unwritten language is to live with the natives until he feels that he has heard and recorded every word and its meaning, using a phonetic alphabet. However, he may easily miss some rare words this way. A faster way is to live with the natives until he feels that he has heard and recorded every *syllable*, using a phonetic alphabet. Typical examples might be "ug," "ba," "pow," "biff," and "pua." He may then have a computer get in these syllables and put out all possible words up to as many syllables as he can afford and deems necessary. Typical examples might be "ugug." "ugba," and "biffpuapow." He may then teach a native the phonetic alphabet and have this native read the list, picking out the meaningful words. Write a program for doing this.

▼7. A philologist trying to decode the ancient Indus Seals decides to analyze the frequency of occurrence of various character pairs. The following figure presents some Indus characters and Roman characters that these character pairs may be arbitrarily associated with for the sake of computer analyses.[8] Write a program that will receive a Roman-coded Indus character-string text and put out the number of occurrences of each occurring character pair.

▼8. Modify the character-recognition procedure to recognize printed characters, which are not all the same width. Assume UNKNOWN is an entire line of print. (Characters may be isolated by noting that they are separated by at least seven blanks.)

9. Modify the character-recognition procedure so that rather than insisting on an exact match, it returns the closest letter in the sense of the procedure DIGNOS in Case Study 13.

10. (Advanced Biology) Write a procedure that will take a red and white blood cell count from an image produced by an optical scanner.

11. (Advanced Physics) Write a procedure that will analyze images of bubble chamber traces produced by an optical scanner, putting out appropriate information if any significant even has taken place.

12. Assume that mug shots of criminals are stored on video tape with framing and magnification adjusted so that the centers of the eyes are at standard locations. Assume also that the various shades of gray are coded as the digits 0 through 9, the whole image being stored as one long character-string. The closeness of match of an unknown and a criminal's image may then be calculated as:

$$\sum_{K=1}^{N} ABS(SUBSTR(UNKNOWN, K, 1) - SUBSTR(CRIMINAL, K, 1))$$

where N is the length of the strings. Write a corresponding subroutine that returns the closeness index.

▼13. Ignoring some punctuation and numerals, we may consider Morse Code to be a bit- or character-string consisting of the following components, arranged in approximate order of decreasing frequency:

space between letters	00	H	1010101	G	1101101
space between words	000	D	110101	B	11010101
E	1	L	10110101	V	10101011
T	11	U	101011	,	110110101011011
A	1011	C	110101101	.	10110101101011
O	11011011	M	11011	J	1011011011
N	1101	P	101101101	K	1101011
I	101	F	10101101	Q	1101101011
R	101101	Y	1101011011	X	110101011
S	10101	W	1011011	Z	110110101

[8] IBM, *Computing Report in Science and Engineering*, Vol. V, No. 3 (White Plains, N.Y:. IBM, June 1969), p. 2.

Write a program that gets in any number of Morse Code strings, putting out the corresponding translations.

14. Same as exercise 13 except translate from ordinary character input to Morse Code output.

15. (Advanced Chemistry) There are fairly regular rules for writing chemical formulas and names such as C2H5OH and ETHYL ALCOHOL. Write a program which translates from one to the other.

16. Modify GRAPH so that the grids are 10-by-10 spaces rather than 10-by-6.

17. Modify GRAPH so that off-scale asterisks are suppressed rather than wrapped around.

▼18. Modify GRAPH so that grid lines are printed with solid dollar signs if they coincide with an axis.

▼19. Write a subroutine that will put out on one graph up to nine functions of the same independent variable, using the symbols 1, 2, etc. rather than asterisks. Assume that YMIN, YMAX, YUNITS, and Y are vectors, the latter given by a subroutine rather than by a function.

▼20. Write a subroutine that will put out the graph of a set of points given by two vectors X and Y.

21. Write a subroutine that will put out the graph of a curve, [X(T), Y(T)] for $T = T_{min}$ to T_{max} where X and Y are given by function procedures.

▼22. Write a subroutine that puts out a line drawing—the graph of straight-line segments between pairs of points given by subscripted variables X and Y.

23. Given the x, y, and z coordinates of a point, its isometric projection is given by

$$u = 0.702(x - y)$$
$$v = 0.405(x + y) - 0.81z$$

For the isometric projection of a brick in Fig. 12-3, $+u$ would be toward the right of the page and $+v$ would be toward the top of the page with the same origin as the x, y, z system. Write a subroutine that calculates U and V vectors from X, Y, and Z vectors, then calls the procedure of exercise 20 to put out the isometric point projection.

24. Same as exercise 23 except call the procedure of exercise 22 to get a line drawing.

25. Rewrite the weather map program assuming the entire map is one characterstring.

26. Devise an output program that displays the results of TEMP in Case Study 12 in a fashion similar to the weather map.

27. Devise a means of also representing temperature on the weather map.

CASE STUDY 15

BEARING SELECTION REVISITED

Structures

All of the elements in an array must have identical attributes. To group data with differing attributes, we may use a different kind of collection called a *structure*.[1] Like an array, the entire structure is given a name; but unlike an array, the individual data items also have distinct *elementary names*. For example, let us reconsider our bearing selection program from Case Study 4. We may group the specifications into the structure portrayed in Fig. 15-1.

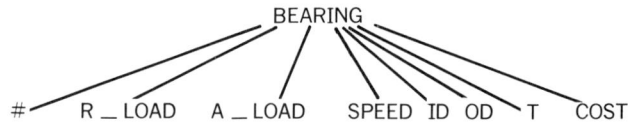

Fig. 15-1 A structure for bearing specifications.

We may refer to an entire structure by using its structure name—for example:

        GET LIST (BEARING);   or
        PUT LIST (BEARING);[2]

Alternatively, we may refer to an individual data item by using its elementary name —for example:

        IF SPEED >= 5000

[1] Student PL does not support structures, and only string data is allowed in SL1 structures.
[2] For the Model 20 and SL1, structures may not be listed in GET and PUT statements.

Sec. 15.1   Declaration of Structures   253

A structure may contain other structures as well as elementary names, enabling us to refer to entire subsets of a structure by a single name. For example, it might be convenient to refer to the two load specifications by the single name **LOAD** and to refer to the three dimension specifications by the single name **DIMENSION**. To do this, we may group the specifications into the structure portrayed in Fig. 15-2. We may then write statements such as "PUT LIST (DIMENSION) ;" to put out all three dimensions.

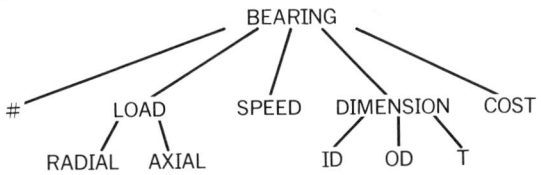

Fig. 15-2 An alternate bearing structure.

Structures which are contained in another structure are called *minor* structures, and structures which are not contained in another are called *major* structures. Minor structures may contain other minor structures, and so on. Note that only elementary names contain data. Structure names serve merely to group these data.

15.1 DECLARATION OF STRUCTURES

PL/I structures have a hierarchical organization similar to a corporate structure with the chairman of the board at the top, down to the lowest echelons. Diagrams such as Figs. 15-1 and 15-2 make this organization clear to us, but we may not include diagrams in a PL/I procedure. The organization is specified in a procedure by assigning *level numbers* to the names and by declaring the names preceded by their level numbers. For example, an appropriate declaration for the structure of Fig. 15-2 is:

```
DECLARE 1  BEARING,
        2  #  FIXED(9),
        2  LOAD,
            3  RADIAL  FIXED(5),
            3  AXIAL   FIXED(5),
        2  SPEED  FIXED(5),
        2  DIMENSION,
            3  ID  FIXED(5,3),
            3  OD  FIXED(5,3),
            3  T   FIXED(5,3),
        2  COST  FIXED(5,2);
```

The indentation is only for legibility. More than one name may be listed per line. Other rules are:

1. The level numbers must be integer constants separated by a space from their names.

2. The level numbers and declaration order must be chosen so that a structure at level n contains all of the names between it and the next name with a level $\leq n$.

3. The major structure name must be at level one, but successive levels need not be numbered consecutively. Some programmers skip level numbers so that they do not need to completely renumber in the event that there are subsequent insertions.

4. The structure declaration is terminated by another name at level one, by a name without a level, or by the end of the DECLARE statement.

5. The level numbers and/or the attributes of the elementary names may be factored.[3]

6. Level numbers are not used in any reference to a structure except its declaration.

According to these rules, an equivalent declaration of the structure BEARING is the one in the following revised version of the procedure BEARNG from Case Study 4:

```
BEARNG:  PROCEDURE   OPTIONS (MAIN);
   DECLARE 1 BEARING, 3 # FIXED(9),
                      3 LOAD, 5 (RADIAL, AXIAL) FIXED(5),
                      3 SPEED FIXED(5),
                      3 DIMENSION, 5 (ID, OD, T) FIXED(5,3),
                      3 COST FIXED(5,2);
      PUT LIST ('BEARING NUMBER', 'RADIAL LOAD', 'AXIAL LOAD',
         'MAXIMUM RPM', 'INSIDE DIAM.', 'OUTSIDE DIAM.', 'THICKNESS',
         'PRICE/UNIT');
   NEXT:   GET LIST (BEARING);
      IF SPEED>=5000 & ID=.625 & T<=.25 THEN PUT SKIP(2) LIST(BEARING);
      GO TO NEXT;
END;
```

15.2 QUALIFIED NAMES

For the previous example, the data for the elementary names must be prepared in the order that these names are listed in the declaration. Similarly, the data for the elementary names will be put out in the order that these names are declared.

---

[3] The SL1 compiler does not allow factoring, a level number greater than four, or skipping of levels. For the Model 20, level numbers must be factored if attributes of the corresponding variables are.

For compilers which support DATA-directed I/O, we may alternatively write:

PUT DATA (BEARING)

This statement would give us output such as:

```
BEARING.#=              57100     BEARING.LOAD.RADIAL=       800
BEARING.SPEED=           6500     BEARING.DIMENSION.ID=    0.625
BEARING.DIMENSION.T=    0.235     BEARING.COST=             6.35;
```

These names are said to be *fully qualified*. The names of successive levels leading up to each data item are separated by periods (decimal points). We may also write qualified names in our procedures.

The same name may occur more than once in a structure provided there is a unique way to qualify each instance. For example, BEARING could have the structure in Fig. 15-3.

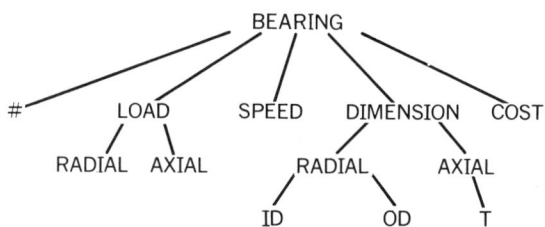

Fig. 15-3 An alternate structure for BEARING.

To distinguish between the radial load and the radial dimensions we may write:

LOAD.RADIAL   or
DIMENSION.RADIAL

Note that we need only partially qualify the names to the extent that they are unique.

15.3 STRUCTURE EXPRESSIONS AND ASSIGNMENT

Major structures, minor structures, and elementary names may be used in expressions and assignment provided all of the operands have the same structuring or are scalar items. For example, suppose that we wish to convert the bearing dimensions from inches to centimeters for the previous example. We may do this with the statement:

DIMENSION = 2.54 * DIMENSION;

Two structures have identical structuring if their diagrams look similar and corresponding names are declared in the same order. Neither the names nor the level numbers of corresponding items need to agree. Corresponding elements may even have different attributes as long as the implied conversion is valid.[4] For example, suppose that MARGIN was declared:

```
1  MARGIN,
  2  X  FLOAT,
  2  Y  FLOAT
```

We could then write:

```
MARGIN = 5E-2;
LOAD   = LOAD - MARGIN;
```

15.4 THE BY NAME OPTION[5]

Consider the problem of combining and analyzing weather reports from a number of different sources. The data reported by typical ships, airplanes, and land stations might be as indicated by the following structures:

```
1  LAND                      1  SHIP                    1  AIRPLANE
  2  TEMPERATURE               2  TEMPERATURE             2  TEMPERATURE
  2  WIND                        3  AIR                   2  WIND
    3  VELOCITY                  3  WATER                   3  VELOCITY
    3  DIRECTION               2  WIND                      3  DIRECTION
  2  HUMIDITY                    3  VELOCITY             2  ALTITUDE
  2  PRESSURE                    3  DIRECTION           2  VISIBILITY
  2  PRECIPITATION             2  PRESSURE
    3  RATE                    2  PRECIPITATION
    3  DAILY_CUMULATIVE          3  RATE
  2  SKY                       2  SKY
                               2  CURRENT
                                 3  VELOCITY
                                 3  DIRECTION
```

SKY could be a character-string variable with typical values of 'SUNNY', 'PARTIALLY CLOUDY', etc. LAND, SHIP, and AIRPLANE have different structurings; so we may not include them in the same ordinary assignment statement. However, we may write:

```
LAND = SHIP, BY NAME;
```

Assignment will take place only for names which are identical in the structures when fully qualified from the second level on. In other words this assignment is

---

[4] For SL1 and the Model 20, only the assignment "structure = structure;" is allowed. Attributes must agree and be either character or arithmetic for Model 20 structure assignment.
[5] Unsupported by the Model 20, SL1, and D compilers.

equivalent to the five element assignments:

```
LAND.WIND.VELOCITY  =  SHIP.WIND.VELOCITY;
LAND.WIND.DIRECTION  =  SHIP.WIND.DIRECTION;
LAND.PRESSURE  =  SHIP.PRESSURE;
LAND.PRECIPITATION.RATE  =  SHIP.PRECIPITATION.RATE;
LAND.SKY  =  SHIP.SKY;
```

Note that SHIP.TEMPERATURE.AIR is not assigned to LAND.TEMPERATURE because the fully-qualified names are not identical from the second level on.

As another example, we may write:

```
LAND  =  .5 * (SHIP + AIRPLANE), BY NAME;
```

Operation and assignment are performed only for elements whose fully-qualified names are identical from the second level on; so this assignment statement is equivalent to:

```
LAND.WIND.VELOCITY  =  .5 * (SHIP.WIND.VELOCITY + AIRPLANE.WIND.VELOCITY);
LAND.WIND.DIRECTION  =  .5 * (SHIP.WIND.DIRECTION + AIRPLANE.WIND.DIRECTION);
```

15.5 THE LIKE ATTRIBUTE[6]

We may use the LIKE attribute to save some writing in the declaration whenever one major or minor structure contains identical names with identical attributes and structuring as another major or minor structure. The general form for this attribute is:

LIKE structure name

As an illustration, the minor structures LAND.WIND, SHIP.WIND, and CURRENT from the previous example contain identical names; so we may declare them as follows:

```
DECLARE 1 LAND,  2 TEMPERATURE FIXED(5,1),
                 2 WIND, 3 (VELOCITY, DIRECTION) FIXED,
                 2 HUMIDITY FIXED,
                 2 PRESSURE FIXED(5,2),
                 2 PRECIPITATION, 3 (RATE, DAILY_CUMULATIVE) FIXED(5,2),
                 2 SKY CHARACTER(20),
         1 SHIP, 2 TEMPERATURE 3 (AIR, WATER) FIXED(5,1),
                 2 WIND LIKE LAND.WIND,
                 2 PRESSURE FIXED(5,2),
                 2 PRECIPITATION, 3 RATE FIXED(5,2),
                 2 SKY CHARACTER(20),
                 2 CURRENT LIKE LAND.WIND,
```

[6] Unsupported by the Model 20, D, and SL1 compilers.

15.6 ARRAYS OF STRUCTURES

Any of the names in a structure may be subscripted.[7] Suppose, for example, that there are 150 land-based weather stations. We may then write:

> DECLARE 1 LAND(150), 2 TEMPERATURE FIXED(5,1), ···

Thus, **LAND** refers to an entire array of weather reports, **LAND(3)** refers to the third report, **PRESSURE** refers to an entire array of pressures, and **PRESSURE(3)** refers to the third pressure. Note that we would have to write **LAND.PRESSURE** and **LAND.PRESSURE(3)** if **SHIP** or **AIRPLANE** were declared in the same procedure.

Suppose that **TEMPERATURE** is also an array containing the last 24 hourly temperatures:

> DECLARE 1 LAND(150), 2 TEMPERATURE(24) FIXED (5,1), ···

Then **LAND(3).TEMPERATURE(5)**, **LAND(3, 5).TEMPERATURE** and **LAND.TEMPERATURE(3, 5)** all refer to the temperature at station three at the fifth hour. Subscripts may be moved to a higher or lower level in a qualified name provided the order is preserved.

The most useful applications of structures are with **RECORD I/O**, which is not discussed in this text, and with list processing, which is discussed in the following case study.

SUMMARY

Structure Declaration

Typical Structure Declaration

```
DECLARE  1  major-name,
         2  minor-name,
            3  elementary-name  attribute-list,
                  .
                  .
                  .
         2  minor-name  LIKE  major  or  minor-name,
                  .
                  .
                  .
```

---

[7] For the Model 20 and D compilers, only elementary names may be subscripted.

The level numbers, which may be factored, must be separated from their names by a blank. The major level must be one, and successive levels must have higher numbers, not necessarily consecutive. Only elementary names, which contain no substructures, may contain data. A major or minor structure may be declared LIKE another major or minor structure. The corresponding names and attributes will then be identical.

The BY NAME Option

Unless the BY NAME option is used, all structures in an assignment statement must have identical structuring. However, the names and attributes need not agree provided any implied conversion is valid. With the BY NAME option, which appears just before the semicolon preceded by a comma, only structure elements having identical fully-qualified names from the second level on, participate.

Structure I/O

Input and output take place in the natural order established in the declaration. Minor structures as well as major may be used in GET, PUT, and assignment statements.

Arrays of Structures

Any of the names in a structure may be subscripted, and the subscripts may be shifted right or left in a qualified name, provided the order is preserved.

EXERCISES

▼1. Do exercise 20 in Case Study 4 using a structure.

2. Do exercise 21 in Case Study 4 using a structure.

3. Do exercise 22 in Case Study 4 using a structure.

▼4. The complete classification of the timber wolf is kingdom *Animal*, phylum *Chordata*, subphylum *Vertebrata*, class *Mammalia*, order *Carnivora*, family *Canidae*, genus *Canis*, species *lupus*.
Show how the animal kingdom may be organized into one giant structure with a character-string description as the data for each species. List only a few representative names.

5. The complete classification of a buttercup flower is kingdom *Plantae*, division *Tracheophyta*, class *Angiospermae*, subclass *Dicotyledones*, order *Ravales*, family *Ranunculaceae*, genus *Ranunculus*, species *acris*. Show how the plant kingdom may be organized into one giant structure with a character-string description as the data for each species.

▼6. Show how the local school administration, government administration, or corporate management may be organized as a structure. Let the titles be the structure names and the employee names be the data.

7. Large systems contain so many parts that it is desirable to catalog them by computer. For example, the parts of a communication satellite might be catalogued into the structure partially indicated below:

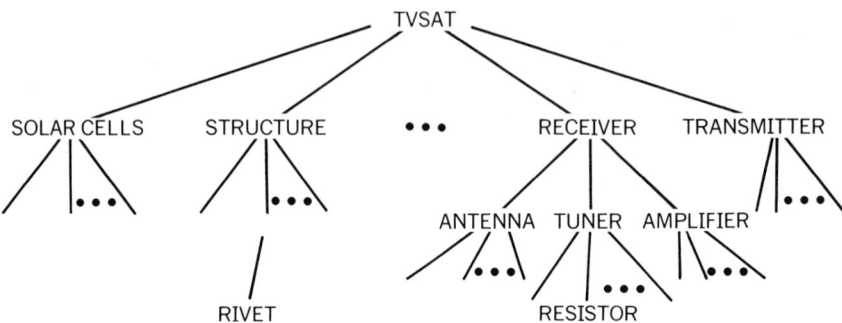

The data could be a character-string containing information such as '#9 1 INCH MILD STEEL' for a rivet and '3700 OHM, 5 WATT, 5%' for a resistor. Show a similar structure and its data in as much detail as possible for one of your household appliances.

8. Suppose we have the following declaration:

```
DECLARE  1  BEAMS,  2  CONCRETE,  3  QUANTITY FIXED,  3  UNIT_COST FIXED(5,2),
              2  STEEL  LIKE  BEAMS.CONCRETE,
         1  COLUMNS  LIKE  BEAMS,
         1  FLOORS,  3  CONCRETE LIKE  BEAMS.CONCRETE,
              3  STEEL,  5  QUANTITY FIXED(5,1),  5  UNIT_COST FIXED(5,2),
              3  WOOD  LIKE  BEAMS.CONCRETE,
         1  TOTAL  LIKE  FLOOR;
```

Show what individual assignments result from the following statements, or indicate that they are invalid.

▼ (a) BEAMS = 1.1 * (BEAMS + COLUMNS);
  (b) TOTAL = COLUMNS + FLOORS;
  (c) TOTAL = 1.1 * (BEAMS + COLUMNS);
  (d) TOTAL.STEEL = BEAMS.STEEL + COLUMNS.STEEL + FLOOR.STEEL;
  (e) TOTAL.STEEL = BEAMS.STEEL + COLUMNS.CONCRETE;
  (f) COST = QUANTITY * UNIT_COST;
▼ (g) TOTAL = BEAMS + COLUMNS + FLOORS, BY NAME;

9. Using the same declaration as in exercise 8, suppose we have the following data card and program segment. Show what the output would look like.

Exercises

```
780   5.86   58   775   48   356.95   5786
```

```
GET  LIST  (BEAMS,  COLUMNS.STEEL,  FLOORS.WOOD.QUANTITY);
PUT  LIST  (COLUMNS.STEEL.QUANTITY,  FLOORS.WOOD.QUANTITY,  BEAMS);
PUT  DATA  (COLUMNS.STEEL.QUANTITY,  FLOORS.WOOD.QUANTITY,  BEAMS);
```

CASE STUDY 16

LIST PROCESSING

POINTER Variables, CONTROLLED and BASED Storage

We have already been exposed to the STATIC and AUTOMATIC storage classes. STATIC variables are allocated before execution is begun, and they remain allocated throughout execution of the program. AUTOMATIC variables are allocated and freed upon entry and exit from the PROCEDURE or BEGIN block in which they are declared. The default is AUTOMATIC.

16.1 CONTROLLED STORAGE

There is also a CONTROLLED storage class which permits greater control over the allocation and freeing of storage. After being declared, storage is allocated with the ALLOCATE statement and freed with the FREE statement.[1] For example, suppose we wish to maximize the potential space for four subscripted variables X, Y, Z, and W, where we first need only X and Y together, then only Y and Z together, then only Z and W together. We may then write:

---

[1] The D, Model 20, and SL1 compilers do not support CONTROLLED storage, the ALLOCATE statement, or the FREE statement. In Student PL all subscripted variables are implicitly CONTROLLED.

```
       DECLARE  (X(*), Y(*), Z(*), W(*)) CONTROLLED;
       GET LIST  (K, L, M, N);
       ALLOCATE  X(K), Y(L);
         .
         .
         .
       FREE X;
       ALLOCATE  Z(M);
         .
         .
         .
       FREE Y;
       ALLOCATE  W(N);
         .
         .
         .
```

We may establish the bounds and any other attributes except storage class in either the **DECLARE** or **ALLOCATE** statements.

The **BEGIN** block will not help in this instance because of the overlapping need for the variables. However, there is an alternative technique using the **DEFINED** attribute which is not discussed in this text.

16.2 THE ALLOCATION FUNCTION

CONTROLLED storage has another useful aspect. Whenever we execute an additional allocation without an intervening freeing, an additional instance of the variable is created without destroying any previous instances. Any reference to a controlled variable will always refer to the most recent instance, but we may always recover previous instances by freeing subsequent instances. It is like a *stack* of similar variables. We may add or remove values from the top with **ALLOCATE** or **FREE** statements, and any reference to the variable always refers to the value at the top of the stack.

This stacking feature provides a convenient alternative to subscripted variables when we need access only to the most recent instance and when we do not wish to decide upon an extent ahead of time. For example, suppose that we wish to simulate a production system where each machine operator works on parts taken from a stack beside him which is added to at irregular intervals. We may keep track of individual parts by keeping a stack of part numbers using a controlled variable **PART#**. Every time we wish to add a part number **PN**, we may execute the statements:

```
       ALLOCATE  PART#;
       PART#  =  PN;
```

Every time we wish to subtract an item, assigning its value to **PN**, we may execute the statements:

IF ⏋ALLOCATION(PART#) THEN GO TO NONE__LEFT;
PN = PART#; FREE PART#;

Here we have introduced the **ALLOCATION** function. It returns '0'B if there are no current allocations of its argument, returning '1'B otherwise. We include it here because presumably the operator acts differently if there are no parts in his stack.

16.3 POINTERS AND THE ADDR FUNCTION

PL/I has a built-in function **ADDR(X)** which returns the machine language address of a variable or the first address in the case of structures and arrays.[2] We may assign this address to a **POINTER** variable—for example:

DECLARE LOC POINTER;
.
.
LOC = ADDR(X);

We may also assign the value of one pointer variable to another:

DECLARE (LOC, WHERE) POINTER;
.
.
WHERE = LOC;

16.4 BASED STORAGE

There is another class of storage called **BASED** that enables us to refer to any of its allocations, not just the most recent. Rather than being stacked, the variable is based on a pointer variable. We then save distinct pointer values for each allocation. To declare based storage, we use the **BASED** attribute followed by parentheses containing the pointer variable upon which the storage is to be based—for example:

---

[2] This address is a four-byte binary number for the F compiler, a three-byte binary number for the D compiler, and a two-byte binary number for the Model 20 compiler. Pointers and the ADDR function are unsupported by SL1 or Student PL.

```
DECLARE 1 WEATHER BASED (PNT), 2 DAY FIXED(6),
         2 DAY FIXED(6),
         2 TEMPERATURE,
           (3 HIGH,
            3 LOW) FIXED(3),
         2 PRECIPITATION FIXED(3,1);
```

PNT is contextually declared to be a pointer.

We may use the ALLOCATE-statement to allocate instances of based variables —for example:

```
ALLOCATE WEATHER;
        or
ALLOCATE WEATHER SET (Q);
```

In the first case, PNT is automatically set to point to the instance; in the latter case, Q, rather than PNT, is set to point to the instance. Q is contextually declared to be a pointer by virtue of its appearance in the SET option.[3]

When used without qualification, for example "PUT LIST (WEATHER);," the instance is the one pointed to by the pointer used in the declaration. To refer to a former instance, we may first assign its pointer to the pointer used in the declaration—for example "PNT = Q; PUT LIST (WEATHER);." Alternatively, we may use the *pointer qualifier* to refer directly to any instance—for example:

```
PUT LIST (Q->WEATHER);
```

The pointer qualifier is formed from an adjacent minus sign and greater-than symbol.[4] This qualifier contextually declares the variable on its left to be a pointer. This notation takes a while to get used to. It may help to think of Q -> WEATHER as WEATHER(Q), but PL/I reserves this notation for subscripted variables.

Whenever we free instances of a based variable, any pointers which point to them become undefined. For example, if we write:

```
FREE WEATHER, Q->WEATHER;
```

then PNT and Q no longer have any valid values. *They are not automatically set to the preceding instances* because based variables are not stacked. Consequently, we must save a distinct pointer value for each allocation.

---

[3] For the D and Model 20 compilers, pointers must be explicitly declared, and BASED variables may be allocated only with the READ and LOCATE statements, which are not discussed in this text. BASED variables are unsupported by SL1 or Student PL.

[4] This operator is written PT in the 48-character set. Operators written in alphabetic characters are reserved identifiers. The D and Model 20 compilers do not support the pointer qualifier.

16.5 THE NULL FUNCTION

By letting one or more elements of a based structure be pointers, we may link together instances of the structure into *lists*. A list is a network of PL/I structures. The structures contain the data and pointer variables which provide the links between the structures. The organization may be linear or rectangular like an array, hierarchical like a PL/I structure, or more general. However, the powerful feature of lists is that their organization may be altered during execution. The organization of arrays or of PL/I structures is fixed in the declaration, and it may not be altered during execution of the procedure.

The manipulation of lists is called *list processing*, and the remainder of this case study is a brief introduction to some applications. For a fuller explanation and additional applications, see the highly-recommended treatment by Knuth.[5] Many of the examples and exercises in this case study are inspired by Knuth's book.

As our first elementary example of list processing, the following program segment builds up a linked linear list of floating-point values:

```
DECLARE  HEAD  POINTER,
      1 X  BASED(LOC), 2 VALUE, 2 LINK POINTER;
    ON ENDFILE (SYSIN) GO TO S10;
    HEAD = NULL;
S5: ALLOCATE X;
    GET LIST (VALUE);
    LINK = HEAD;
    HEAD = LOC;
    GO TO S5;
S10: ...
```

NULL is a built-in function that returns a special pointer value outside the implementation's valid range of addresses. It provides positive identification of the end of the list. Also, we may initialize the pointer of a based variable to NULL and test for it later to determine if any instances have been allocated. (The ALLOCATION function may not be used with based variables.)

We may graphically represent this linear list by the diagram in Fig. 16-1. One element of each instance contains the floating-point value, and the other element

---

[5] Donald E. Knuth, *The Art of Computer Programming, Vol. 1* (Reading, Mass.: Addison-Wesley, 1968), Chap. 2. Note that Knuth uses the symbols ← for assignment, ⇐ for allocation, and WEATHER(Q) for Q -> WEATHER.

BASED variables for the D and Model 20 compilers are intended only for efficient I/O because the ALLOCATE and FREE statements are unsupported. CONTROLLED variables for Student PL are intended only for adjustable bounds because structures and pointers are unsupported. Nevertheless, list processing may be accomplished less elegantly and less efficiently for these compilers and SL1 by using unbased subscripted variables with indirect addresses as described in Sec. 10.10. All of the list processing examples in this case study may be adapted to this technique.

Sec. 16.5 The Null Function

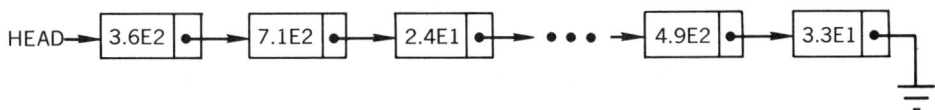

Fig. 16-1 A linear linked list.

contains the link to the next instance. The link to NULL is indicated by the electrical ground symbol.

As an example of the manipulation of this list, the following program segment sums the values in it:

```
        SUM  =  0;
        NEXT =  HEAD;
MORE:   IF NEXT = NULL THEN GO TO FINISHED;
        SUM  =  SUM + NEXT->VALUE;
        NEXT =  NEXT->LINK;
        GO TO MORE;
FINISHED: ...
```

Note that we may point to an element of a based structure just as we may point to the entire structure.

So far, we have done nothing that could not be done more efficiently with a singly-subscripted variable. However, suppose that our values are arranged in ascending order, and we wish to get in a value, then merge it in its proper order. We may do this as follows:

```
        ALLOCATE X;
        GET LIST (VALUE);
        LAST, NEXT = HEAD;
AGAIN:  IF NEXT = NULL THEN GO TO INSERT;
        IF VALUE <- NEXT->VALUE THEN GO TO INSERT;
        LAST = NEXT;
        NEXT = NEXT->LINK;
        GO TO AGAIN;
INSERT: LINK = NEXT;
        IF NEXT = HEAD THEN HEAD = LOC; ELSE LAST->LINK = LOC;
```

After finding the right position, all that we have to do is reset two links; whereas with a subscripted variable we would have to relocate all elements beyond the inserted one. As another example where a linked linear list is preferable to a singly-subscripted variable, suppose that we wish to join two linear lists, A and B, and that we have pointers pointing to the tail of A and the head of B. We may do this as follows:

```
        TAILA->LINK = HEADB;
```

We could find the tail of list A by stepping through from the head end if we did not have the foresight to set TAILA when the list was created.

16.6 QUEUE EXAMPLE

Suppose that in our production system, a worker takes his parts from a *queue* rather than a stack, meaning items are added at one end and subtracted at the other. We may simulate this by joining instances onto the tail end of a linear list of linked, based structures and subtracting them from the head end. For example, suppose we have the following declaration:

```
DECLARE  (HEAD,  TAIL)  POINTER,
    1  PART  BASED(WHERE),  2  PART#  FIXED(9),  2  LINK  POINTER;
```

If we have established a linked list with **HEAD** and **TAIL** pointing to the corresponding ends, then we may add an instance with part number **PN** at the tail end by the following statements:

```
ALLOCATE  PART;
PART#  =  PN;
LINK  =  NULL;
TAIL->LINK  =  WHERE;
TAIL  =  WHERE;
```

To add to an empty list, we may use the same statements with "TAIL -> LINK = WHERE;" replaced with "HEAD = WHERE;."

To subtract a part from the head of the list, assigning its part number to **PN**, we may execute the statements:

```
IF  HEAD  =  NULL  THEN  GO  TO  NONE_LEFT;
WHERE  =  HEAD;
HEAD  =  LINK;
PN  =  PART#;
FREE  PART;
```

16.7 SUBORDINATE PROCEDURES FOR LIST PROCESSING

Linked lists provide an efficient means for operating upon sparse vectors and matrices where most of the elements are zeros. For example, we may represent a sparse vector by linking together instances of the following structure:

```
1  X  BASED  (XPNT),
    2  VALUE  FLOAT,
    2  SUBSCRIPT  FIXED  BINARY,
    2  LINK  POINTER
```

Only instances with nonzero VALUE are included. As an example of the manipulation of such vectors, the following function procedure returns the inner product of two such vectors:

```
VVPROD:   PROCEDURE (HEADX, HEADY);   /* THIS FUNCTION PROCEDURE RETURNS
          THE INNER PRODUCT OF TWO SPARSE VECTORS STORED AS LINKED
          STRUCTURES SIMILAR TO THOSE BELOW. */
       DECLARE (HEADX, HEADY) POINTER,
          1 X BASED(XPNT), 2 VALUE FLOAT, 2 SUBSCRIPT FIXED BINARY,
             2 LINK POINTER,
          1 Y BASED(YPNT) LIKE X;
       SUM = 0;   XPNT = HEADX;   YPNT = HEADY;
S10:   IF XPNT = NULL | YPNT = NULL   THEN RETURN (SUM);
          /* FINISHED WHEN EITHER IS NULL */
       IF X.SUBSCRIPT = Y.SUBSCRIPT   THEN DO;
          SUM = SUM + X.VALUE*Y.VALUE;   XPNT = X.LINK;  YPNT = Y.LINK;
          GO TO S10;   END;
       IF X.SUBSCRIPT > Y.SUBSCRIPT   THEN YPNT = Y.LINK;
                                      ELSE XPNT = X.LINK;
       GO TO S10;
END;
```

Note that instead of being based structures, the parameters are *pointers* to the based structures. Also, there is no allocation within VVPROD. Consequently, the structure declarations within it are merely templates which describe the structures based on XPNT and YPNT so that the proper relative locations of the value, the subscript, and the link are established.

16.8 MULTIPLE LINKS

Did Shakespeare really write all of the plays and sonnets that are attributed to him? Some scholars believe that someone else wrote some of them. To help find out, we may have a computer generate a *verbal index* for each play and sonnet. A verbal index is an alphabetical list of all the words used and the number of times each occurs. Perhaps a comparison of these indices would reveal two different styles.

We may generate a verbal index by getting in and storing the words in a play in a subscripted variable, then sorting and combining redundant entries. However, this technique would involve a great many comparisons and interchanges. A faster technique is to build a *verbal index tree*, and then put out the contents in alphabetical order. For example, Fig. 16-2 shows a verbal index tree of the following lines:

> *Double, double toil and trouble;*
> *Fire burn and cauldron bubble.*

This diagram is called a tree because it resembles an upside-down branching tree. More specifically, this diagram is a *binary tree* because each branch divides

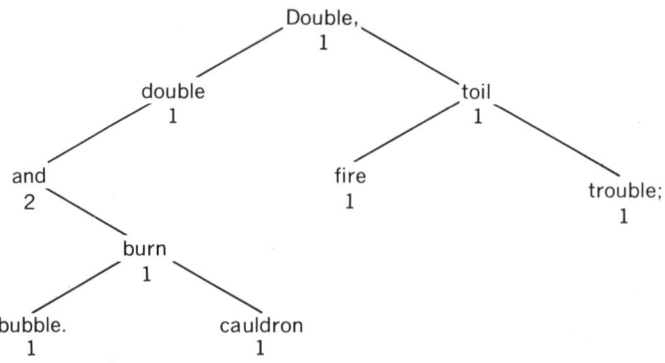

Fig. 16-2 An example of a verbal index tree.

into no more than two branches at each joint or *node*. A distinct word and the number of its occurrences are listed at each node, and the node occupied by "Double," is called the *root*. Note that adjacent punctuation is considered part of a word and judged according to collating sequence; so "double" is different from "Double,." The flow chart in Fig. 16-3 shows how to build this tree, and Fig. 16-4 shows some successive stages of its development. To thoroughly understand the flow chart, it is advisable to complete the missing stages and try a few other passages.

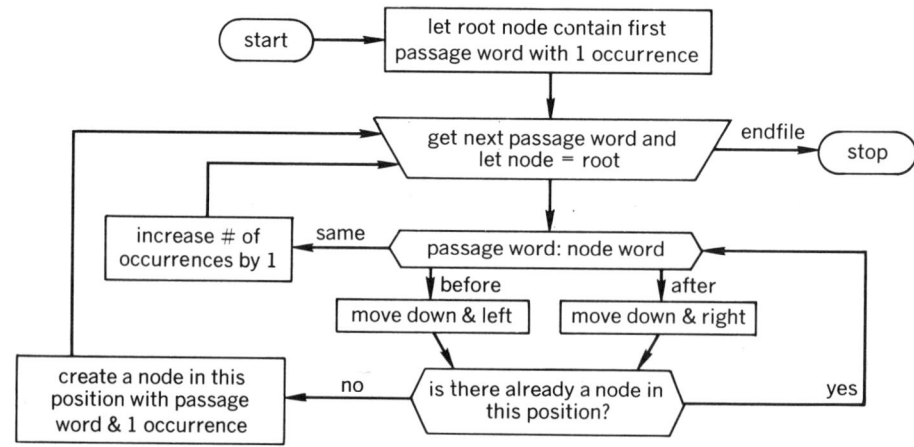

Fig. 16-3 Flow chart for building a verbal index tree.

Note that our tree is quite different from a PL/I structure which can have data only at the terminal nodes and cannot grow during a program. We may write a program which builds this tree using based structures with two links such as:

```
1  X  BASED(XPNT),
   2  VALUE    CHARACTER(16),
   2  COUNT    FIXED  BINARY,
   2  (LEFT, RIGHT) POINTER
```

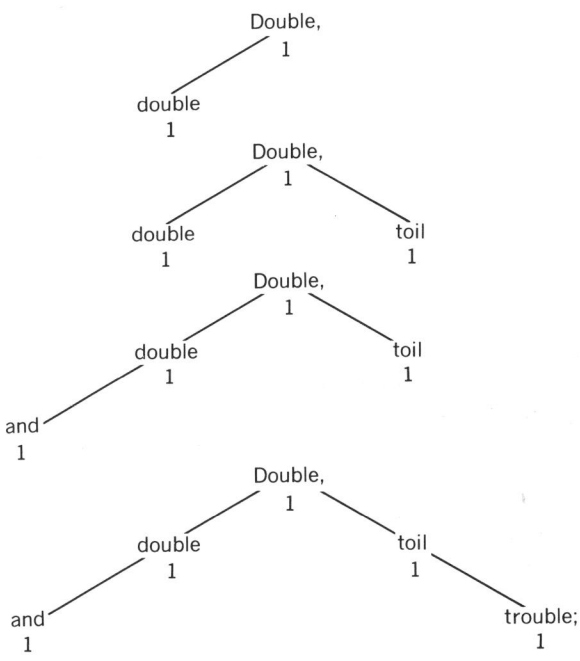

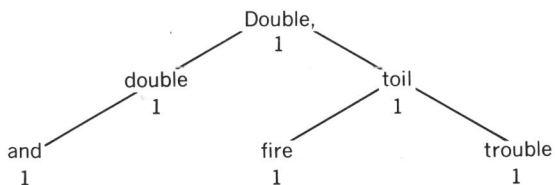

Fig. 16-4 Successive stages of a verbal index tree.

We may set the LEFT and/or RIGHT pointer to NULL wherever there are no branches. Accordingly, the following program segment gets in a list of words and forms their verbal index tree:

```
INDEX:  PROCEDURE   OPTIONS (MAIN);
    DECLARE (ROOT, PARENT) POINTER, WORD CHARACTER(16),
        1 X BASED(XPNT), 2 VALUE CHARACTER(16), 2 COUNT FIXED BINARY,
        2 (LEFT, RIGHT) POINTER;

NEW:    PROCEDURE;
    ALLOCATE X;   VALUE = WORD;   COUNT = 1;  LEFT, RIGHT = NULL;
    END;
ON ENDFILE (SYSIN)   GO TO S30;
```

```
*   ESTABLISH FIRST NODE:   */

    GET LIST (WORD);   CALL NEW;   ROOT = XPNT;

        /*   ESTABLISH REMAINING NODES:   */
    S10:   GET LIST (WORD);
           XPNT = ROOT;
    S20:   IF WORD = VALUE   THEN DO;   COUNT = COUNT + 1;   GO TO S10;   END;
           IF WORD < VALUE   THEN DO;
              IF LEFT ¬= NULL   THEN GO TO S20;
              PARENT = XPNT;
              CALL NEW;   PARENT->LEFT = XPNT;
              GO TO S10;
              END;
           IF RIGHT ¬= NULL   THEN GO TO S20;
           PARENT = XPNT;
           CALL NEW;   PARENT->RIGHT = XPNT;
           GO TO S10;
```

We have made use of an internal procedure, described in Case Study 5, to avoid redundant programming.

After forming the tree, we may automatically put the words in alphabetical order by a *postorder traversal:*

Starting at the root, move through the tree, always exploring left subtrees before right and putting out the contents of a node after its entire left subtree has been explored. Exploration of a null subtree consists of merely checking to find out that it is null.

For example, a postorder traversal of the tree shown in Fig. 16-5 gives the result A H G I B C F K J D E. (The null links are indicated by the electrical ground symbol.) This concept is somewhat difficult; so it is advisable to verify that it gives alphabetical order for Fig. 16-2 and to try exercises 1 and 2 at the end of this case study before proceeding further.

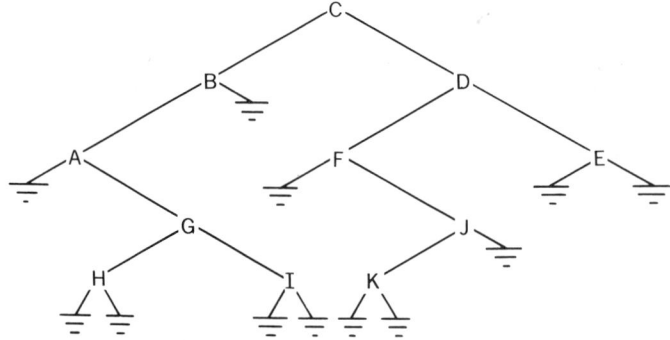

Fig. 16-5 A binary tree.

To implement this postorder traversal, we must keep a stack of pointers from the root so that we may find our way back after exploring subtrees. Accordingly, we may complete our procedure **INDEX** with the following statements:

```
              DECLARE PREDECESSOR CONTROLLED POINTER;
        S30:  XPNT = ROOT;
        S40:  IF XPNT = NULL   THEN GO TO S50;
              ALLOCATE PREDECESSOR;   PREDECESSOR = XPNT;
              XPNT = LEFT;
              GO TO S40;

        S50:  IF ¬ALLOCATION(PREDECESSOR)    THEN STOP;
              XPNT = PREDECESSOR;   FREE PREDECESSOR;
              PUT SKIP LIST (VALUE, COUNT);
              XPNT = RIGHT;
              GO TO S40;
        END;
```

It is advisable to check through these statements by hand for our example in Fig. 16-2.

16.9 ARRAYS OF POINTERS

Zoologists have found that a group of chickens often settles into a unique *pecking order*, and sociologists are interested in this as a model of human behavior. Two chickens do not usually peck each other. One of the pair is generally dominant. Also, the group may usually be ranked into a pecking order, meaning that for any three chickens I, J, and K, if chicken I pecks on chicken J and chicken J pecks on chicken K, then chicken I pecks on chicken K. To save observation time, we would like to be able to determine the pecking order, or determine that there is none, without waiting for every chicken to peck or be pecked by every other chicken. For example, suppose that we have observed the dominance relations among nine chickens as indicated in Fig. 16-6, where an arrow from node I to node J means that chicken I pecks chicken J.

The linked *network* in Fig. 16-6 is more general than a tree because there may be more than two paths from one node to another. A pecking order is impossible if we may make a closed path or *circuit*, following the directions of arrows. Fig. 16-6 contains no circuits, but it is not complete enough to establish a complete

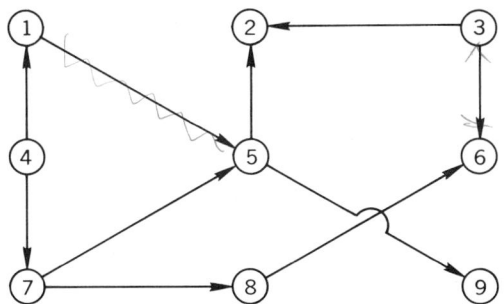

Fig. 16-6   Dominance relationships.

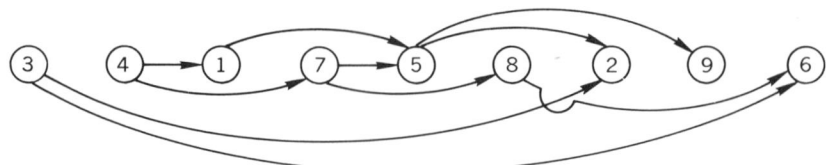

Fig. 16-7 A partial pecking order.

pecking order. Nevertheless, we may establish a *partial ordering* such as is shown in Fig. 16-7. A partial ordering means that for all known relations, the dominated chicken is listed after the dominating one. A partial ordering would help the observer determine where to encourage further confrontations. When the number of chickens is large, we need a systematic way to find partial orderings because the graph becomes hopelessly tangled. We may start with a table such as is shown below, then for each relation "J pecks K", we list K as an inferior of J and increase K's number of superiors by one. The reader should verify the indicated entries.

data: 1 → 5, 4 → 1, 4 → 7, 7 → 5, 7 → 8, 8 → 5, 3 → 6, 3 → 2,
5 → 2, 5 → 9

output queue

chicken #	1	2	3	4	5	6	7	8	9
# of superiors	Ø1	Ø1 2	0	0	Ø1 2	Ø1 2	Ø1	Ø1	Ø1
inferiors	5		6 2	1 7	2 9		5 8	6	

Next, we may list all chickens without superiors (#3 and #4) in an output queue. No chicken dominates the one at the head of this queue; so we may list him first in our partial ordering. Consequently, we may cross out his number, and we may think of this as removing him from the chicken yard. Accordingly, the chickens in his inferior list may have their numbers of superiors reduced by one. Whenever a chicken's number of superiors becomes zero, the chicken may be put at the tail of the output queue because he is dominated by none remaining in the yard. After this, we may put out the next chicken at the head of the queue and reduce its inferiors' numbers of superiors by one, putting any that become zero into the output queue. We may repeat this process until there is none left in the output queue. There is a circuit if some are still left in the chicken yard.

This process has been started in the table below. It is left for the reader to verify that the method gives the partial ordering of Fig. 16-7.

output queue	3̸ 4 1 7

chicken #	1	2	3	4	5	6	7	8	9
# of superiors	Ø̷X0	Ø̷X2̷1	0	0	Ø̷X2	Ø̷X2̷1	Ø̷X0	Ø1	Ø1
inferiors	5		6 2	1 7	2 9		5 8	6	

To program this method, we may store the queue and each of the inferior lists as a linear list of based structures. We may then have a subscripted pointer variable HEAD__INFERIOR(K) which points to the head of chicken K's inferior list. We may also store the number of superiors as a subscripted variable #__OF__SUPERIORS(K). Accordingly, we may implement our technique as follows:

```
PECK:   PROCEDURE OPTIONS (MAIN);  /*  THIS PROCEDURE PUTS OUT THE
        PECKING ORDER, GIVEN LIST-DIRECTED DATA SPECIFYING THE NUMBER OF
        CHICKENS FOLLOWED BY ANY NUMBER OF PAIRS J, K -- MEANING CHICKEN
        J PECKS CHICKEN K.   */
    DECLARE (HEAD_OUT, TAIL_OUT) POINTER,
        1 OUT BASED(OUTPNT), 2 ID, 2 LINK POINTER,
        1 INFERIOR BASED(INFPNT) LIKE OUT,
        S LABEL;
    GET LIST (N);
    BEGIN;
        DECLARE #_OF_SUPERIORS(N) FIXED BINARY INITIAL ((N)0),
            HEAD_INFERIOR(N) POINTER;

        /*  INPUT STAGE:  */
        ON ENDFILE (SYSIN)  GO TO S20;
        S = FIRST;  HEAD_OUT = NULL;  HEAD_INFERIOR = NULL;  I = 0;
S10:    GET LIST (J,K);
        #_OF_SUPERIORS(K) = #_OF_SUPERIORS(K) + 1;
        ALLOCATE INFERIOR;  INFERIOR.ID = K;
        INFERIOR.LINK = HEAD_INFERIOR(J);
        HEAD_INFERIOR(J) = INFPNT;
        GO TO S10;

        /*  INITIALIZE OUTPUT QUEUE:  */
S20:    I = I + 1;   IF I > N  THEN GO TO S30;
        IF #_OF_SUPERIORS(I) ¬= 0   THEN GO TO S20;
        ALLOCATE OUT;  OUT.ID = I;  OUT.LINK = NULL;
        GO TO S;
FIRST:  HEAD_OUT, TAIL_OUT = OUTPNT;   S = NOT_FIRST;
        GO TO S20;
NOT_FIRST:   TAIL_OUT->OUT.LINK = OUTPNT;   TAIL_OUT = OUTPNT;
        GO TO S20;
```

```
                /* PUT OUT THE DOMINANT REMAINING ID: */
        S30:    IF HEAD_OUT = NULL   THEN STOP;
                OUTPNT = HEAD_OUT;   HEAD_OUT = OUT.LINK;   I = OUT.ID;
                PUT LIST (I);    FREE OUT;

                /*  REDUCE THE #_OF_SUPERIORS OF THE PUT OUT CHICKEN'S
                    INFERIORS:  */
                INFPNT = HEAD_INFERIOR(I);
        S40:    IF INFPNT = NULL    THEN GO TO S30;
                #_OF_SUPERIORS(INFERIOR.ID) = #_OF_SUPERIORS(INFERIOR.ID) - 1;
                IF #_OF_SUPERIORS(INFERIOR.ID) = 0
                    THEN DO;  /*   PUT IN OUTPUT QUEUE  */
                        ALLOCATE OUT SET (TAIL_OUT->OUT.LINK);
                        OUT.ID = INFERIOR.ID;   OUT.LINK = NULL;
                        TAIL_OUT = TAIL_OUT->OUT.LINK;
                    END;
                INFPNT = INFERIOR.LINK;
                GO TO S40;
        END;
END;
```

Note that we have used a label variable, first discussed in exercise 10 in Case Study 3.

SUMMARY

Storage Classes

The BASED and CONTROLLED storage class attributes are alternatives to STATIC and AUTOMATIC.

The ALLOCATE and FREE Statements

Instances of based and controlled variables are created by the ALLOCATE statement and destroyed by the FREE statement. Instances of controlled variables are stacked, with any reference denoting the top instance on the stack. Instances of based variables are based on a pointer variable, and any instance may be denoted by pointing to it—e.g. TAIL -> B. If unspecified, the pointer is assumed to be that declared with the BASED attribute—e.g., BPNT for the declaration 1 B BASED(BPNT), 2 VALUE, 2 LINK POINTER.

Built-in Functions

The ALLOCATION function returns '1'B if there are any current instances of a controlled variable, returning a '0'B otherwise.

The ADDR function returns the address of a variable.

The NULL function returns a special address outside the valid range. It may be used to indicate terminal links.

EXERCISES

▼1. Write out the nodes of the following tree in postorder:

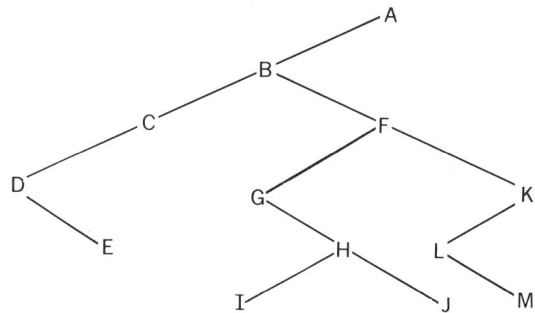

2. Write out the nodes of the following nonalphabetic tree in postorder:

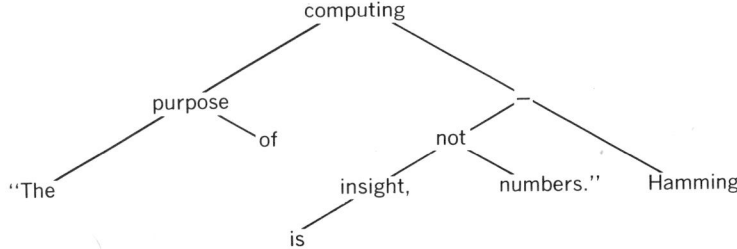

3. Write a function procedure that returns the number of instances in a list such as that of Fig. 16-1.

4. Write a subroutine that merges a value into a list arranged in ascending order.

▼5. Write a subroutine that builds the vector sum of two sparse vectors, where the vectors are stored as linked lists.

6. Write a function procedure that returns the length of a sparse vector stored as a linked list, where the length is defined as the square root of the sum of the squared values.

▼7. Linear lists such as in Fig. 16-1 have the disadvantage that we may not reach every node from an arbitrary point in the list. We may overcome this disadvantage by linking the last node back to the first instead of to NULL, thus making a *circular list* as follows:

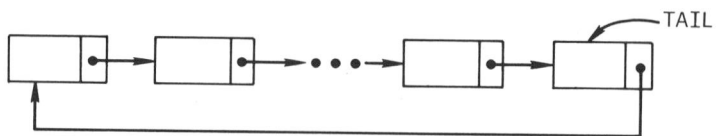

Write a subroutine that inserts the value V at the "tail" of such a list.

8. Same as exercise 7 except write a subroutine that inserts a value V at the "head" of such a list.

9. Same as exercise 7 except write a function that deletes the node at the head, returning its value.

▼10. To permit easy movement in either direction, we may make a doubly-linked circular list as follows:

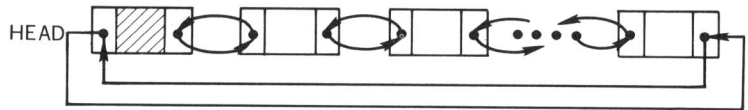

Write a subroutine that inserts a value V at an arbitrary position pointed to by a pointer variable HERE. If no data is stored in the head node, an empty list may be represented by a solitary head node.

11. Same as exercise 10 except write a subroutine that deletes the node pointed to by HERE.

▼12. We may represent a sparse matrix by a doubly-linked list of instances of the structure 1 X BASED(XPNT), 2 VALUE, 2 I, 2 J, 2 RIGHT POINTER, 2 DOWN POINTER, where I is the row, J is the column, RIGHT points to the next element in the same row, and DOWN points to the next element in the same column. The last item in each row may have RIGHT point back to the first item in the row or to null—whichever seems more convenient. Similarly, the last item in each column may point back to the first item in the column or to null. The first item in each column may be pointed to by a subscripted pointer HEAD__ROW; and the first item in each column may be pointed to by a subscripted pointer HEAD__COL. Write a function procedure that returns the sum of the diagonal values or *trace* of a square matrix stored in this fashion.

13. For a sparse matrix stored as in exercise 12 above, write a function procedure that returns the infinity-norm as defined in exercise 18 in Case Study 11.

14. In a general tree, each node may have any number of offspring—not just two. However, we may represent any tree as a binary tree by the following technique:

Link together the offspring of each node.
Erase all but the leftmost link from each parent to its offspring.
Rotate the diagram clockwise so that the right and left links both tend downwards.

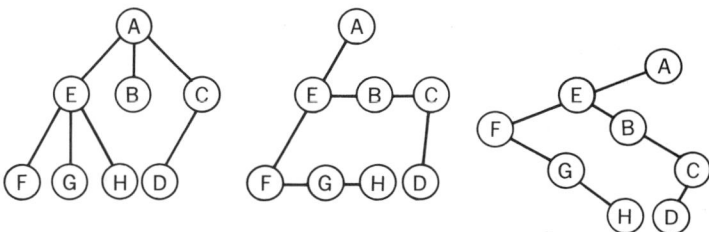

Describe the reverse process whereby the original tree may be recovered from its binary representation.

15. For some applications it is desirable to put out the contents of a binary tree in *preorder* rather than postorder. Starting at the root, we move through the tree, always exploring left subtrees before right and putting out the contents of a node *before* its left subtree has been explored. Most people find this order more natural than postorder. For the tree of exercise 1, it gives A B C D E F G H I J K L M. What does it give for the tree in Fig. 16-5? Write a general-purpose subroutine that puts out the contents of a binary tree in preorder.

▼16. For some applications it is desirable to put out the contents of a binary tree in *endorder*. Starting at the root, we move through the tree, always exploring left subtrees before right, and putting out the contents of a node after both its left and right subtrees have been fully explored. For the tree of exercise 1 it gives E D C I J H G M L K F B A. What does it give for the tree in Fig 16-5? Write a general-purpose subroutine that puts out the contents of a binary tree in endorder.

17. Personalize the pecking order program so that the input and output involve names such as Henry, Claudia, Henrietta, Henny-Penny, and Chicken-Little rather than chicken numbers.

▼18. A planar maze is one whose paths may be represented on a sheet of paper as a network without any crossing lines, meaning there are no tunnels or bridges. If the start and finish of the maze could also be joined by an additional imaginary line that crosses no other—for example, if both are on the outside—then the following rule will get us through the maze: *At a junction, always choose the extreme right-hand path.* Devise a linked structure representation of such a maze and a program for generating it and finding the way through. (This method is more efficient than the matrix iterative method described in Case Study 13 when we wish to know only if there is a path between two particular nodes.)

19. The method of exercise 18 is usually of no help to a lost spelunker because caves are often nonplanar, or the "start" (where he first realizes he is lost) often cannot be joined to the finish (which is any entrance) by an imaginary line that crosses no others. However, the following more general method by Tarry will always work: *Never go twice in the same direction along any branch, and take unexplored paths before ones which have been explored in the opposite direction.* In a cave we may insure this by making an arrow at the ends of each branch in the direction we are traversing it, and checking for previous arrows before taking a branch. In a program we may accomplish this by changing the settings of one-bit variables that indicate

which ways each branch has been traversed. Devise a linked structure representation of a maze together with a program for generating it and finding the way through according to Tarry's method.

20. We are often interested in finding the shortest, quickest, or least expensive path between two points in a network. One of the most efficient ways of doing this is a method due to George Dantzig. This method is most easily explained by an analogy due to Ira Pohl: Imagine a family of amoebae moving from the starting point at uniform speed, dividing as necessary whenever a new node is reached. The first amoeba to reach the finish has reached it in the minimum possible time. In programming this method, it is helpful to keep a list of conquered nodes together with the time it has taken to conquer them. We may then compare the sums of these times plus the branch times to connected unconquered nodes to determine which node is conquered next. Write a program implementing this technique.

▼21. Ford's iterative method for determining the quickest path is as follows: The time required to reach the starting node is set to zero, and the time required to reach all other nodes is tentatively set to infinity. We then repetitively sweep through all of the nodes, reducing their tentative times whenever the time of a neighbor plus the time of the branch between them constitutes a reduction. The times are the minimums when none of them changes for one entire sweep. Write a program implementing this technique.

▼22. Suppose that a utility company wishes to connect N towns at minimum cost, and the company has calculated the cost of directly connecting each town to every other. The minimum cost way to connect the towns may be found by the following method by Kruskal: Start with the cheapest connection; then successively add the cheapest remaining connection that does not form a circuit until $N - 1$ connections have been used. At this point all N towns will be joined by a minimum cost tree of $N - 1$ branches. Assume that the input consists of a series of number pairs—J, K—arranged in nondecreasing order of the direct cost of linking town J with town K. The output may consist of similar pairs for the links that have been accepted. To check for circuits, it is helpful to have an integer vector PARENT(K), initially 0, that gives the node which is the parent of node K. To see if a prospective branch forms a circuit, both of its ends may be traced back to see if they have the same nonzero parent.[6]

▼23. The flow graph shown below is a *cascade* flow graph because all signals flow in one direction. For cascade flow graphs, the *gain* is the sum of the branch products in every path from the start to the finish. This gain is ($abc + dc + ae$) in the graph below. Write a general-purpose program that gets in the data for a cascade flow graph, then calculates and puts out the gain.

---

[6] For additional list-processing applications related to those in exercises 18 through 22, see Claude Berge, *The Theory of Graphs* (New York: John Wiley & Sons, Inc., 1962); and L.R. Ford, Jr., and D.R. Fulkerson, *Flows in Networks* (Princeton, N.J.: Princeton University Press, 1962).
For additional list-processing applications related to those in exercises 23 and 24, see Charles S. Loren, *Flowgraphs for the Modeling of Linear Systems* (New York: McGraw-Hill Book Company, 1964).

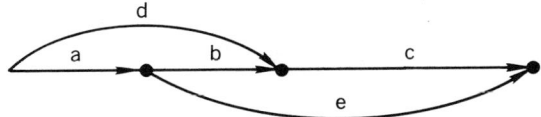

24. The graph below depicts a *discrete Markov system*. The nodes are possible states of the system, and the branches denote possible state transitions with the transition probabilities beside them. If the system is in state J we may calculate the probability of reaching state K in exactly M steps as the sum of all M-branch products leading from state J to state K. For example, the two-step probability of going from state 2 to state 5 is 0.4*1 + 0.3*0.5 = 0.55. Write a program that builds a Markov graph of linked structures from input data, then uses it to get any number of values for J, K, and M, calculating and putting out the corresponding probability.

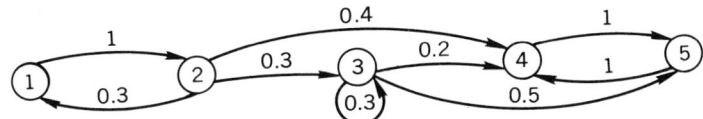

CASE STUDY 17

COMPLEX ARITHMETIC

Complex Variables[1]

In Case Study 5 we found the complex roots of a quadratic equation without resorting to complex variables. However, PL/I does have the capability for complex arithmetic. We may denote an imaginary constant by simply writing an "I" immediately after an arithmetic constant—for example:

$$3I \quad \text{or}$$
$$-8.673E-3I$$

We may denote a complex constant by writing a real constant $\pm$ an imaginary constant—for example:

$$8 + 3I \quad \text{or}$$
$$9 - 8.673E-3I$$

This is also the form for complex input and output data.

17.1 THE COMPLEX ATTRIBUTE

We may use the COMPLEX attribute to make a variable complex—for example:

DECLARE C COMPLEX

---

[1] Unsupported by the D, Model 20, SL1, and Student PL compilers. For these compilers it is a good exercise to write a package of subroutines for complex arithmetic.

Sec. 17.1 The Complex Attribute

Complex values may be used in expressions. With the exception of integer powers, real values are converted to complex with an imaginary part of zero when combined with complex values. Similarly, the imaginary part is set to zero when a real value is assigned to a complex variable. For example, if C is complex and R is real, then the following statements are valid:

```
C = C/(5 + 1I);
R = C**5 + C**R;
C = 1.5E9;
C = EXP(C) + ABS(C);
```

The "1" is necessary in the first example—it may not be implied. The first term in the second example is evaluated by repetitive multiplication, but the second term is evaluated by logarithms and exponentials unless R is declared to be fixed-point integer. The imaginary part is dropped when the sum is assigned to R. The imaginary part of C becomes zero in the third example. The fourth example illustrates that we may use complex arguments for built-in functions.

The built-in functions which may be used with complex arguments are implementation dependent.[2] In addition to the ordinary built-in functions, there are several intended especially for complex arithmetic. They are:

```
REAL(c)
IMAG(c)
CONJ(c)
COMPLEX(x,y)
```

The REAL function returns the real part of its complex argument; the IMAG function returns the imaginary part if its complex argument; the CONJ function returns the complex conjugate of its complex argument; and the COMPLEX function returns the complex value formed from its two real arguments. For example, if C has the value 3 + 5I, X has the value 15.6, and Y has the value 12.9, then:

```
REAL(C) is 3
IMAG(C) is 5
CONJ(C) is 3 - 5I
COMPLEX(X,Y) is 15.6 + 12.9I
```

Note that we may not write:

```
X + YI
```

We may append "I" only to constants.

REAL and IMAG may be used as pseudo-variables to assign the value of a real expression to either part of a complex variable. COMPLEX may be used as

---

[2] With the F compiler, we may use ABS ADD BINARY, DECIMAL, DIVIDE, FIXED, FLOAT, MULTIPLY, PRECISION, ROUND, SIN, COS, TAN, ATAN(x), SINH, COSH, TANH, ATANH, LOG, SORT, PROD, and SUM.

a pseudo-variable to assign the real and imaginary parts of a complex expression to two real variables. For example:

```
REAL  (C)    =  3.14159;
IMAG  (C)    =  4.71432;
COMPLEX (X, Y) = LOG (5E2 + 18I);
```

17.2 POLYNOMIAL ROOTS EXAMPLE

As an example of complex variables, let us use the standard *IBM Scientific Subroutine Package* procedure for finding the roots of a polynomial.

An excerpt from the instruction manual is:[3]

```
PURPOSE:
PRTC calculates all roots of a given complex
polynomial.
Usage:
CALL  PRTC  (C, N);
C(N) — COMPLEX  BINARY  FLOAT  [(53)]
       Given coefficient vector of normalized
       polynomial
```
$$P(Z) = Z^N + C_1 Z^{N-1} + \cdots + C_N$$
```
       Resultant N complex roots of given polynomial.
N —    BINARY  FIXED
       Given dimension of coeffecnt vector.
       N is also the degree of the polynomial and
       the number of roots to be calculated.
```

The square brackets around the precision of C mean that there is an optional double-precision version.

The following program uses this subroutine to find the roots of any number of polynomials:

```
ROOTS:   PROCEDURE OPTIONS(MAIN);
         DECLARE C(100) COMPLEX BINARY;
    IN:  GET LIST (N, (C(K) DO K = 1 TO N));
         PUT SKIP LIST ('COEFFICIENTS:');
         PUT SKIP DATA ((C(K) DO K = 1 TO N));
         CALL PRTC(C, N);
         PUT SKIP LIST ('ROOTS:');
         PUT SKIP DATA ((C(K) DO K = 1 TO N));
         GO TO IN;
    END;
```

---

[3] Reprinted by permission from *H20-0554-0—System/360 Scientific Subroutine Package (PL/I)*. © 1968 by International Business Machines Corporation.

If our coefficients are real, we simply punch zeros for the imaginary part in the input data. With the F compiler, BINARY and DECIMAL floating-point numbers are actually the same; so there is no harm in simply declaring C(100) COMPLEX with the "DECIMAL" default.

17.3 TRANSFER FUNCTION EXAMPLE

In automatic control theory, we frequently deal with a rational transfer function that is a ratio of two polynomials of a complex variable—

$$T(S) = \frac{A(M)S^M + A(M-1)S^{M-1} + \cdots A(0)S^0}{B(N)S^N + B(N-1)S^{N-1} + \cdots B(0)S^0}$$

where the As and Bs are real.

The following complex function procedure uses Horner's rule, described in exercise 15 in Case Study 2, to evaluate a transfer function.

```
T:  PROCEDURE (M, N, A, B, S) COMPLEX;
    DECLARE (S, NUM, DENOM) COMPLEX FLOAT, A(*), B(*);
    ON ZERODIVIDE NUM = 1E70;  ON OVERFLOW NUM = 1E70;
    NUM = A(M);
    DO K = M-1 TO 0 BY -1;
        NUM = NUM*S + A(K);
    END;
    DENOM = B(N);
    DO K = N-1 TO 0 BY -1;
        DENOM = DENOM*S + B(K);
    END;
    NUM = NUM/DENOM;
    RETURN(NUM);
END;
```

We are often interested in knowing the *amplitude* or *gain* which is the absolute value of this function; together with the *phase* which is the arctangent of the imaginary over the real part:

```
TF    = T(S);
AMP   = ABS(TF);
PHASE = ATAN(IMAG(TF), REAL(TF));
```

17.4 CONFORMAL MAPPING EXAMPLE

A real function of a real variable maps points and intervals of one axis onto points and intervals of another as is shown in Fig. 17-1. More conventionally this is depicted as in Fig. 17-2, but Fig. 17-1 is better for our purposes.

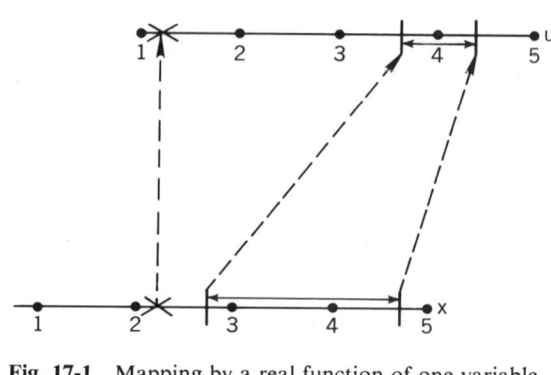

Fig. 17-1 Mapping by a real function of one variable.

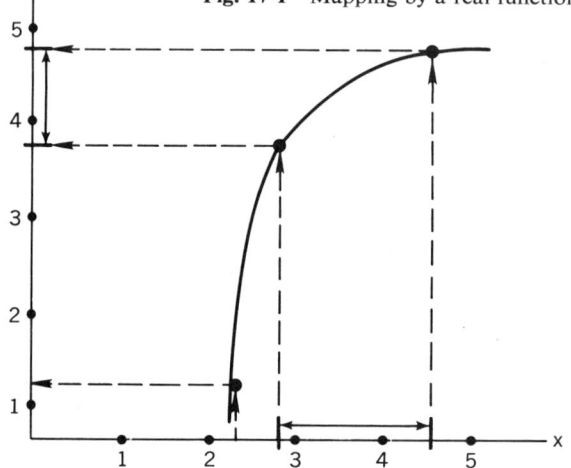

Fig. 17-2 Conventional portrayal of a real function of one variable.

A complex function of a complex variable, symbolically denoted $w = f(z)$, maps points, curves, and regions of one plane onto points, curves, and regions of another as is shown in Fig. 17-3.

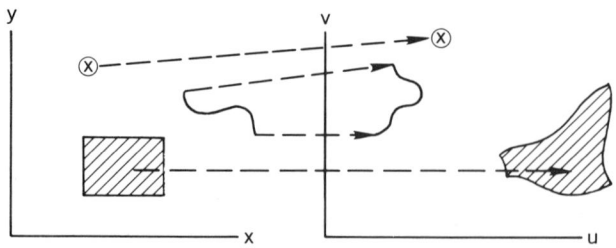

Fig. 17-3 Mapping by a complex function of a complex variable

This can be seen more easily if we break $f(z)$ into its real and imaginary parts. For example, take $w = z^2 = (x + iy)^2 = x^2 + 2ixy - y^2 = (x^2 - y^2) + i(2xy)$

$= u + iv$. We can thus consider this mapping as being comprised of two *real* functions of two *real* variables:

$$u = x^2 - y^2$$
$$v = 2xy$$

Take, for example, the z plane point $x = 1$, $y = 1$; it maps into the w plane point $u = 0$, $v = 2$. Similarly, if we take any locus of points in the z plane, they will map into a locus in the w plane.

Any algebraic complex function such as z^2, z^3, z^n, $1/(z^2 + z^3 + 8)$, etc. can be broken into its real and imaginary parts by using the simple rules of arithmetic with complex numbers. Transcendental complex functions such as $\sin(z)$ can similarly be manipulated by using their Taylor series equivalents, e.g.,

$$\sin(z) = z - \frac{z^3}{2!} + \frac{z^5}{5!} \cdots$$

The mappings which result from such complex functions have a *conformal* property which make them extremely useful for the solution of many two-dimensional field problems.

Many problems in engineering and science reduce to that of solving Laplace's partial differential equation in a given two-dimensional region subject to a given distribution of T on the boundary:

$$\frac{\partial^2 T}{\partial x^2} + \frac{\partial^2 T}{\partial y^2} = 0$$

For certain simple regions such as a unit circle or a semi-infinite half plane, the solution is already known for any boundary distribution. For other less simple regions, such as an infinite strip or a concentric annulus, the solution is known provided the boundary distribution is constant along certain arcs. When our region and boundary distribution do not fall into one of these two categories, we may make x and y into the complex variable $z = x + iy$; then we may search for a mapping function $w = f(z)$ which maps our region and its boundary distribution into a case for which the solution is known. If we can find such a mapping function, the remarkable fact is that

$$T(z) = \tau(w)$$

where τ is the solution for the simple region and $w = f(z)$.

In other words, if a mapped function obeys Laplace's equation in a conformally mapped region, then the given function obeys Laplaces's equation in the given region.[4]

---

[4] Conformal mapping may also be used for Poisson's equation, $\partial^2 T/\partial x^2 + \partial^2 T/\partial y^2 = S(x, y)$, and for the biharmonic equation, $\partial^4 T/\partial x^4 + \partial^4 T/\partial x^2 \partial y^2 + \partial^4 T/\partial y^4 = 0$.

To assist us in our search for the right mapping function, there are several rather general-purpose mapping functions, and there are catalogs of specific mapping functions. We may illustrate the technique with the following example.

Suppose we have the eccentric annular region shown in Fig. 17-4 with constant $T = T_1$ imposed on the inside boundary and $T = T_2$ imposed on the outside boundary. Arcs of constant temperature map into arcs of the same constant temperature, and we know or can easily derive the solution for a *concentric* annulus with two constant temperatures on the two boundaries, so let us search for a mapping from an eccentric to a concentric annulus. We look in a catalog and find that there is such a mapping, and it is given by the function

$$w = \frac{z - a}{az - 1}$$

where

$$a = \frac{1 + x_1 x_2 + \sqrt{(1 - x_1^2)(1 - x_2^2)}}{x_1 + x_2}$$

$$R_0 = \frac{1 - x_1 x_2 + \sqrt{(1 - x_1^2)(1 - x_2^2)}}{x_1 - x_2}$$

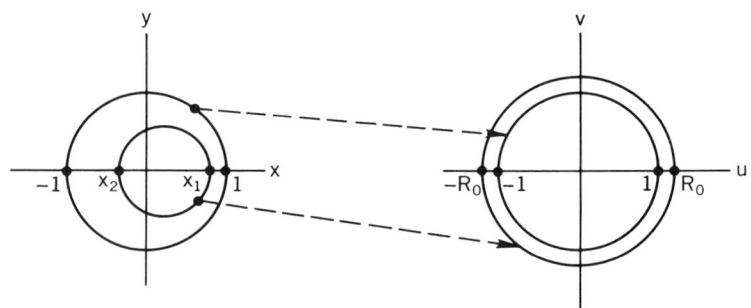

Fig. 17-4 Conformal mapping of an annular region.

The mapping turns the annulus inside out, so on the concentric annulus we have T_1 on the outside and T_2 on the inside. The known or easily derived solution for this case is

$$\tau(w) = T_2 + (T_1 - T_2) \log_2 |w/R_0|$$

To find the temperature at any point z, we simply evaluate $w(z)$, then evaluate the formula:

```
ECCEN:  PROCEDURE OPTIONS (MAIN);
        DECLARE Z COMPLEX;
        GET LIST (X1, X2, T1, T2);
        X1X2 = X1*X2;   S = 1 + SQRT((1 - X1*X1)*(1 - X2*X2));
        A = (S + X1X2)/(X1 + X2);   RO = (S - X1X2)/(X1 - X2);
    IN: GET LIST (Z);
        PUT SKIP LIST (Z, T2 + (T1-T2)*LOG(ABS((Z-A)/RO*(A*Z-1))));
        GO TO IN;
END;
```

SUMMARY

Complex Variables

Complex variables are declared with the attribute COMPLEX.

Complex Constants

An imaginary constant is written by writing an "I" immediately after a real constant—for example, 3I or −8.673E−3I.

A complex constant or data are written as a real constant ±an imaginary constant.

Complex Expressions

With the exception of integer powers, real values are converted to complex when combined with them.

The imaginary part is set to zero when real values are assigned to complex variables.

The imaginary part is truncated when complex values are assigned to real variables.

Built-in Functions and Pseudo-Variables

REAL(c), IMAG(c), CONJ(c), and COMPLEX(x, y) are built-in functions for complex arithmetic. In addition, an implementation-dependent selection of the other built-in functions such as LOG and ABS may be used with complex arguments. REAL, IMAG, and COMPLEX, may also be used as pseudo-variables.

EXERCISES

1. Modify ROOTS so that it works for *unnormalized* polynomials.
2. Rewrite the subroutine QUAD of Case Study 5 using complex numbers.
3. (Difficult) Most numerical analysis books present one or more general techniques for finding all of the roots of a polynomial. Find out one such method and write a general-purpose subroutine for it.
▼4. Most approximation formulas for the standard transcendental functions may be used for complex arguments as well as real. Accordingly, do exercise 17 in Case Study 7 for a complex argument.

5. Rewrite exercise 17 in Case Study 7 for a complex argument.
6. Rewrite exercise 19 in Case Study 7 for a complex argument.
7. Rewrite exercise 20 in Case Study 7 for a complex argument.
8. Rewrite exercise 21 in Case Study 7 for a complex argument.
9. Rewrite exercise 22 in Case Study 7 for a complex argument.
10. Rewrite exercise 23 in Case Study 7 for a complex argument.
11. Rewrite exercise 24 in Case Study 7 for a complex argument.
▼12. Many vector and matrix operations, such as inner products, norms, and Gaussian elimination, are valid for complex elements as well as for real elements. Rewrite exercise 6 in Case Study 10 accordingly.
▼13. Rewrite exercise 7 in Case Study 10 for complex elements.
14. Rewrite exercise 12 in Case Study 10 for complex elements.
15. Rewrite exercise 13 in Case Study 10 for complex elements.
16. Rewrite MVPROD of Case Study 11 for complex elements.
17. Rewrite MMPROD of Case Study 11 for complex elements.
18. Rewrite GAUSS of Case Study 12 for complex elements.
19. Rewrite exercise 16 in Case Study 11 for complex arguments.
20. Rewrite exercise 17 in Case Study 11 for complex arguments.
21. Rewrite exercise 18 in Case Study 11 for complex arguments.
22. Rewrite exercise 19 in Case Study 11 for complex arguments.
▼23. In many applications, complex matrices are *Hermetian:*

$$A(J, I) = CONJ(A(I, J))$$

For such matrices, we may store only the lower triangular portion in compacted form since the upper triangular elements are easily derived from the lower ones. Write a function procedure that finds the sum of all the elements in a Hermetian matrix stored in compacted form.

24. Rewrite SYMPRD of Case Study 12 for complex elements.
25. Rewrite FINNRM of Case Study 12 for complex elements.
26. Rewrite exercise 20 in Case Study 12 for complex elements.
27. Rewrite exercise 21 in Case Study 12 for complex elements.
28. Transfer functions are sometimes expressed in a partial fraction expansion. If none of the *poles* of the function are multiple, this has the form

$$T = \frac{C(1)}{S - R(1)} + \frac{C(2)}{S - R(2)} + \frac{C(3)}{S - R(3)} + \cdots \frac{C(N)}{S - R(N)}$$

where S, C, and R are complex. Write a corresponding function procedure for T.

▼29. Suppose that we wish to solve Laplace's equation for a semi-infinite strip with $T = 0$ on the sides and $T = 1$ on the end. As is shown in the figure, this region may be mapped into an infinite strip with $T = 1$ on one side and $T = 0$ on the other by the transformation

$$w = \log\left[\frac{\sin z - 1}{\sin z + 1}\right]$$

The solution for the infinite strip is known to be simply $T = v/\pi$. Write a program that puts out T for any number of points in the semi-infinite strip.

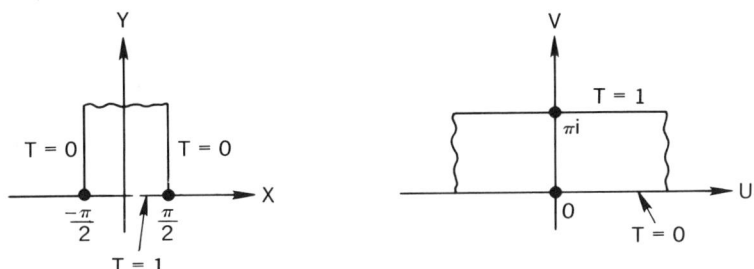

▼30. Using polar coordinates $w = re^{i\theta}$, the general solution to Laplace's equation for the unit circle is given by Poisson's formula:

$$\tau(r, \theta) = \int_{-\pi}^{\pi} \frac{\bar{\tau}(\phi)\, d\phi}{1 + r[r - 2\cos(\phi - \theta)]}$$

where $\bar{\tau}(\phi)$ is the boundary distribution as a function of angle ϕ. Therefore, Laplace's equation may be solved by numerical integration for an arbitary boundary distribution provided we know how to map the region into the unit circle. Write a program that solves Laplace's equation in the exterior of an ellipse with foci at $x = \pm 1$ and a major radius of "a" for arbitrary boundary conditions. This region is mapped into the unit circle by the function

$$w = \frac{z - \sqrt{z^2 - 1}}{a - \sqrt{a^2 - 1}}$$

▼31. Sometimes we are given a normal derivative of T on the boundary rather than T itself:

$$g(z) = \frac{\partial T}{\partial n}$$

The corresponding derivative on a transformed boundary will be

$$\gamma(w) = \left|\frac{dw}{dz}\right| g(z)$$

For the unit circle, the formula analogous to Poisson's is

$$\tau(r, \theta) = \int_{-\pi}^{\pi} \log\{1 + r[r - 2\cos(\phi - \theta)]\}\, \gamma(\phi)\, d\phi$$

Adapt exercise 30 accordingly.

CASE STUDY 18

LINEAR EQUATIONS REVISITED

A Package of Subroutines

The subroutine **GAUSS** developed in Case Study 12 is inadequate as a general-purpose library procedure. The intent was merely to provide enough familiarity to use a standard library procedure. However, some libraries do not have a good linear equations subroutine. Other libraries, such as the IBM Scientific Subroutine Package, have such sophisticated linear equations subroutines that Case Study 12 provides insufficient background to use them easily. Also, programmers who have a frequent demand for this application may wish to develop their own subroutines for greatest efficiency in their special circumstances. Consequently, this case study will develop the subject more thoroughly, borrowing heavily upon ideas in the highly recommended book by George Forsythe and Cleve B. Moler, *Computer Solutions of Linear Algebraic Systems* (Englewood Cliffs, N. J.: Prentice-Hall, Inc., 1967).

One of the drawbacks of our procedure **GAUSS** in Case Study 12 is that it destroys the original **A** matrix and **B** vector; so copies must be explicitly made if the original values are needed later in the program. On the other hand, twice the storage is needed if **A** and **B** are not destroyed. Fortunately, it is an easy matter to write a subroutine that may be used either way. We may set up a separate parameter **U** for the upper triangular matrix that is derived from **A**. We may then use the same argument for both parameters when we do not need to save the original values. Consequently, in subsequent discussions, any reference to **A** will denote an original value, and any reference to **U** will denote a modified value even though both may be stored in the same location.

18.1 THE LU DECOMPOSITION

When a physical system described by linear equations is given a different input or loading vector, only the right-hand constants change. The elimination will give the same upper triangular matrix; so it seems a shame to repeat it merely for the sake of the accompanying forward substitution. Exercise 8 in Case Study 12 suggests solving for several loading vectors simultaneously, but this reduces the maximum size system that may be solved, and we often wish to solve for additional loading vectors after seeing the result of the first.

Fortunately, we may avoid repeating the elimination if we save the factors used in the forward substitution. We may then perform a forward substitution upon any new B vector without having to rederive the same upper triangular matrix.

For example, we first multiplied B(1) by A(2, 1)/A(1, 1), then subtracted from B(2). Let us call this factor L(2, 1). Next, we multiplied B(1) by A(3, 1)/A(1, 1) and subtracted from B(3). Let us call this factor L(3, 1), and so on up to L(N, 1). Then we multiplied B(2) by U(3, 2)/U(2, 2) and subtracted from B(3) when we began eliminating the second column of the lower triangle. We may call this factor L(3, 2). Continuing in this fashion we will get a *lower triangular* L matrix which we may regard as the *forward substitution* matrix. To save space we may store L in the zero lower triangular portion of U by using the same name for both L and U —say LU (which we declare FLOAT). Conceptually, however, they are different matrices. If we consider L to have implied ones along its diagonal, it turns out that A equals the matrix product L·U; so this scheme is often called the LU decomposition of A.

Finding the LU decomposition of A is no more work than our original elimination. In fact, there is no need to include any B vectors during the decomposition. It is no more work to apply the forward substitution separately than it is to apply it while performing the elimination, and it is more flexible to apply these operations separately.

Another way to avoid repeating an elimination is to evaluate a matrix called the *inverse* of A, then multiply any number of B vectors by this inverse to find the solutions. However, this multiplication takes just as long as forward plus backward substitution, and the inverse takes three times as long to derive as the LU decomposition. Consequently, calculating the inverse is an inefficient way to solve equations. However, the inverse is occasionally desired for its own sake, so we will learn how to derive it later.

Accordingly, the following two subroutines perform the decomposition and the substitutions:

```
GAUSLU:   PROCEDURE (N, A, LU);   /*   THIS PROCEDURE FINDS THE LU
          DECOMPOSITION OF AN N BY N MATRIX 'A' USING GAUSS' METHOD.  TO
          DOUBLE THE AVAILABLE SPACE, USE THE SAME ARGUMENT FOR 'A' AND
          AND LU IF 'A' IS NOT SUBSEQUENTLY NEEDED.   */
```

```
      DECLARE A(*,*), LU(*,*) FLOAT;
      IF ADDR(A) = ADDR(LU)  THEN GO TO S10;
      DO I = 1 TO N;  /*  INITIALIZE LU IF DIFFERENT FROM A   */
         DO J = 1 TO N;
            LU(I,J) = A(I,J);
         END;
      END;
S10:  DO J = 1 TO N-1;  /*  SET COLUMN  */
      JP1 = J+1;  PIVOT = LU(J,J);
      DO I = JP1 TO N;
         F, LU(I,J) = LU(I,J)/PIVOT;  /*  RECORD MULTIPLE  */
            DO K = JP1 TO N;  /* SUBTRACT MULTIPLE OF JROW FROM IROW */
               LU(I,K) = LU(I,K) - F*LU(J,K);
            END;
         END;
      END;
END;

GSSUB:  PROCEDURE (N, LU, B, X);  /*  GIVEN THE LU DECOMPOSITION OF AN
        N BY N MATRIX A, THIS SUBROUTINE SOLVES THE LINEAR EQUATIONS
        AX = B USING FORWARD AND BACKWARD SUBSTITUTION.  USE THE SAME
        ARGUMENT FOR B AND X WHEN SPACE IS CRITICAL AND B IS NOT NEEDED
        AFTER THE SOLUTION.  */
   DECLARE LU(*,*) FLOAT, B(*), X(*), SUM FLOAT(16)  /* PRECISION HERE
        AND IN THE MULTIPLY FUNCTION ARE IMPLEMENTATION DEPENDENT  */;

   /*  FORWARD SUBSTITUTION:  */
   DO I = 1 TO N;
      SUM = 0;
      DO J = 1 TO I-1;
         SUM = SUM + MULTIPLY(LU(I,J), X(J), 16);
      END;
      X(I) = B(I) - SUM;
   END;

   /*  BACKWARD SUBSTITUTION:  */
   DO I = N TO 1 BY -1;
      SUM = 0;
      DO J = I+1 TO N;
         SUM = SUM + MULTIPLY (LU(I,J), X(J), 16);
      END;
      X(I) = (X(I) - SUM)/LU(I,I);
   END;
END;
```

GAUSLU introduces the built-in function **ADDR** that returns the address of its argument or the first address in the case of structures or arrays. Another notable feature is the check for a zero multiplier before wasting the time of multiplying a whole row.

GSSUB uses the **MULTIPLY** function to accumulate double-precision inner products. This substantially reduces the cumulative chopoff error with only a negligible increase in storage requirements.

We could save time by initializing the forward sum to B(I), but this might degrade the accuracy because B(I) often has a different order of magnitude than the individual terms in the sum. For a similar reason, the backward sum is not initialized to X(I).

To promote understanding, it is a good idea to apply **GAUSLU** then **GSSUB** by hand to the following numerical example. This example has been especially designed to require only simple positive integer arithmetic. The answer may be checked by matrix-vector multiplication:

$$\begin{bmatrix} 2 & 1 & 2 & 3 \\ 4 & 4 & 5 & 8 \\ 2 & 7 & 6 & 11 \\ 6 & 5 & 8 & 16 \end{bmatrix} \begin{bmatrix} X(1) \\ X(2) \\ X(3) \\ X(4) \end{bmatrix} = \begin{bmatrix} 11 \\ 30 \\ 39 \\ 48 \end{bmatrix}$$

18.2 DOOLITTLE'S METHOD

The elements of LU are actually inner products, but this fact is disguised by the fact that these products are not completely formed one at a time. Consequently it is impossible to accumulate them in double precision without paying the penalty of storing all of LU in double precision. However, Doolittle devised a means of deriving the same LU decomposition wherein each element is completely derived one at a time, facilitating double-precision accumulation. Briefly, Doolittle's method is as follows:

1. The first row of LU is the same as A.

2. The remainder of the first column is the first column of A divided by the diagonal element LU(1, 1).

3. The remainder of the next row is the same as A minus an inner product of previously calculated elements of LU above and to the left.

4. The remainder of the next column is the same as A minus an inner product of previously calculated elements above and to the left—the whole quantity divided by the diagonal element in the same column of LU.

5. Steps 3 and 4 are repeated until all elements have been derived.

The order in which the elements are calculated is illustrated below for a 4-by-4 matrix:

1st	2nd	3rd	4th
5th	8th	9th	10th
6th	11th	13th	14th
7th	12th	15th	16th

The method may be stated more precisely by the following procedure:

```
DOOLU:   PROCEDURE (N, A, LU);   /*  THIS PROCEDURE FINDS THE LU
         DECOMPOSITION OF AN N BY N MATRIX 'A' USING DOOLITTLE'S METHOD.
         TO DOUBLE THE AVAILABLE SPACE, USE THE SAME ARGUMENT FOR 'A' AND
         LU IF 'A' IS NOT SUBSEQUENTLY NEEDED.    */
```

```
    DECLARE A(*,*), LU(*,*) FLOAT, SUM FLOAT(16); /*  PRECISION HERE AND
        IN THE MULTIPLY FUNCTION ARE IMPLEMENTATION DEPENDENT     */
    DO I = 1 TO N;

        DO J = 1 TO N;
            SUM = A(I,J);
            DO K = 1 TO I-1;
                SUM = SUM - MULTIPLY (LU(I,K), LU(K,J), 16);
            END;
            LU(I,J) = SUM;
        END;

        /*   TRANSFORM COLUMN:    */
        PIVOT = LU(I,I);
        DO J = I+1 TO N;
            SUM = A(J,I);
            DO K = 1 TO I-1;
                SUM' = SUM - MULTIPLY (LU(J,K), LU(K,I), 16);
            END;
            LU(J,I) = SUM/PIVOT;
        END;
    END;
END;
```

To promote understanding, it is a good idea to apply this procedure by hand to the same example used for GAUSLU to see that it gives the same decomposition.

On a computer, Gauss's and Doolittle's methods are approximately the same speed, requiring time on the order of $N^3T/3$ where N is the number of equations, and T is the time of one multiplication plus one subtraction plus one DO cycle. By hand, Doolittle's method is considerably faster because it requires no recording of intermediate results. Without double-precision accumulation both methods give identical results.

18.3 CROUT'S METHOD

Another method by *Crout* is similar to Doolittle's except it begins with the first column, and the rows instead of the columns are divided by the pivots. This method gives a different LU decomposition; so it must be used with a different forward–backward substitution scheme. Briefly, the method is as follows:

1. The first column of LU is the same as A.

2. The remainder of the first row is the first row of A divided by the diagonal element LU(1, 1).

3. The remainder of the next column is the same as A minus an inner product of previously calculated elements of LU above and to the left.

4. The remainder of the next row is the same as A minus an inner product of previously calculated elements above and to the left—the whole quantity divided by the diagonal element in the same row of LU.

5. Steps 3 and 4 are repeated until all elements have been derived.

The order in which the elements are calculated is illustrated below for a 4-by-4 matrix:

1st	5th	6th	7th
2nd	8th	11th	12th
3rd	9th	13th	15th
4th	10th	14th	16th

Crout's and Doolittle's methods are often confused, and they are both often called Gauss's method.

Crout's decomposition and the corresponding forward–backward substitution may be stated more precisely by the following procedures:

```
CRTLU:  PROCEDURE (N, A, LU);  /*  THIS PROCEDURE FINDS THE LU
    DECOMPOSTION OF AN N BY N MATRIX 'A' USING CROUT'S METHOD.
    FOLLOW BY CRTSUB TO SOLVE LINEAR EQUATIONS.  TO DOUBLE THE
    AVAILABLE SPACE, USE THE SAME ARGUMENT FOR 'A' AND LU IF 'A'
    IS NOT SUBSEQUENTLY NEEDED.  */
  DECLARE A(*,*), LU(*,*) FLOAT, SUM FLOAT(16);  /*  PRECISION HERE
    AND IN THE MULTIPLY FUNCTION ARE IMPLEMENTATION DEPENDENT  */
  DO J = 1 TO N;

    /*  TRANSFORM COLUMN:  */
    DO I = J TO N;
      SUM = A(I,J);
      DO K = 1 TO J-1;
        SUM = SUM - MULTIPLY (LU(I,K), LU(K,J), 16);
      END;
      LU(I,J) = SUM;
    END;

    /*  TRANSFORM ROW:  */
    PIVOT = LU(J,J);
    DO I = J+1 TO N;
      SUM = A(J,I);
      DO K = 1 TO J-1;
        SUM = SUM - MULTIPLY (LU(J,K), LU(K,I), 16);
      END;
      LU(J,I) = SUM/PIVOT;
    END;
  END;
END;

CRTSUB:  PROCEDURE (N, LU, B, X);  /*  GIVEN CROUT'S LU DECOMPOSITION
    OF AN N BY N MATRIX 'A', THIS SUBROUTINE SOLVES THE LINEAR
    EQUATIONS AX = B USING FORWARD AND BACKWARD SUBSTITUTION.  USE
    THE SAME ARGUMENT FOR B AND X WHEN SPACE IS CRITICAL AND WHEN B
    IS NOT NEEDED AFTER THE SOLUTION.  */
  DECLARE LU(*,*) FLOAT, B(*), X(*), SUM FLOAT(16)  /*  PRECISION HERE
    AND IN THE MULTIPLY FUNCTION ARE IMPLEMENTATION DEPENDENT  */;

  /*  FORWARD SUBSTITUTION:  */
  DO I = 1 TO N;
    SUM = 0;
    DO J = 1 TO I-1;
      SUM = SUM + MULTIPLY (LU(I,J), X(J), 16);
    END;
    X(I) = (B(I) - SUM)/LU(I,I);
  END;
```

```
    /*    BACKWARD SUBSTITUTION:    */
    DO I = N TO 1 BY -1;
       SUM = 0;
       DO J = I+1 TO N;
          SUM = SUM + MULTIPLY (LU(I,J), X(J), 16);
       END;
       X(I) = X(I) - SUM;
    END;
END;
```

As before, it is worthwhile to apply these two procedures by hand to the following numerical example which requires only positive integer arithmetic:

$$\begin{bmatrix} 2 & 4 & 2 & 6 \\ 1 & 4 & 7 & 5 \\ 2 & 5 & 6 & 8 \\ 1 & 4 & 9 & 10 \end{bmatrix} \begin{bmatrix} X(1) \\ X(2) \\ X(3) \\ X(4) \end{bmatrix} = \begin{bmatrix} 28 \\ 30 \\ 38 \\ 37 \end{bmatrix}$$

Crout's method has essentially the same speed and accuracy as Doolittle's. The inner looping may be simplified for both methods if they are modified to alternate between corresponding rows and columns in the order illustrated below for a 4-by-4 matrix:

1st	5th	7th	9th		1st	2nd	3rd	4th
2nd	6th	11th	13th		5th	6th	8th	10th
3rd	8th	12th	15th		7th	11th	12th	14th
4th	10th	14th	16th		9th	13th	15th	16th

Order for modified   *Order for modified*
Crout's method   *Doolittle's method*

For the modified Crout's method, the subroutine may be written as follows:

```
MCRTLU:  PROCEDURE (N, A, LU);  /*  THIS PROCEDURE FINDS THE LU
         DECOMPOSITION OF AN N BY N MATRIX 'A' USING A MODIFICATION OF
         CROUT'S METHOD.  FOLLOW BY CRTSUB TO SOLVE LINEAR EQUATIONS.
         TO DOUBLE THE AVAILABLE SPACE, USE THE SAME ARGUMENT FOR 'A' AND
         LU IF 'A' IS NOT SUBSEQUENTLY NEEDED.  */
   DECLARE A(*,*), LU(*,*) FLOAT, SUM FLOAT(16);

   /*    INITIALIZE FIRST COLUMN IF DIFFERENT FROM A:  */
   IF ADDR(LU) = ADDR(A)  THEN DO J = 1 TO N;  LU(I,J) = A(I,J);  END;

   /*    DERIVE REMAINDER OF LU:  */
   DO I = 1 TO N-1;
      IP1 = I+1;   PIVOT = LU(I,I);
      DO J = IP1 TO N;
         SUM = A(I,J);
         DO K = 1 TO I-1;   /*  TRANSFORM ROW ELEMENT  */
            SUM = SUM - MULTIPLY (LU(I,K), LU(K,J), 16);
         END;
         LU(I,J) = SUM/PIVOT;
```

```
            SUM = A(J,IP1);
            DO K = 1 TO I;    /*  TRANSFORM COLUMN ELEMENT  */
               SUM = SUM - MULTIPLY (LU(J,K), LU(K,IP1), 16);
            END;
            LU(J,IP1) = SUM;
         END;
      END;
   END;
```

The inner loops may be simplified even further if we form the sums for corresponding row and column elements simultaneously:

```
         DO J = IP1 TO N;
           SUMROW = A(I, J);
           SUMCOL = A(J, IP1);
           DO K = 1 TO I-1;
              SUMROW = SUMROW + MULTIPLY(LU(I, K), LU(K, J), 16);
              SUMCOL = SUMCOL + MULTIPLY(LU(J, K), LU(K, IP1), 16);
           END;
           LU(I, J) = SUMROW/PIVOT;
           LU(J, IP1) = SUMCOL - MULTIPLY(LU(J, I), LU(I, IP1), 16);
         END;
```

However, this second modification may execute slower than the first modification for some implementations, depending on the use of registers.

If there are names for these modifications, the author is unaware of them.

18.4 SCALED PARTIAL PIVOTING

All of the decomposition schemes we have written so far have the serious drawback that they will terminate from **ZERODIVIDE** if any of the diagonal elements happen to be zero. Actually, even a relatively small pivot causes a serious loss of accuracy because it makes the terms involving the pivot row large compared with the coefficients from which they are algebraically subtracted. This causes many digits of these coefficients not to participate, so their accuracy is greatly degenerated.

However, we do not have to pivot with the diagonal element. We may use any element in the same column whose row has not already been used for pivoting. At each stage let us then choose from these the element that has the largest magnitude relative to the remaining elements in its row. This strategy is called *scaled partial pivoting*, and if in spite of it we are trapped into using a zero pivot, then the matrix is singular. The *rank* is the number of pivots we have used before this happens.

Nothing can be done about magnified experimental errors. Double-precision storage of A, LU, B, or X is useless unless A and B are more accurate than their single-precision representations. Inaccurate solutions due to ill-conditioning properly reflect erratic physical systems. However, we shall see that *iterative improve-*

ment can reduce the magnified chopoff error, but it is worth doing only if the magnified chopoff error is greater than the magnified experimental error.

Scaled partial pivoting does not appreciably increase the execution time, but it does complicate the programming. We end up with an LU decomposition that is scrambled or permuted, as is shown in Fig. 18-1. However, L is still suitable for

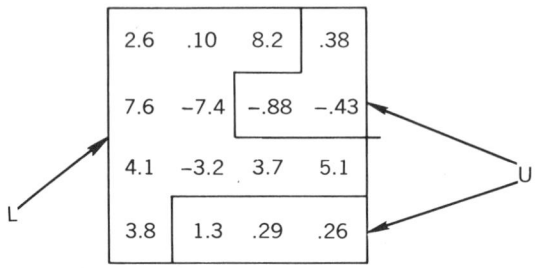

Fig. 18-1 A permuted LU decomposition.

a forward substitution because it will have exactly one row with one element, one row with two elements, and so on up to N. Similarly, U is still suitable for back substitution. However, we must keep a record of the order in which the pivots were chosen so that we do not use any row as a pivot twice and so that we may perform the substitutions in proper order. We may do this by initializing the values of a *permutation* vector to equal their own subscript, PERM(I) = I, then interchanging PERM(J) with PERM(K) whenever the Jth row is taken as the Kth pivot. For example, if we take the third row as the first pivot, then at this stage PERM(1) = 3, PERM(2) = 2, PERM(3) = 1, PERM(4) = 4, ... PERM(N) = N. We may then use the same DO loops that we did without scaled partial pivoting, but using PERM(I), PERM(J), and PERM(K) wherever I, J, and K, respectively, are used as row subscripts. This technique bears a great resemblance to the section on indirect sorting in Case Study 10. Alternatively, we may interchange rows to bring the largest relative pivots to the diagonal, keeping track of the interchanges. Which of the two schemes is more efficient depends upon the implementation. However, it seems probable that indirect addressing is the best approach for Gauss's method and that interchange is the best approach for Crout's or Doolittle's method. Accordingly, we use interchange, leaving the other as an exercise.

With scaled partial pivoting, Crout's method or its modification has a slight advantage over Doolittle's because the latter must temporarily store the whole column in double precision until the best pivot is found to divide it by, or else suffer single-precision divisions. With scaled partial pivoting and a few more refinements, MCRTLU and CRTSUB become:

```
(NOUNDERFLOW):   MCRTLU:   PROCEDURE (N, A, LU, PERM, IER);
     /* THIS PROCEDURE FINDS THE LU DECOMPOSITION OF AN           1   5   7   9
        N BY N MATRIX USING A MODIFICATION OF CROUT'S METHOD      2   6  11  13
        WITH SCALED PARTIAL PIVOTING AND DOUBLE PRECISION         3   8  12  15
        ACCUMULATION OF THE INNER PRODUCTS.  TO REDUCE            4  10  14  16
        THE NUMBER OF DO LOOPS, THE ORDER OF CROUT'S CALCU-
        LATIONS HAS BEEN MODIFIED AS IS INDICATED FOR A 4 BY 4 MATRIX
        ABOVE.  IF A IS NO LONGER NEEDED AFTER DECOMPOSITION, USE THE
        SAME ARGUMENT FOR THE PARAMETERS A AND LU TO PERMIT LARGER
        COEFFICIENT MATRICES.  */
```

```
      DECLARE  /* PARAMETERS */
          N              /* GIVEN:  # OF ROWS AND COLUMNS USED IN A & LU */,
          A(*,*)         /* GIVEN:  MATRIX TO BE DECOMPOSED */,
          LU(*,*) FLOAT  /* RESULT:  LU DECOMPOSITION OF A PERMUTED VERSION
                                    OF A */,
          PERM(*) FIXED BINARY /* RESULT:  PERMUTATION VECTOR INDICATING
                                    ORDER IN WHICH ROWS OF A WERE TAKEN */,
          IER            /* RESULT:  IER = -1 IF ROW 1 HAS ALL ZEROS;
                                    IER = I IF A IS SINGULAR OR NEAR-SINGULAR WITH
                                    RANK I;  IER = 0 IF ALL IS WELL. */;

      DECLARE  /* INTERNAL VARIABLES: */
          WEIGHT(N)      /* ROW WEIGHTS USED TO HELP DETERMINE PIVOTS */,
          (SUMROW, SUMCOL, PIVOT) FLOAT(16)  /* DOUBLE PRECISION INNER
                                    PRODUCTS AND PIVOT.  PRECISION HERE AND IN THE
                                    MULTIPLY FUNCTION ARE IMPLEMENTATION
                                    DEPENDENT */,
          NEXT LABEL INITIAL(DIFFERENT);

      /* INITIALIZE PERM AND LU;  FIND ROW WEIGHTS AND FIRST PIVOTAL
         ROW: */
      ON ZERODIVIDE  GO TO ZEROROW;
      IF ADDR(LU) = ADDR(A)  THEN NEXT = SAME;
      BIGGEST, IER = 0;
      DO I = 1 TO N;
          PERM(I) = I;  ROWMAX = 0;  J = N;
       DIFFERENT:  LU(I,J) = A(I,J);
       SAME:  AB = ABS(A(I,J));  ROWMAX = MAX(ROWMAX, AB);
          J = J-1;
          IF J > 0  THEN GO TO NEXT;
          WEIGHT(I) = ROWMAX;  AB = AB/ROWMAX;
          IF AB > BIGGEST  THEN DO;  IROW = I;  BIGGEST = AB;  END;
      END;
      PIVOT = A(IROW,1);

         /* DERIVE REMAINDER OF LU KEEPING TRACK OF INTERCHANGES: */
         ON ZERODIVIDE  GO TO SINGULAR;  ON OVERFLOW  GO TO SINGULAR;
         DO I = 1 TO N-1;
             IF IROW ¬= I  THEN DO;
                 K = PERM(I);  PERM(I) = PERM(IROW);  PERM(IROW) = K;
                 AB=WEIGHT(I);  WEIGHT(I)=WEIGHT(IROW);  WEIGHT(IROW)=AB;
                 DO J = 1 TO N;
                     AB = LU(I,J);  LU(I,J) = LU(IROW,J);  LU(IROW,J) = AB;
                 END;  END;
             BIGGEST = 0;  IPLUS1 = I+1;
             DO J = IPLUS1 TO N;
                 SUMROW, SUMCOL = 0;
                 DO K = 1 TO I-1;
                     SUMROW = SUMROW + MULTIPLY(LU(I,K), LU(K,J), 16);
                     SUMCOL = SUMCOL + MULTIPLY(LU(J,K), LU(K,IPLUS1), 16);
                 END;
                 LU(I,J) = (LU(I,J) - SUMROW)/PIVOT;
                 AB, LU(J,IPLUS1) = LU(J,IPLUS1) -
                     (SUMCOL + MULTIPLY(LU(J,K), LU(K,IPLUS1), 16));
                 AB = ABS(AB)/WEIGHT(J);
                 IF AB > BIGGEST  THEN DO;  IROW = J;  BIGGEST = AB;  END;
             END;
             PIVOT = LU(IROW,IPLUS1);
         END;  IF PIVOT ¬= 0  THEN RETURN;
      SINGULAR:  IER = I;  RETURN;
      ZEROROW:  IER = -1;
      END;
```

```
        /* INITIALIZE PERM AND LU; FIND ROW WEIGHTS AND FIRST PIVOTAL
           ROW: */
        ON ZERODIVIDE   GO TO ZEROROW;
        IF ADDR(LU) = ADDR(A)   THEN NEXT = SAME;
        BIGGEST, COND, IER = 0;
        DO I = 1 TO N;
            PERM(I) = I;   ROWMAX = 0;   J = N;
            DIFFERENT:   LU(I,J) = A(I,J);
            SAME:   AB = ABS(A(I,J));   ROWMAX = MAX(ROWMAX, AB);
              J = J-1;
              IF J > 0   THEN GO TO NEXT;
              WEIGHT(I) = ROWMAX;   AB = AB/ROWMAX;
              IF AB > BIGGEST   THEN DO;   IROW = I;   BIGGEST = AB;   END;
        END;
        PIVOT = A(IROW,1);
        /* DERIVE REMAINDER OF LU, UPDATING COND AND KEEPING TRACK OF
           ROW INTERCHANGES: */
        ON ZERODIVIDE   GO TO SINGULAR;   ON OVERFLOW   GO TO SINGULAR;
        DO I = 1 TO N-1;
            IF IROW ¬= I   THEN DO;
                K = PERM(I);   PERM(I) = PERM(IROW);   PERM(IROW) = K;
                AB=WEIGHT(I);   WEIGHT(I)=WEIGHT(IROW);   WEIGHT(IROW)=AB;
                DO J = 1 TO N;
                    AB = LU(I,J);   LU(I,J) = LU(IROW,J);   LU(IROW,J) = AB;
                END;   END;
            BIGGEST = 0;   IPLUS1 = I+1;
            DO J = IPLUS1 TO N;
                SUMROW = 0;   SUMCOL = 0;
                DO K = 1 TO I-1;
                    SUMROW = SUMROW + MULTIPLY(LU(I,K), LU(K,J), 16);
                    SUMCOL = SUMCOL + MULTIPLY(LU(J,K), LU(K,IPLUS1), 15);
                END;
                LU(I,J) = (LU(I,J) - SUMROW)/PIVOT;
                SAVE = LU(J,IPLUS1);
                AB = SAVE - (SUMCOL + MULTIPLY(LU(J,K), LU(K,IPLUS1), 15));
                LU(J,IPLUS1) = AB;   AB = ABS(AB)/WEIGHT(J);
                IF AB > BIGGEST   THEN DO;   IROW = J;   BIGGEST = AB;
                    ORIGINAL = SAVE;   END;
            END;
            PIVOT = LU(IROW,IPLUS1);
            COND = MAX(COND, ABS(ORIGINAL/PIVOT));
        END;
        RETURN;
    ZEROROW:   IER = -1;   RETURN;
    SINGULAR:   IER = I;
END;

(NOUNDERFLOW):   CRTSUB:   PROCEDURE (N, LU, PERM, B, X);
    /* THIS PROCEDURE FINDS THE SOLUTION X TO THE SIMULTANEOUS EQUATIONS
       L*U*X = A'*X = B', WHICH IS THE SAME AS THE SOLUTION TO A*X = B.
       A' AND B' ARE PERMUTED VERSIONS OF A AND B ACCORDING TO THE PER-
       MUTATION VECTOR PERM.  L AND U ARE STORED IN THE MATRIX LU.  LU
       AND PERM MUST BE DERIVED FROM THE MATRIX A BY THE PROCEDURE
       MCRTLU.  DO NOT USE THE SAME ARGUMENT FOR B AND X. */
    DECLARE   /* PARAMETERS: */
        N               /* GIVEN:   NUMBER OF ROWS AND COLUMNS IN LU */,
        LU(*,*) FLOAT   /* GIVEN:   DECOMPOSITION OF 'A' */,
        PERM(*) FIXED BINARY /* GIVEN:   PERMUTATION VECTOR INDICATING
                            THE ORDER IN WHICH THE ELEMENTS OF B MUST BE
                            USED */,
        B(*)            /* GIVEN:   VECTOR OF CONSTANT TERMS FOR WHICH THE
                            EQUATIONS A*X = B ARE BEING SOLVED */,
        X(*)            /* RESULT:   VECTOR OF SOLUTIONS */;
```

```
    DECLARE /* INTERNAL VARIABLES: */
       SUM FLOAT(16) /* DOUBLE PRECISION INNER PRODUCT.  PRECISION HERE
                        AND IN THE MULTIPLY FUNCTION ARE IMPLEMENTATION
                        DEPENDENT. */;
    DO I = 1 TO N;
       SUM = 0;
       DO J = 1 TO I-1;
          SUM = SUM + MULTIPLY (LU(I,J), X(J), 16);
       END;
       X(I) = (B(PERM(I)) - SUM)/LU(I,I);
    END;
    DO I = N-1 TO 1 BY -1;
       SUM = 0;
       DO J = I+1 TO N;
          SUM = SUM + MULTIPLY (LU(I,J), X(J), 16);
       END;
    END;
END;
```

The **NOUNDERFLOW** condition prefix is discussed in exercise 29 in Case Study 7.

MCRTLU and CRTSUB provide a powerful, flexible combination that can be used in many different ways. For example, the flow chart in Fig. 18-2 indicates how we could use them to solve for any number of B vectors and save a copy of LU and PERM for possible future use:

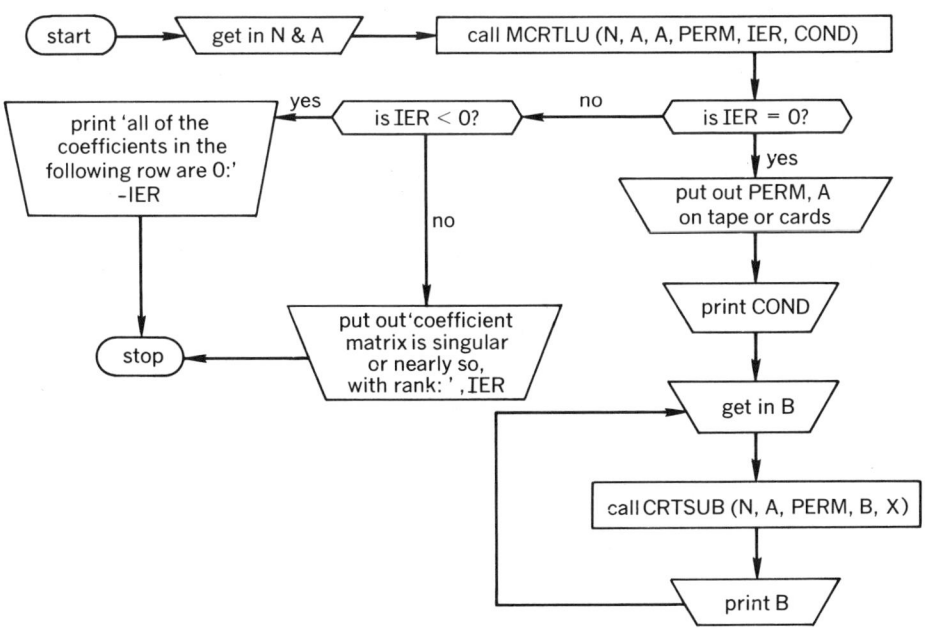

Fig. 18-2 Flow chart for one use of MCRTLU and CRTSUB.

18.5 ITERATIVE REFINEMENT

After calculating a solution X, we may multiply it by A to find out how much this product differs from B due to chopoff error:

$$R = B - A \cdot X$$

This R vector is called the residual vector, and it is usually small compared with B even if the matrix is ill-conditioned. R would be zero if our computer had infinite precision. We may now find the solution to the system

$$A \cdot C = R$$

If C were exact, then X + C would be the exact solution to our original equations because

$$A \cdot (X + C) = A \cdot X + A \cdot C = A \cdot X + R = B$$

We may regard C as a correction to X. Actually, C is not exact, but since R is small compared with B, C is small compared to X; so the absolute errors in C will be small compared with those in X. Moreover, we may reduce the errors even further by assigning X + C to X and repeating the whole process—deriving a second residual and a second correction. We may iterate in this fashion until the relative corrections become less than some specified relative tolerance.

It is important to note, however, that iterative refinement may increase the residuals while it decreases the error in the solution; and often we are interested in finding an X such that A·X is as close as possible to B rather than an X which is as close as possible to the solution of A·X = B.

Iterative refinement is usually not necessary for procedures which use double-precision accumulation of inner products. When necessary, it should take less than 25 per cent of the time required for decomposition. The following procedure implements this technique, using **CRTSUB** to perform the solutions:

```
(NOUNDERFLOW):  CRTMPV:  PROCEDURE (N,A,LU,PERM,B,X,RELTOL,DIGITS);

    /* THIS PROCEDURE USES ITERATIVE REFINEMENT TO REFINE THE SOLUTION
       OF THE LINEAR EQUATIONS 'A*X = B' UNTIL THE MAXIMUM RELATIVE
       CHANGE IN TWO SUCCESSIVE ITERATES HAS AN ABSOLUTE VALUE LESS THAN
       RELTOL.  IF THIS ACCURACY CANNOT BE ACHIEVED IN 16 ITERATIONS,
       RELTOL IS SET TO THE TOLERANCE THAT HAS BEEN ACHIEVED.  THIS PRO-
       CEDURE MAY BE USEFUL WHEN ACCURATE SOLUTIONS ARE MORE IMPORTANT
       THAN SMALL RESIDUALS AND N*1E-6 IS GREATER THAN THE RELATIVE
       EXPERIMENTAL ERRORS IN THE ELEMENTS OF 'A' AND 'B'.  OTHERWISE,
       ROUNDOFF ERROR IS NEGLIGIBLE COMPARED TO THE NATURAL MAGNIFICA-
       TION OF EXPERIMENTAL ERRORS, AND USE OF CRTMPV TO GET AN
       ACCURATE MATHEMATICAL SOLUTION TO A PHYSICALLY INACCURATE SYSTEM
       IS MISLEADING.   */

    /* NOTE:  THIS PROCEDURE REQUIRES THE ORIGINAL 'A' AND 'B' AND THE
       PROCEDURE CRTSUB.  */
```

Sec. 18.5 *Iterative Refinement*

```
DECLARE   /* PARAMETERS: */
   N                    /* GIVEN:  THE NUMBER OF EQUATIONS AND UNKNOWNS */,
   A(*,*)               /* GIVEN:  THE COEFFICIENTS OF THE UNKNOWNS */,
   LU(*,*) FLOAT        /* GIVEN:  THE LU DECOMPOSITION OF PERMUTED 'A' */,
   PERM(*) FIXED BINARY /* GIVEN:  THE ORDER OF THE PIVOTAL ROWS */,
   B(*)                 /* GIVEN:  THE CONSTANT TERMS OF THE EQUATIONS */,
   X(*)                 /* GIVEN-RESULT:  THE SOLUTION WHICH IS REFINED */,
   RELTOL               /* GIVEN-RESULT:  AN INDICATION OF THE RELATIVE
                           ERROR WHICH IS DESIRED FOR X.  IF THIS ERROR IS
                           NOT ACHIEVED WITHIN 16 ITERATIONS, THEN RELTOL
                           IS SET TO AN INDICATION OF WHAT ACCURACY WAS
                           ACHIEVED. */,
   DIGITS               /* RESULT:  APPROXIMATE NUMBER OF SIGNIFICANT
                           DIGITS IN UNREFINED SOLUTION.  */;

DECLARE   /* INTERNAL VARIABLES: */
   R(N)              /* THE RESIDUALS 'B - A*X' */,
   DX(N)             /* THE CHANGES IN THE SOLUTIONS */,
   SUM FLOAT(16)     /* IMPLEMENTATION DEPENDENT DOUBLE PRECISION INNER
                        PRODUCT */,
   NORMX FLOAT       /* THE MAXIMUM ABSOLUTE COMPONENT OF X */,
   NORMDX FLOAT      /* THE MAXIMUM ABSOLUTE CHANGE IN AN X COMPONENT*/;
NORMX = 0;
DO I = 1 TO N;
   NORMX = MAX (NORMX, ABS(X(I)));
END;
IF NORMX = 0 THEN DO;  DIGITS = 6;  RETURN;  END;

DO ITER = 1 TO 16;
   DO I = 1 TO N;
      SUM = 0;
      DO J = 1 TO N;
         SUM = SUM + MULTIPLY(A(I,J), X(J), 16);
      END;
      R(I) = B(I) - SUM;
   END;
   CALL CRTSUB(N, LU, PERM, R, DX);
   NORMDX = 0;
   DO I = 1 TO N;
      NORMDX = MAX (NORMDX, ABS(DX(I)));
      X(I) = X(I) + DX(I);
   END;
   IF ITER = 1 THEN DIGITS = -LOG10 (MAX (1E-6, NORMDX/NORMX));
   IF NORMDX <= RELTOL*NORMX  THEN RETURN;
END;
RELTOL = NORMDX/NORMX;
END;
```

The subroutines we have developed here are quite powerful, and although they do not need to be thoroughly understood to be used, they require too much instruction for the novice. Consequently, it would be nice to combine them into an easy-to-use package with all of the sophistication transparent to the user. To avoid misuse, a package such as this should be as accurate as possible, it should not destroy the original A or B, and it should automatically put out an appropriate message and terminate in case of singularity or a zero row. This package may be slower, more wasteful of space, and less flexible than necessary; but this is the price of a foolproof, easy-to-use subroutine:

```
LINEAR:  PROCEDURE (N, A, B, X);
   /* THIS PROCEDURE SOLVES THE GENERAL SET OF SIMULTANEOUS LINEAR
         EQUATIONS:     A(1,1)*X(1) + A(1,2)*X(2) + ... A(1,N)*X(N) = B(1)
                        A(2,1)*X(1) + A(2,2)*X(2) + ... A(2,N)*X(N) = B(2)
                            .              .                .           .
                            .              .                .           .
                            .              .                .           .
                        A(N,1)*X(1) + A(N,2)*X(2) + ... A(N,N)*X(N) = B(N)
```

WHERE
THE XS ARE THE UNKNOWNS, THE AS ARE THEIR COEFFICIENTS AND THE
BS ARE CONSTANTS. THE AS AND BS ARE NOT ALTERED BY THIS PRO-
CEDURE. WHEN A ROW OF COEFFICIENTS ARE ALL ZEROS OR WHEN THE
MATRIX OF COEFFICIENTS IS SINSULAR OR VERY NEARLY SO, AN APPROP-
RIATE MESSAGE IS PRINTED AND THE ERROR CONDITION IS RAISED
BECAUSE A SOLUTION IS IMPOSSIBLE. OTHERWISE, THE SOLUTION WILL
BE ACCURATE TO DEFAULT PRECISION. THIS PROCEDURE IS DESIGNED TO
BE ESPECIALLY ACCURATE, FOOLPROOF, AND EASY TO USE. HOWEVER, THE
THREE PROCEDURES THAT IT CALLS MAY BE USED DIRECTLY FOR GREATER
EFFICIENCY IF ANY OF THE FOLLOWING CONDITIONS PREVAIL:
  1. THE SAME AS ARE TO BE USED, EITHER WITHIN ONE PROGRAM OR
     OVER A PERIOD OF TIME, WITH MORE THAN ONE SET OF BS.
  2. A MAGNIFIED RELATIVE ROUNDOFF ERROR OF GREATER THAN 1E-6 IS
     ACCEPTABLE.
  3. THE COEFFICIENTS WILL NOT ALL FIT IN THE MEMORY, BUT HALF
     THEIR NUMBER WOULD, AND ALTERATION OF THE COEFFICIENTS
     DURING THE SOLUTION IS ACCEPTABLE.
  4. SMALL RESIDUALS ARE REQUIRED RATHER THAN ACCURATE
     SOLUTIONS, WHERE THE ITH RESIDUAL IS B(I) - A(I,1)*X(1) -
     A(I,2)*X(2) - ... A(I,N)*X(N).
FOR DETAILS SEE DAVID R. STOUTEMYER, 'PL/I PROGRAMMING FOR
ENGINEERING AND SCIENCE', (ENGLEWOOD CLIFFS, N.J.: PRENTICE
HALL, INC., 1971), CASE STUDY 18. */

/* NOTE: THIS PROCEDURE REQUIRES THE PROCEDURES MCRTLU, CRTSUB, AND
CRTMPV; SO BE SURE THEY ARE INCLUDED EITHER WITHIN THIS PROC-
EDURE OR EXTERNALLY. */

DECLARE   /* PARAMETERS: */
   N   /* GIVEN: THE NUMBER OF EQUATIONS AND UNKNOWNS */,
   A(*,*)   /* GIVEN: THE COEFFICIENTS OF THE UNKNOWNS */,
   B(*)   /* GIVEN: THE CONSTANT TERMS OF THE EQUATIONS */,
   X(*)   /* RESULT: THE SOLUTIONS */;

   DECLARE /* INTERNAL VARIABLES: */
      PERM(N) FIXED BINARY /* ORDER IN WHICH PIVOTAL ROWS ARE TAKEN */,
      LU(N,N) FLOAT   /* LU DECOMPOSITION OF PERMUTED A */;
   CALL MCRTLU (N, A, LU, PERM, IER);
   IF IER < 0 THEN DO; PUT LIST
      ('ALL OF THE COEFFICIENTS IN THE FOLLOWING ROW ARE ZEROS', -IER);
      SIGNAL ERROR; END;
   IF IER > 0 THEN DO; PUT LIST
      ('COEFFICIENT MATRIX IS SINGULAR OR NEARLY SO WITH RANK:', IER);
      SIGNAL ERROR; END;
   CALL CRTSUB (N, LU, PERM, B, X);
   RELTOL = 1E-6;
   CALL CRTMPV (N, A, LU, PERM, B, X, RELTOL, DIGITS);
   IF RELTOL > 1E-6 THEN DO; PUT LIST
      ('ITERATIVE IMPROVEMENT DID NOT CONVERGE. COEFFICIENT MATRIX IS
SINGULAR OR NEARLY SO'); SIGNAL ERROR; END;
END;

18.6 COMPUTING THE DETERMINANT

The determinant of A turns out to be simply the diagonal product of LU. The diagonal elements of LU correspond to the pivot elements of LU, and rows would have to be interchanged to bring them to the diagonal. Each such interchange multiplies the determinant by -1; so

$$\text{DET}(A) = (-1)^K \sum_{I=1}^{N} \text{LU}(\text{PERM}(I), I)$$

where K is the number of interchanges required to bring the pivots to the diagonal. Actually, this number is not unique, but it will always be either odd or even for a given permutation. This number may be determined by counting the number of interchanges while sorting **PERM** into increasing order. The pivot product is then multiplied by -1 if this number is odd. Note that although a zero determinant implies a singular matrix, a small magnitude determinant does not necessarily imply an ill-conditioned matrix, and a large determinant does not necessarily imply a well-conditioned matrix.

18.7 COMPUTING THE INVERSE

The inverse matrix A^{-1} is defined as the matrix whose product with A gives the identity matrix "I" with all zeros except for "ones" along the diagonal:

$$\begin{matrix} 1 & 0 & 0 & \cdots & 0 \\ 0 & 1 & 0 & \cdots & 0 \\ 0 & 0 & 1 & \cdots & 0 \\ \cdot & \cdot & \cdot & & \cdot \\ \cdot & \cdot & \cdot & & \cdot \\ \cdot & \cdot & \cdot & & \cdot \\ 0 & 0 & 0 & \cdots & 1 \end{matrix}$$

We can see that each column of the inverse matrix is given by the solution to the linear equations with the corresponding column of I as B. Thus, we may use the following subroutine to derive the inverse:

```
CRTINV:   PROCEDURE (N, LU, PERM, AINV);   /* THIS PROCEDURE FINDS THE
          INVERSE AF A MATRIX GIVEN ITS PERMUTED LU DECOMPOSITION AS
          CALCULATED BY MCRTLU. */
          DECLARE LU(*,*) FLOAT, PERM(*) FIXED BINARY, AINV(*,*), X(N), B(N);
          DO J = 1 TO N;
             B = 0;   B(J) = 1;
             CALL CRTSUB (N, LU, PERM, B, X);
             DO I = 1 TO N;
                AINV(I,J) = X(I);
             END;
          END;
END;
```

We may regard the identity matrix as a sort of generalized "one," and we may regard the inverse matrix as a sort of generalized reciprocal:

$$A^{-1} = \frac{I}{A}$$

This inverse does not exist when A is singular; so we may regard a singular matrix as a sort of generalized zero. Once we have the inverse, we may obtain the solution for any B vector by the product

$$X = A^{-1}B$$

However, inversion followed by multiplication is less accurate and more time-consuming than direct use of the decomposition because more steps are involved. As we have pointed out, multiplication takes essentially the same time as forward plus backward substitution and the inverse takes a total of three times as long as decomposition.

18.8 SYMMETRIC, POSITIVE DEFINITE MATRICES

When a matrix is symmetric, the decomposition time may be cut almost in half, because then for Crout's decomposition,

$$U(I, J) = L(J, I)/L(I, I)$$

However, to take advantage of this, we must sacrifice our pivoting strategy because it destroys any symmetry. This sacrifice is too dangerous to risk unless we know in advance that the matrix is also *positive definite*. The diagonal elements are good pivot choices for such matrices, and for many applications we know in advance that the matrix is this type. When this is so, we may cut the space as well as the work in half by using a different decomposition due to *Cholesky*. LU as well as A is symmetric for this decomposition. Because of this, U is said to be the *transpose* of L:

$$U = L^T$$

The elements of L may be calculated by the following formulas:

$$L(I, J) = \frac{A(I, J) - \sum_{K=1}^{J-1} L(I, K) * L(J, K)}{L(J, J)} \quad \text{for } J = 1 \text{ to } I - 1$$

$$L(I, I) = \sqrt{A(I, I) - \sum_{J=1}^{I-1} L(I, J)^2}$$

We may evaluate the elements in compacted natural order indicated by the pattern below for a 4-by-4 example:

```
1st
2nd   3rd
4th   5th   6th
7th   8th   9th   10th
```

To avoid the square roots, Wilkinson suggests an alternate decomposition for positive definite symmetric matrices into a diagonal matrix D, a lower triangular

Sec. 18.8 Symmetric, Positive Definite Matrices

matrix with unit diagonal L, and its transpose L^T:[1]

$$A = L \cdot D \cdot L^T$$

Using implied unit diagonals, if we store D, L, and L^T in the same square matrix LDLT, the elements may be derived by the following formulas:

$$\left. \begin{array}{l} T(J) = A(I, J) - \sum_{K=1}^{J-1} T(K)*LDLT(J, K) \\ LDLT(I, J) = T(J)/LDLT(J, J) \end{array} \right\} \text{ for } J = 1 \text{ to } I - 1$$

$$LDLT(I, I) = A(I, I) - \sum_{J=1}^{I-1} T(J)*LDLT(I, J)$$

where T is a temporary storage vector.

To take advantage of the storage savings, we must store A and LDLT in compacted form as is described in Case Study 12. When this is done, the corresponding decomposition and forward–backward substitution subroutines are:

```
WKLDLT: PROCEDURE (N, A, LDLT);  /* THIS PROCEDURE USES WILKENSON'S
   METHOD TO FIND THE LDLT DECOMPOSITION OF AN N BY N POSITIVE DEF-
   INATE SYMMETRIC MATRIX A.  THE LOWER TRIANGULAR ELEMENTS OF A AND
   LDLT ARE STORED IN COMPACTED FORM:  TO SAVE ADDITIONAL SPACE, USE
   THE SAME ARGUMENT FOR A AND LDLT IF A IS NO LONGER NEEDED AFTER
   DECOMPOSITION.  */
DECLARE A(*), LDLT(*) FLOAT, TEMP(N), SUM FLOAT(16);
M = 0;
DO I = 1 TO N;
   J = 1;
   S10:  M = M+1;
   SUM = A(M);
   DO K = 1 TO J-1;
      SUM = SUM - MULTIPLY(TEMP(K), LDLT(.5*J*(J-1) + K), 16);
   END;
   TEMP(J) = SUM;
   IF J = I  THEN GO TO S20;
   LDLT(M) = SUM/LDLT(.5*J*(J+1));
   J = J+1;
   GO TO S10;
   S20:  M = M+1;  SUM = A(M);  JO = .5*J*(J-1);
      SUM = SUM - MULTIPLY (TEMP(K), LDLT(JO+K), 16);
      LDLT(M) = SUM/LDLT(JO+J);
END;
END;

WLKSUB: PROCEDURE (N, LDLT, B, X);  /* THIS PROCEDURE SOLVES THE
   LINEAR EQUATIONS AX = B FOR POSITIVE DEFINATE, SYMMETRIC A,
   GIVEN ITS LDLT DECOMPOSITION STORED IN COMPACTED FORM.  TO
   SAVE SPACE, USE THE SAME ARGUMENT FOR B AND X IF B IS NO LONGER
   NEEDED AFTER THE SOLUTION.  */
DECLARE LDLT(*) FLOAT, B(*), X(*) SUM FLOAT(16);
```

---

[1] Anthony Ralston and Herbert S. Wilf, *Mathematical Methods for Digital Computers Vol. 2* (New York: John Wiley & Sons, Inc., 1967), p. 72.

```
    /* FORWARD SUBSTITUTION:   */
    M = 1;
    DO I = 1 TO N;
       SUM = 0;
       DO J = 1 TO I-1;
          SUM = SUM + MULTIPLY (LDLT(M), X(J), 16);
          M = M + 1;
       END;
       X(I) = B(I) - SUM;
       M = M + 1;
    END;

    /* DIAGONAL DIVISION:   */
    M = 0;
    DO I = 1 TO N;
       M = M + 1;
       X(I) = X(I)/LDLT(M);
    END;

    /* BACK SUBSTITUTION:   */
    DO I = N-1 TO 1 BY -1;
       SUM = 0;
       DO J = I+1 TO N;
          SUM = SUM + MULTIPLY (LDLT(.5*J*(J-1)+I), X(J), 16);
       END;
       X(I) = X(I) - SUM;
    END;
END;
```

EXERCISES

1. Run versions of MCRTLU with indirect addressing and with row interchanges to see which is faster on your implementation.

2. Run versions of MCRTLU with both modifications to see which is faster on your implementation.

3. Write subroutines for Gaussian decomposition and substitution with scaled partial pivoting.

▼4. Adapt Wilkinson's method for tridiagonal matrices, as described in exercise 25 in Case Study 12.

5. If determinants are required frequently, it is a good idea to include a result parameter in the decomposition which has the value ± 1 according to whether the number of interchanges is odd or even. Modify MCRTLU accordingly and write a companion procedure DET which evaluates the determinant using the results of MCRTLU.

▼6. The following procedure uses Gauss-Jordan elimination, which is described in exercise 9 in Case Study 12, to invert a matrix in its own space—thus permitting the inversion of larger matrices. Modify the procedure to include scaled partial pivoting.

```
GJINV:  PROCEDURE (N, A);
    /* A IS REPLACED BY ITS INVERSE SHIFTED 1 COLUMN TO THE RIGHT */
    DECLARE A(*,*);

    /* SET N+1'TH COLUMN TO N'TH COLUMN OF IDENTITY MATRIX: */
    NP1 = N+1;   A(N,NP1) = 1;
    DO I = 1 TO N-1;  A(I,NP1) = 0;   END;

    DO J = N TO 1 BY -1;
        PIVOT = A(J,J);
        /* DIVIDE PIVOT ROW BY PIVOT: */
        DO K = 1 TO NP1;  A(J,K) = A(J,K)/PIVOT;   END;

        /* SUBTRACT MULTIPLE OF PIVOT ROW FROM EACH OF THE OTHER ROWS: */
        DO I = 1 TO J-1, J+1 TO N;
            F = A(I,J);
            IF F ¬= 0   THEN DO K = 1 TO N-1;
                A(I,K) = A(I,K) - F*A(J,K);   END;
        END;

        /* CORRECT 2 ELEMENTS TO GET J-1'TH COLUMN OF IDENTITY MATRIX IN
           COLUMN J: */
        A(J,J) = 0;   A(J-1,J) = 1;
    END;
END;
```

7. Write a subroutine for Cholesky's decomposition and substitution using matrices stored in compacted form.

▼8. Adapt Wilkinson's method for band structured matrices as described in exercise 26 in Case Study 12.

APPENDIX

Answers to Selected Exercises

```
/* 1-1: */
ENERGY: PROCEDURE  OPTIONS (MAIN);
    E = 0.0001*0.0025/2;
    PUT DATA (E);
END;

/* 1-4: */
SOLVE: PROCEDURE  OPTIONS (MAIN);
    B = (.062 + 18.6*57.3)/17.6;
    PUT DATA (B);
END;

/* 1-6:
REACT AND H2004 ARE VALID FOR THE F COMPILER.
*/

/* 1-7:
1.68E-4 IS VALID, REPRESENTING .000168;   76E3.5 IS INVALID.
*/

/* 1-9:
DECLARE (MAGNITUDE, INTERVAL, LAG) FLOAT, (PERCENTILE, SCORE) FIXED;
*/

/* 1-10:
FMAGNITUDE, FINTERVAL, FLAG, IPERCENTILE, ISCORE
*/

/* 1-14:
A*(B+C)
*/
```

/* 1-15:   A. $\dfrac{a + 2 \times 10^{14}}{b^{-n}}$ B.   MEANINGLESS

```
*/

/* 1-16:
3.6E7 + X/(Y - Z**-5/(6 + Q))
*/

 /* 2-1: */
T:  PROCEDURE (S);
    TERM = (1800 - S)/133;
    RETURN (61.4/(TERM + 1/TERM));
END;

TORQUE: PROCEDURE  OPTIONS (MAIN);
    T1 = T(0E0);    T2 = T(450E0);
    T3 = T(1710E0); T4 = T(1800E0);
    PUT DATA (T1, T2, T3, T4);
END;

/* 2-3: */
WAVE_L: PROCEDURE (N, M);
    RETURN (911.8/(1E0/(N*N) - 1/(M*M)));
END;
SPCTRM: PROCEDURE  OPTIONS (MAIN);
    N = 2;   M1 = 3;   M2 = 4;   M3 = 5;
    W1 = WAVE_L(N, M1);   W2 = WAVE_L(N, M2);   W3 = WAVE_L(N,M3);
    PUT DATA (W1, W2, W3);
END;
```

```
/* 2-6:  C = Y INTERCEPT = .8;   B = SLOPE AT Y INTERCEPT = 1.7;
   TO PASS THRU 4'TH POINT,  2.5 = A*4.4**2 + 1.7*4.4 + .8;  SO
   A = (2.5 - .8 -1.7*4.4)/4.4**2 = -.3; */

EXPER:  PROCEDURE (X);
   RETURN (-.3*X*X + 1.7*X +.8);
END;

/* 2-7: */
L:  PROCEDURE (N, M) FLOAT;
   RETURN (911.8/(1E0/(N*N) - 1/(M*M)));
END;

SPCTRM:  PROCEDURE OPTIONS (MAIN);
   DECLARE L ENTRY (FIXED BINARY, FIXED BINARY) RETURNS (FLOAT);
   S1 = L(2,3);   S2 = L(2,4);   S3 = L(2,5);
   PUT DATA (S1, S2, S3);
END;

/* 2-8 N:
Q = 1/TAND(A);
*/

/* 2-9:  WHEN X IS NEAR ZERO, EXP(X) AND EXP(-X) ARE BOTH NEARLY 1
   CAUSING A LARGE LOSS OF ACCURACY IF THEIR DIFFERENCE IS TAKEN.  */

/* 2-11: */
ASIND:  PROCEDURE (Y);
   RETURN (ATAND(Y, SQRT(1-Y*Y)));
END;

/* 2-14: */
SINE:  PROCEDURE (A);
   ASQ = A*A;
   RETURN (((.007603*ASQ - .16596)*ASQ + .999892)*A);
END;

BESSEL:  PROCEDURE (X);
   XSQ = X*X;
   RETURN((-6.09*XSQ + 345.5)/((XSQ + 2.56)*XSQ + 345.5));
END;

/* 2-15: */
BESSEL:  PROCEDURE (X);
   XSQ = X+X;
   RETURN (-6.09/(XSQ - 56.3 + 3550/(XSQ + 56.9)));
END;

/* 2-16: */
Y:  PROCEDURE (X);
   P = X*(X+1)+1;  /* 3 OCCURRENCES OF Q+1 GIVES ADDITIONAL SAVING  */
   RETURN(((P+X)*P + 1)*P);
END;

/* 3-3   D & F COMPILERS:
   K = 2   A = 6.00000E+00   M = 4   K = 2   A = 8.00000E+00   M = 4
   K = 2   A = 1.00000E+00   M = 4   K = 3   A = 6.00000E+00   M = 9
   ENDFILE MESSAGE                                            */
```

```
/* 3-7:   */
DISTRT:   PROCEDURE   OPTIONS (MAIN);
   /* FOR ANY NUMBER OF TRIANGLES, THIS PROCEDURE EVALUATES THE
      PERCENT DISTORTION SA IN ANGLE A DUE TO PERCENT DISTORTIONS
      OF SX, SY, AND SZ IN THE 3 SIDES WHICH ARE OPPOSITE ANGLES
      A, B, AND RESPECTIVELY.   DATA FOR EACH TRIANGLE SHOULD CONSIST
      OF SX, SY, SZ, C, AND B -- THE LATTER TWO IN DEGREES. */
      PUT PAGE LIST ('SX', 'SY', 'SZ', 'C', 'B', 'SA');
   IN:   GET LIST (SX, SY, SZ, C, B);
      SA = (SX - SY)/TAND(C) + (SX - SZ)/TAND(B);
      SA = (SX - SY)/TNAD(C) + (SX - SZ)/TAND(B);
      PUT SKIP LIST (SX, SY, SZ, C, B, SZ);
      GO TO IN;
END;

/* 3-10:   DECLARE X LABEL;
         X = A;
      COMMON:
         ...
         GO TO X;
      A:
         ...
         X = B;   GO TO COMMON;
      B:
         ...
         X = A;   GO TO COMMON;         */

/* 4-1: */
ABS:   PROCEDURE (X);
   IF X < 0   THEN RETURN(-X);
   RETURN(X);
END;

/* 4-4: */
```

ENTER WITH X → N = X → IS N <= 0? → N = N - 1
                        NO ↓   YES
                        RETURN WITH N

```
FLOOR:   PROCEDURE (X);
   N = X;
   IF N >= 0   THEN RETURN (N);
   IF N ¬= X   THEN N = N-1;
   RETURN (N);
END;

/* 4-7: */
THEMOD:   PROCEDURE (A, B);
   X = A/B;
   N = FLOOR (X);
   RETURN ((X-N)*B);
END;

/* 4-10:   A*MAX(0,T-T0);         */

/* 4-13:   A*MOD(T,P),   A*MOD(-T,P)   */

/* 4-15:   FCN(T1 + MOD(T-T1, T2-T1))   */

/* 4-16:   A1*MAX(T-T1) + A2*MAX(T-T2) + A3*MAX(T-T3)   */

/* 4-17:   MIN (LAMBDAMAX, MAX(-LAMBDAMAX, L*I))   */

/* 4-23: */
```

```
NQUAD:   PROCEDURE (X, Y);
     IF X >= 0   THEN GO TO ONEORFOUR;
     IF Y >= 0   THEN RETURN (2);
     RETURN (3);
   ONEORFOUR:  IF Y >= 0   THEN RETURN (1);
     RETURN (4);
END;

/* 4-25: */
SINE:   PROCEDURE (A);
     ADUP = A;  PI = 3.14159;   TWOPI = 6.28319;   PINEG = -3.14159;
   S10:   IF ADUP < PI   THEN GO TO S20;
     ADUP = ADUP - TWOPI;   GO TO S10;
   S20:   IF ADUP < PINEG  THEN GO TO S30;
     ADUP = ADUP + TWOPI;   GO TO S20;
   S30:   IF ADUP > 1.57076   THEN ADUP = PI - ADUP;
     IF ADUP < -1.57080   THEN ADUP = PINEG - ADUP;
     ASQ = ADUP*ADUP;
     RETURN(((.007603*ASQ - .16596)*ASQ + .999892)*ADUP);
END;

/* 4-30: */
IN:   PROCEDURE (X, Y);
   IF Y<=.5*X+2 & Y>=-.5*X+3 & Y>=2*X-8   THEN RETURN(1);
   RETURN (0);
END;

/* 4-33: */
IN:   PROCEDURE (X, Y);
   IF Y>=0 & X>=0 & (Y<=5-2*X | Y<=3-.5*X)   THEN RETURN (1);
   RETURN (0);
END;

/* 4-34: */
IN:   PROCEDURE (X, Y, Z);
   IF 14*Z<=27-5*X+17*Y & 2*Z<=6+6*X-4*Y & 2*Z>=17-5*X+5*Y &
      4*Z>=90-14*X-8*Y   THEN RETURN (1);
   RETURN(0);
END;

/* 5-3: */
TWOEQN:   PROCEDURE (A, B, C, P, Q, R, X, Y);
   D = A*Q - P*B;
   X = (C*Q - R*B)/D;
   Y = (A*R - P*C)/D;
END;

/* 5-5: */
AAS:   PROCEDURE (A, B, X, C, Y, A);
   COEF = X/SIND(A);
   Y = COEF*SIND(B);
   C = 180 - A - B;
   Z = COEF*SIND(C);
END;

/* 5-6: */
TRINGL:  PROCEDURE  OPTIONS (MAIN);
   IN:  GET LIST (A, B, X);
      CALL AAS (A,B,X,C,Y,Z);
      PUT SKIP LIST (A,B,X,C,Y,Z);
   GO TO IN;
END;
```

```
/* 5-9: */
SSS:    PROCEDURE (X, Y, Z, A, B, C);
    C = ACOSD (((X*X + Y*Y - Z*Z)/(2*X*Y));
    B = ACOSD (((X*X + Z*Z - Y*Y)/(2*X*Z));
    A = 180 - B - C;
END;

/* 5-10: */
TRNGL:  PROCEDURE OPTIONS (MAIN);
        PUT PAGE LIST ('X', 'Y', 'Z', 'A', 'B', 'C');
    IN: GET LIST (X, Y, Z);
        CALL SSS (X, Y, Z, A, B, C);
        PUT SKIP LIST (X, Y, Z, A, B, C);
        GO TO IN;
END;

/* 5-13: */
SSA:    PROCEDURE (X,Y,A,B1,C1,Z1,B2,C2,Z2);
    COEF = SIND(A)/X;   Q = COEF*Y;
    B1, B2 = ATAND(Q, SQRT(1-Q*Q));
    C1, C2 = 180 - A - B1;
    Z1, Z2 = SIND(C1)/COEF;
    IF X > Y   THEN RETURN;
    B2 = 180 - B1;
    C2 = 180 - A - B2;
    Z2 = SIND(C2)/COEF;
END;


/* 5-14: */
TRINGL: PROCEDURE OPTIONS (MAIN);
    S1: GET LIST (N, Q1, Q2, Q3);
        IF N ¬= 1   THEN GO TO S2;
        CALL AAS(Q1,Q2,Q3,C,Y,Z); PUT SKIP LIST (Q1,Q2,C,Q3,Y,Z); RETURN;
    S2: IF N ¬= 2   THEN GO TO S3;
        CALL ASA(Q1,Q2,Q3,X,C,Y); PUT SKIP LIST (Q1,Q3,C,X,Y,Q2); RETURN;
    S3: IF N ¬= 3   THEN GO TO S4;
        CALL SSS(Q1,Q2,Q3,A,B,C); PUT SKIP LIST (A,B,C,Q1,Q2,Q3); RETURN;
    S4: IF N ¬= 4   THEN GO TO S5;
        CALL SAS(Q1,Q2,Q3,A,Z,B); PUT SKIP LIST (A,B,Q2,Q1,Q3,Z); RETURN;
    S5:   CALL SSA(Q1,Q2,Q3,B1,C1,Z1,B2,C2,Z2);
        PUT SKIP LIST(Q3,B1,C1,Q2,Q3,Z1);
        IF B2 ¬= B1   THEN PUT LIST (B2, C2, Z2);
        GO TO S1;
END;

/* 5-18: */
TRNSLT: PROCEDURE (X, Y, A, FXY, FXX, FYY);
    FXY = FXY + A*X*Y;
    FXX = FXX + A*X*X;
    FYY = FYY + A*Y*Y;
END;

/* 5-20: */
SECTN:  PROCEDURE OPTIONS (MAIN);
        PUT PAGE LIST ('IXY', 'IXX', 'IYY');
    IN: GET LIST (A, ALPHA, X, Y, FXY, FXX, FYY);
        CALL TRNSLT(X, Y, A, FXY, FXX, FYY);
        CALL ROTATE(ALPHA, FXY, FXX, FYY);
        PUT SKIP LIST (FXY, FXX, FYY);
    GO TO IN;
END;
```

```
/* 5-23:  USING 'RT2 = -B - RT1' MIGHT CAUSE THE SUBRACTION OF TWO
   NEARLY EQUAL QUANTITIES.  IT IS IMPOSSIBLE FOR QUAD TO RAISE THE
   ZERODIVIDE CONDITION BECAUSE IF THE LARGER ROOT IS ZERO, THEN THE
   SMALLER ROOT, HENCE THE DISCRIMINANT, MUST ALSO BE ZERO. THIS
   CAUSES TRANSFER TO S10 WHERE NO DIVISION IS PERFORMED.*/

/* 6-2: */
HMEAN:  PROCEDURE  OPTIONS (MAIN);
        SUM, N = 0;
        ON ENDFILE (SYSIN)  GO TO S10;
    S5: GET LIST (X);
        SUM = SUM + 1/X;   N = N + 1;
        GO TO S5;
   S10: AVG = N/SUM;   PUT PAGE DATA (AVG);
END;

/* 6-4: */
ATANMN:  PROCEDURE  OPTIONS (MAIN);
        ON ENDFILE (SYSIN)  GO TO S20;
        A,N = 0;
   S10:GET LIST (X);
        A = A + ATAN(X);
        N = N + 1;
        GO TO S10;
    S20:   PUT LIST (TAN(A/N));
END;  /* ARITHMETIC = 1'ST POWER, HARMONIC = -1 POWER, GEOMETRIC=LOG */

/* 6-6: */
INTEG:  PROCEDURE  OPTIONS (MAIN);
        ON ENDFILE (SYSIN)  GO TO OUT;
        AREA = 0;
    S10:    GET LIST (W, H);
        AREA = AREA + W*H;
        GO TO S10;
    OUT:   PUT DATA (AREA);
END;

/* 6-7:
            A.    (C - D)(U + V)              F.    COS (A)/10
*/

/* 6-8:
NO.  WHEN THE DISCRIMINANT HAS A LARGE RELATIVE ERROR, IT WILL BE
SMALL RELATIVE TO THE OTHER TERM, -.5B
*/

/* 6-9 (D & F COMPILERS): */
TWOEQN:  PROCEDURE (A, B, C, P, Q, R, X, Y);
    DECLARE F FLOAT(16);
    F = DIVIDE(P, A, 16);
    Y, F = (R - F*C)/(Q - F*B);
    X = (C - B*F)/A;
END;

/* 6-15:  SIN(THETA)**2/(1+COS(THETA))   */

/* 6-16: */
PLYGN:  PROCEDURE  OPTIONS (MAIN);
        ON ENDFILE (SYSIN)  GO TO S10;
        GET LIST (X1, Y1);   AREA = 0;
    S5:  GET LIST (X2, Y2);
        AREA = AREA + X1*Y2 - X2*Y1;
        X1 = X2;   Y1 = Y2;
        GO TO S5;
    S10:   PUT LIST (.5*AREA);
END;
```

```
/* 6-20: */
AREA:   PROCEDURE (A, B, N);
    DX = (B-A)/N;   SUM = Y(A) + Y(B);   K = 1;
  S10:  SUM = SUM + 4*Y(A+K*DX);
    K = K+1;
    IF K = N   THEN RETURN(SUM*DX/3);
    SUM = SUM + 2*Y(A+K*DX);
    K = K + 1;
    GO TO S10;
END;

/* 6-23: */
FACTRL:   PROCEDURE (N);
    IF N > 7   THEN DO;
      X = N;   RETURN (2.50663*EXP((X+.5)*LOG(X)-X*
          (((-139/51840/X + 1/288)/X + 1/12)/X + 1));   END;
    NFACT = 1;   K = 2;
  S10:  IF K > N   THEN RETURN (NFACT);
    NFACT = NFACT*K;
    K = K + 1;
    GO TO S10;
END;

/* 6-24: */
GAMMA:   PROCEDURE (X);
    Q = X-1;   FACTOR = 1;
  S10:  IF Q >= 0   THEN GO TO S20;
    Q = Q+1;   FACTOR = FACTOR/Q;   GO TO S10;
  S20:  IF Q <= 1   THEN GO TO S30;
    FACTOR = FACTOR*Q;   Q = Q-1;   GO TO S20;
  S30:  RETURN (FACTOR*
        (((((-.101068*X+.42455)*X-.699859)*X+.951236)*X-.574865)*X+1));
END;

/* 7-2:
            0 COS 0 - .5 = -.5       PI/4 COS PI/4 - .5 = +.055
*/

FCN:   PROCEDURE (THETA);
    RETURN (THETA*COS(THETA) - .5);
END;

R_RT:   PROCEDURE   OPTIONS   (MAIN);
    DECLARE FCN ENTRY;
    PUT LIST (RGLFLS(0E0, .786E0, -.5E0, .055E0, 1E-4, FCN));
END;


/* 7-5: */
MOSTRT:   PROCEDURE (XLEFT, XRIGHT, RELTOL, F);
    DECLARE F ENTRY;
    ON ZERODIVIDE;   ON OVERFLOW;   ON UNDERFLOW;
    PUT PAGE LIST ('ROOTS:');
    IF XLEFT < 0   THEN FACTOR = .875;   ELSE FACTOR = 1.125;
    IF SIGN(XLEFT)=SIGN(XRIGHT) THEN XRT=XRIGHT;   ELSE XRT=-1E-10;
    X1 = XLEFT;
  S5:   Y1 = F(X1);   X2 = FACTOR*X1;   Y2 = F(X2);
  S10:  IF SIGN(Y2) ¬= SIGN(Y1)   THEN DO;
        CALL SECANT((X1),(Y1),(X2),(Y2),RELTOL,F,X,IER);
        IF IER = 0   THEN PUT SKIP LIST (X);
      END;
```

```
              X3 = FACTOR*X2;   Y3 = F(X3);
              IF (Y2>0 & Y2<=Y1 & Y2<Y3) | (Y2<0 & Y2>Y1 & Y2>=Y3)   THEN DO;
                 CALL SECANT((X1),(Y1),(X2),(Y2),RELTOL,F,X,IER);
                 IF IER=0   THEN PUT SKIP LIST (X);   END;
              X1 = X2;   Y1 = Y2;   X2 = X3;   Y2 = Y3;
              IF X3 < XRT   THEN GO TO S10;
              IF XRT = XRIGHT   THEN RETURN;
              XRT = XRIGHT;   FACTOR = 1.125;   X1 = 1E-10;
              GO TO S5;
END;

/* 7-9: */
RESUB:   PROCEDURE (X1, RELTOL, G, X, IER);
         DECLARE G ENTRY;
         KOUNT, IER = 0;
   ITERATE:   X = G(X1);
         IF ABS(X-X1) <= ABS(RELTOL*X)   THEN RETURN;
         X1 = X;   KOUNT = KOUNT + 1;
         IF KOUNT < 100   THEN GO TO ITERATE;
         IER= 1;
END;

/* 7-12: */
MAXDCT:   PROCEDURE (X1, X4, RELTOL, F, X3, Y3);
          DECLARE F ENTRY;
          DX = X4 - X1;
   S10:   DX = .53125*DX;   X3 = X1 + DX;   X2 = X4 - DX;
          Y2 = F(X2);   Y3 = F(X3);
          IF ABS(DX) <= ABS(RELTOL*X3)   THEN GO TO S20;
          IF Y3 > Y2   THEN X1 = X2;   ELSE X4 = X3;
          GO TO S10;
   S20:   IF Y3 > Y2   THEN RETURN;
          Y3 = Y2;   X3 = X2;
END;

/* 7-15: */
GOLDEN:   PROCEDURE (X1, X4, RELTOL, F, X3, Y3);
          DECLARE F ENTRY;
          DX = .617*(X4 - X1);   X2 = X4 - DX;   Y2 = F(X2);
   S10:   X3 = X1 + DX;   Y3 = F(X3);
   S20:   IF ABS(DX) <= ABS(RELTOL*X3)   THEN GO TO S30;
          IF RELTOL*(Y3 - Y2) > 0   THEN DO;   X1 = X2;   Y2 = Y3;
              DX = .617*DX;   GO TO S10;   END;
          X4 = X3;   X3 = X2;   Y3 = Y2;   DX = .617*DX;
          X2 = X4 - DX;   Y2 = F(X2);   GO TO S20;
   S30:   IF RELTOL*(Y3 - Y2) > 0   THEN RETURN;
          Y3 = Y2;   X3 = X2;
END;

/* 7-18: */
DILOG:   PROCEDURE (X, DINEW, IER);
   /* THIS PROCEDURE CALCULATES DINEW = DILOG(X).   IER IS
      SET TO 1 IF THE SERIES HAS NOT CONVERGED IN 20 TERMS.
      OTHERWISE IER IS SET TO 0.   */
   IER = 0;   K = 1;   DIOLD, TERM, ONEMX = 1 - X;
   NEXT:   TERM = TERM*ONEMX;
       DINEW = DIOLD + TERM/(K*K);
       IF DINEW = DIOLD   THEN RETURN;
       DIOLD = DINEW;
       K = K + 1;
       IF D < 20   THEN GO TO NEXT;
       IER = 1;
END;
```

```
/* 7-21: */
SINSIN:  PROCEDURE (X, RESULT, IER);
    K = 1;   IER = 0;   RESULT = X*X;   XONPI = X/3.14159;
  NEXT:  COEF = 1 - (XONPI/K)**4;
      IF COEF = 1   THEN RETURN;
      RESULT = COEF*RESULT;   K= K+1;
      IF K < 40   THEN GO TO NEXT;
      IER = 1;
END;

/* 7-23: */
EXPINT:  PROCEDURE (N, X, EXPOLD, ABERROR);
      C = N;   TERM, EXPOLD, EXPNEW = EXP(-X)/X;
      ABERROR = ABS(TERM);
  S10:  TERM = - TERM*C/X;
      ABNEW = ABS(TERM);
      IF ABNEW >= ABERROR   THEN RETURN;
      ABERROR = ABNEW;   EXPOLD = EXPNEW;
      EXPNEW = EXPOLD + TERM;
      IF EXPNEW = EXPOLD   THEN RETURN;
      C= C + 1;
      GO TO S10;
END;

/* 7-24: */
EXPINT:  PROCEDURE (N, X, RELTOL, CNVGNTNEW);
      M = 0;   CNVGNTOLD = X;
  S10:  CNVGNTNEW = X;   J = M;
  S20:  CNVGNTNEW = X + (N+J)/(1 + (J+1)/CNVGNTNEW);
      J = J - 1;
      IF J >= 0   THEN GO TO S20;
      P = ABS(CNVGNTNEW - CNVGNTOLD);   Q = ABS(CNVGNTNEW);
      IF P <= RELTOL*Q   THEN GO TO S30;
      CNVGNTOLD = CNVGNTNEW;   M = M + 1;
      IF M < 16   THEN GO TO S10;
      RELTOL = P/Q;
      S30:   CNVGNTNEW = EXP(-X)/CNVGNT;
END;

/* 7-29:  IN BLZNO, RGLFLS, SCNT, SECANT, AND MOSTRT, UNDERFLOW AND
    OVERFLOW ARE THE RESULT OF FUNCTION EVALUATION;   SO THE CORRES-
    PONDING NULL ON STATEMENTS COULD NOT BE EFFECTIVELY REPLACED WITH
    CONDITION PREFIXES.   */

/* 7-30:  NONE OF THE PROCEDURES IN THIS CASE STUDY COULD MAKE GOOD
    USE OF AN ON ERROR STATEMENT. */

/* 7-31: */
RESUB:  PROCEDURE (X1, RELTOL, G, X);
      DECLARE G ENTRY, IER EXTERNAL;
      KOUNT, IER = 0;
   ITERATE:  X = G(X1);
      IF ABS(X-X1) <= ABS(RELTOL*X)   THEN RETURN;
      X1 = X; KOUNT = KOUNT + 1;
      IF KOUNT < 100   THEN GO TO ITERATE;
      IER = 1;
END;
```

```
/* 7-32: */
HAV:   PROCEDURE (X, HAVNEW, LOC);
       DECLARE LOC LABEL;
       XSQ = X*X;  K = 4; HAVOLD, TERM = .25*XSQ;
  NEXT:  TERM = -TERM*XSQ/(K*(K-1));
       HAVNEW = HAVOLD + TERM;
       IF HAVNEW = HAVOLD  THEN RETURN;
       HAVOLD = HAVNEW;  K = K + 2;
       IF K < 20  THEN GO TO NEXT;
       GO TO LOC;
END;

HAVUSE:   PROCEDURE OPTIONS (MAIN);
       DECLARE ERROR LABEL;
   IN:  GET LIST (X);
       CALL HAV (X, Y, ERROR);
       PUT SKIP LIST (X, Y);
       GO TO IN;
   ERROR:  PUT SKIP LIST (X, 'THIS ANGLE IS TOO LARGE');
       GO TO IN;
END;

/* 8-1: */
EFFNCY:   PROCEDURE  OPTIONS (MAIN);
   PUT PAGE LIST ('GAS TURBINE EFFICIENCY WITH AIR:');
   PUT SKIP(2) LIST ('PRESSURE RATIO', 'EFFICIENCY');
   DO R = .1 TO 6 BY .1;
       PUT SKIP LIST (R, 1-R**.285);
   END;
END;

/* 8-5: */
AREA:   PROCEDURE (A, B, N, F);
   DECLARE F ENTRY;
   DX = (B-A)/N;   SUM = .5*(F(A)+F(B));
   DO K = 1 TO N-1;
       SUM = SUM + F(A + K*DX);
   END;
   RETURN (SUM*DX);
END;

/* 8-7: */
NEWTON:   PROCEDURE (RELTOL, FCN, X, IER);
   DECLARE FCN ENTRY;
   ON UNDERFLOW;
   IER = 0;
   DO K = 1 TO 20;
       DIFF = FCN(XOLD);
       X = XOLD - DIFF;
       IF ABS(DIFF) <= ABS(RELTOL*X)  THEN RETURN;
   END;
   IER = 1;
END;
```

Appendix

```
/* 8-8: */
BESSEL:  PROCEDURE (X, BESNEW, IER);    /* THIS PROCEDURE
     CALCULATES BESNEW = ZERO'TH ORDER BESSEL FUNCTION OF X.  IER IS
     SET TO 1 IF CONVERGENCE IS NOT ACHIEVED WITHIN 10 TERMS.  OTHER-
     WISE,  IER IS SET TO ZERO.   */
  IER = 0;   XSQ = X*X;    BESOLD, TERM = 1;
  DO K = 2 TO 20 BY 2;
     TERM = -.25*TERM*XSQ/(K*(K-1))**2;
     BESNEW = BESOLD + TERM;
     IF BESNEW = BESOLD  THEN RETURN;
     BESOLD = BESNEW;
  END;
  IER = 1;
END;

/* 8-12: */
T:  PROCEDURE (S);
   TERM = (1800 - S)/133;
   RETURN (61.4/(TERM + 1/TERM));
END;

TORQUE:  PROCEDURE  OPTIONS (MAIN);
   DO S = 0, 450, 1710, 1800;
      PUT SKIP LIST (S, T(S));
   END;
END;

/* 8-16: */
ENTRPY:  PROCEDURE  OPTIONS (MAIN);
   DECLARE (V FIXED(3,1), T FIXED(5,0), GAMMA FIXED(3,1)) DECIMAL;
   PUT PAGE LIST ('NORMALIZED ENTROPY TABLE');
   DO V = 1 TO 8 BY .5;
      PUT SKIP(2) DATA (V);   Q = LOG(V);
      PUT SKIP LIST ('GAMMA:','1.1','1.2','1.3','1.4','1.5','1.6');
      PUT SKIP LIST ('TEMPERATURE');
      DO  T = 100 TO 1000 BY 25;
         PUT SKIP LIST (T);   P = LOG(T);
         DO GAMMA = 1.1 TO 1.6 BY .1;
            PUT LIST (Q + P/(GAMMA - 1));
         END;
      END;
   END;
END;

/* 8-18: */
REACTN:  PROCEDURE  OPTIONS (MAIN);
   ANOTHER:  RC = 1;
      GET LIST (M);
      DO I = 1 TO M;
         GET LIST (A,K);  RC = RC*A**K;
      END;
      GET LIST (N);
      DO J = 1 TO N;
         GET LIST (B, L);   RC = RC/B**L;
      END;
      PUT SKIP LIST (RC);
      GO TO ANOTHER;
END;
```

```
/* 8-19: */
TRIODE:    PROCEDURE OTRIONS (MAIN);
    DECLARE (EA, EG, I) FIXED BINARY;
    PUT PAGE LIST ('TRIODE CURRENT FOR C = 40 AND MU = 92');
    PUT SKIP(2) LIST ('GRID VOLTAGE:', '-20','-15','-10','-5','0');
    PUT SKIP LIST ('ANODE VOLTAGE');
    DO EA = 50 TO 400 BY 10;
        PUT SKIP LIST (EA);
        TERM = EA/92;
        DO EG = -20 TO 0 BY 5;
            Q = EG + TERM;
            IF Q > 0
                THEN DO;
                    I = 40*Q**1.5;
                    IF I<=55 & I*EA<=1600   THEN PUT LIST (I);
                                            ELSE PUT LIST (' ');
                END;
                ELSE PUT LIST (' ');
        END;
    END;
END;


/* 8-22: */   TABLE4:   PROCEDURE   OPTIONS (MAIN);
    DECLARE ((THETA_I, THETA_R) FIXED(6,4), (R, RLEFT, RRIGHT)
        FIXED(3,2)) DECIMAL;
    PUT PAGE LIST ('ANGLE OF REFRACTION IN DEGREES');
    RLEFT = 1.1;
S10:    RRIGHT = MIN (RLEFT+.3, 2.0);
    PUT SKIP(2) LIST ('INDEX:');
    DO R = RLEFT TO RRIGHT BY .1;
        PUT LIST (R);
    END;
    PUT SKIP LIST ('INCIDENT ANGLE');
    DO THETA_I = .1 TO 10 BY .1, 11 TO 89;
        PUT SKIP LIST (THETA_I);
        SINE = SIND(THETA_I);
        DO R = RLEFT TO RRIGHT BY .1;
            U = SINE/R;   THETA_R = ATAND(U/SQRT(1-U*U));
            PUT LIST (THETA_R);
        END;
    END;
    RLEFT = RLEFT + .4;
    IF RRIGHT < 2.0   THEN GO TO S10;
END;

/* 8-23: */
B:   PROCEDURE (N, K);
    ANSW = 1;
    DO J = N-K+1 TO N;
        ANSW = ANSW*J;
    END;
    DO J = 2 TO K;
        ANSW = ANSW/J;
    END;
    RETURN (ANSW);
END;
```

```
/* 8-25: */
TABLE:  PROCEDURE   OPTIONS (MAIN);
   DECLARE (THETA_I, THETA_R) FIXED (7,4);
   PUT PAGE LIST ('REFRACTION TABLE IN DEGREES FOR INDEX OF 1.52:');
   PUT SKIP(2) LIST ('INCIDENT ANGLE', 'REFRACTED ANGLE');
   R = 1.52;
   ON ENDPAGE (SYSPRINT)  PUT PAGE LIST ('INCIDENT ANGLE',
      'REFRACTED ANGLE');
   DO THETA_I = .5 TO 89.5 BY .5;
      U = SIND(THETA_I)/R;   THETA_R = ATAND(U/SQRT(1-U*U));
      PUT SKIP LIST (THETA_I, THETA_R);
   END;
END;


/* 9-4: */
CG:  PROCEDURE (N, W, X, Y, Z, XBAR, YBAR, ZBAR);
   DECLARE W(*), X(*), Y(*), Z(*);
   XBAR, YBAR, ZBAR, WTOT = 0;
   DO J = 1 TO N;
      XBAR = XBAR + W(J)*X(J);
      YBAR = YBAR + W(J)*Y(J);
      ZBAR = ZBAR + W(J)*Z(J);
      WTOT = WTOT + W(J);
   END;
   XBAR = XBAR/WTOT;
   YBAR = YBAR/WTOT;
   ZBAR = ZBAR/WTOT;
   RETURN;
END;

MANYCG:  PROCEDURE   OPTIONS (MAIN);
      DECLARE W(2500), X(2500), Y(2500), Z(2500);
   S10:   GET LIST (N, (X(K), Y(K), Z(K), W(K) DO K = 1 TO N));
      CALL CG (N, W, X, Y, Z, XBAR, YBAR, ZBAR);
      PUT SKIP DATA (XBAR, YBAR, ZBAR);
      GO TO S10;
END;

/* 9-6: */
LINES:  PROCEDURE   OPTIONS (MAIN);
   DECLARE X(5000), Y(5000);
   GET LIST (N, (X(K), Y(K) DO K = 1 TO N));
   CALL LINE (N, X, Y, A, B);
   GET LIST (N, (X(K), Y(K) DO K = I TO N));
   CALL LINE (N, X, Y, S, T);
   CALL TWOEQN(-A,1E0,B,-S,1E0,T,U,V);
   PUT LIST ('X AND Y COORDINATES OF INTERSECTION ARE:',U,V);
END;

/* 9-9: */
ESAKI:   PROCEDURE (V);
   DECLARE AMPS(0:15) INITIAL(0,.09,.13,.15,.14,.12,.09,.06,.04,.02,
      .02,.03,.04,.06,.09,.17) STATIC;
   K = MAX (0, MIN(20*V, 14));
   RETURN (AMPS(K) + 20*(V-.05*K)*(AMPS(K+1) - AMPS(K)));
END;
```

```
/* 9-11: */
ESAKI: PROCEDURE (V);
   DECLARE AMPS(0:15)   INITIAL (0,.09,.13,.15,.14,.12,.09,.06,.04,
      .02,.02,.03,.04,.06,.09,.17) STATIC;
   R= 20*V;   K = R;
   IF K = R   THEN RETURN (AMPS(K));
   RETURN (AMPS(K) + 20*(V-.05*K)*(AMPS(K+1)-AMPS(K)));
END;  /* THIS TECHNIQUE TAKES LONGER ON THE AVERAGE   */

/* 9-14: */
INTERP:   PROCEDURE (N,XIN,XMIN,DX,Y,YOUT,IER);
   DECLARE Y(*);
   IER = 1;   K = (XIN-XMIN)/DX;
   IF K<0 | K>N-2   THEN RETURN;
   YOUT = Y(K) + (XIN-XMIN-K*DX)*(Y(K+1)-Y(K))/DX;
   IER = 0;
END;

/* 9-18: */
F:   PROCEDURE (N, XIN, X, Y);   /* THIS GENERAL PURPOSE FUNCTION
      PROCEDURE EVALUATES ANY EMPIRICAL OR MATHEMATICAL FUNCTION BY
      QUADRATIC INTERPOLATION FOR ARBITRARILY SPACED VALUES OF THE
      INDEPENDENT VARIABLE.  EXTRAPOLATION IS USED FOR VALUES OF XIN
      OUTSIDE THE RANGE OF X.  */
   DECLARE N      /* THE NUMBER OF TABULATED POINTS  */,
           XIN    /* THE INDEPENDENT VARIABLE  */,
           X(*), Y(*)   /* THE THE TABULATED VALUES OF THE INDEPENDENT
                           AND DEPENDENT VARIABLE  */;
      DO J = 3 TO N-1;
         IF XIN <= X(J)   THEN GO TO S10;
      END;
   S10:   D1 = (Y(J-1)-Y(J-2))/(X(J-1)-X(J-2));
      D2 = (Y(J)-Y(J-1))/(X(J)-X(J-1));       RETURN
         (Y(J-2)+(XIN-X(J-2))*(D1+(XIN-X(J-1))*(D2-D1)/(X(J)-X(J-2))));
END;

/* 9-20: */
SMOOTH:   PROCEDURE (N,X);
   DECLARE X(*);
   DO J = 2 TO N-1;
      X(J) = (X(J-1) + X(J) + X(J+1))/3;
   END;
END;

/* 9-22: */
DERIV1:   PROCEDURE (N, DX, Y, DYONDX);
   DECLARE Y(*), DYONDX(*);
   COEF = .5/DX;
   DO J = 2 TO N-1;
      DYONDX (J) = COEF*(Y(J+1)- Y(J-1));
   END;
END;
```

```
/* 10-1: */
SORT:   PROCEDURE (N, X);
   DECLARE X(*);
   DO J = 1 TO N-1;
      K_OF_XMIN = J;   XMIN = X(K_OF_XMIN);
      DO K = J+1 TO N;
         IF X(K) < XMIN  THEN DO;  K_OF_XMIN = K;  XMIN = X(K);  END;
      END;
      X(K_OF_XMIN) = X(J);   X(J) = XMIN;
   END;
END;

/* 10-4: */
BUBBLE:   PROCEDURE (N,X);
     DECLARE X(*);
     KMAX = N-1;
  S10:  KLAST = 1;
     DO K = 1 TO KMAX;
        IF X(K+1) > X(K) THEN DO; XMAX = X(K+1);
           X(K+1) = X(K); X(K) = XMAX; KLAST = K; END;
     END;
     KMAX = KLAST - 1;
     IF KMAX > 0 THEN GO TO S10;
END;

/* 10-6:
DOT = SUM(A*B);
*/

/* 10-7: */
DOT:   PROCEDURE (N, A, B);
   DECLARE A(*), B(*);
   SUM = 0;
   DO J = 1 TO N;
      SUM = SUM + A(J)*B(J);
   END;
   RETURN (SUM);
END;

/* 10_14:   Y = C + SUM(A*SIN(K*T/3.14*P)) + B*COS(K*T/(3.14*P)));    */

/* 10-15: */
Y: PROCEDURE (T, P, C, N, A, B);
   DECLARE A(*), B(*);
      TOTAL = T/(3.14159*P);
      C1, CKM1 = COS(TOTAL);   SKM1 = SIN(TOTAL);
      CK = 2*CKM1*CKM1;   SK = 2*SKM1*CKM1;
      TOTAL = C + A(1)*SKM1 + B1*CKM1;
      K = 2;
   S10:   TOTAL = TOTAL + A(K)*SK + B(K)+CK;
      IF K >= N  THEN RETURN (TOTAL);
      CKP2 = 2*C1*CK - CKM1;
      SKP1 = 2*C1*SK - SKM1;
      K = K + 1;
      CKM1 = CK;    CK = CKP1;
      SKM1 = SK;    SK = SKP1;
      GO TO S10;
END;
```

```
/* 10-17: */
GENCNT:   PROCEDURE OPTIONS (MAIN);
   GET LIST (N);
  BEGIN;
       DECLARE A(N), B(N), C(N), D(N);
      IN: GET LIST (A, B, C);
         D= (A+B+C)/3;
         PUT SKIP LIST (D);
         GO TO IN;
   END;
END;

/* 10-18: */
FIT:   PROCEDURE OPTIONS (MAIN);
   S10: GET LIST (N);
      BEGIN;
         DECLARE X(N), Y(N);
         GET LIST ((X(K), Y(K) DO K = 1 TO N));
         CALL LINE (N, X, Y, A, B);
         STDEV= SQRT (SUM((Y-A*X-B)**2)/N);
         PUT SKIP LIST (A, B, STDEV);
      END;
      GO TO S10;
END;

/* 10-20: PARAMETERS MUST BE DECLARED WITH EITHER CONSTANT OR ASTERISK
   BOUNDS FOR THE F COMPILER.  ALSO 'DECLARE X(N)' WOULD BE LESS
   FLEXIBLE THAN 'DECLARE X(*)' BECAUSE THE LATTER PERMITS THE USER TO
   USE EITHER A CONSTANT OR VARIABLE EXTENT IN THE CALLING PROGRAM.   */

/* 10-23: */
PLYADD:   PROCEDURE (M, N, A, B, C);
   DECLARE A(*), B(*), C(*);
   MINMN = MIN (M, N);
   DO K = 0 TO MINMN;
      C(K) = A(K)+B(K);
   END;
   IF MINMN = M
      THEN DO K = M+1 TO N;
         C(K) = B(K);
      END;
      ELSE DO K = N+1 TO M;
         C(K) = A(K);
      END;
END;

/* 10-25: */
PLYDIV:   PROCEDURE (M, N, A, B, Q);
   DECLARE A(*), B(*), Q(*)   /* LOWER BOUNDS PRESUMED ZERO */;
   D = B(N);
   DO K = M-N TO 0 BY -1;
      F, Q(K) = A(N+K)/D;
      DO J = K TO K+N-1;
         A(J) = A(J) - F*B(J-K);
      END;
   END;
END;

/* 10-27: */
PLYDIF:   PROCEDURE (M, A, B);
   DECLARE A(*), B(*) /* LOWER BOUNDS PRESUMED ZERO */;
   DO K = 1 TO M;
      B(K-1) = K*A(K);
   END;
END;
```

```
/* 11-1: */
MAXRN:  PROCEDURE  OPTIONS (MAIN);
   DECLARE RAIN (12,7);
   GET LIST (RAIN);   N = 7;
   DO I = 1 TO 12;
      PUT SKIP LIST (I, VMAX(N,RAIN(I,*)));
   END;
END;

/* 11-3: */
MAXRN:  PROCEDURE OPTIONS (MAIN);
   DECLARE RAIN(12,7), CROSSMAX (12);
   GET LIST (RAIN);
   CROSSMAX = RAIN(*,1);
   DO J = 2 TO 7;
      CROSSMAX = MAX(CROSSMAX, RAIN(*,J));
   END;
   PUT LIST(CROSSMAX);
END;

/* 11-5: */
PRECIP: PROCEDURE OPTIONS (MAIN);
   DECLARE (RAIN, SNOW, SLEET, HAIL, PRECIPITATION) (12,7);
   GET LIST (((RAIN(I,J), SNOW(I,J), SLEET(I,J), HAIL(I,J) DO J = 1
             7) DO I = 1 TO 12));
   PRECIPITATION = RAIN + .1*SNOW + .9*SLEET + .7*HAIL;
   CALL MTXOUT (BINARY(12), BINARY(7), PRECIPITATION);
END;

/* 11-7: */
INOUT: PROCEDURE OPTIONS (MAIN);
   DECLARE RAINS(1984:2000, 12, 7);
   GET LIST ((((RAINS(I,J,K)   DO I = 1984 TO 2000) DO J = 1 TO 12)
             DO K = 1 TO 7));
 M = 12;   N = 7;
   DO I = 1984 TO 2000;
      PUT PAGE LIST ('YEAR = ', I);
      CALL MTXOUT (M,N, RAINS(I, *, *));
   END;
END;

/* 11-9: */
VOLUME:  PROCEDURE OPTIONS (MAIN);
   DECLARE RAIN(12,7), AREA(7), VOL(12);
   GET LIST(RAIN, AREA);
   M =12;   N = 7;
   CALL MVPROD(M,7,RAIN, AREA, VOL);
   PUT DATA (VOL);
END;

/* 11-12: */
OILMTX:  PROCEDURE OPTIONS (MAIN);
   DECLARE AMOUNT(57,57), CRUDE(57,57), COMPONENT(57,57);
   GET LIST (M, N, L, ((AMOUNT(I,J) DO J = 1 TO N) DO I = 1 TO M),
          ((CRUDE(J,K)   DO J = 1 TO N) DO K = 1 TO L)); /* CRUDE IN
          COLUMN ORDER     */
   CALL MMPROD(M, N, L, AMOUNT, CRUDE, COMPONENT);
   CALL MTXOUT(N, L, COMPONENT);
END;
```

```
/* 11-15: */
INERTA:  PROCEDURE OPTIONS (MAIN);
     DECLARE  (A, T, TA, ATA)(3,3);
     MNL = 3;
S10: GET LIST (A, T);
     CALL MMPROD(MNL, MNL, MNL, T, A, TA);
     CALL MMPROD(MNL, MNL, MNL, A, TA, ATA);
     CALL MTXOUT(MNL, MNL, ATA);
     GO TO S10;
END;

/* 11-16: */
DIAG:  PROCEDURE (N, A);
   DECLARE A(*,*);
   PDT = 1;
   DO I = 1 TO N;
      PDT = PDT*A(I,I);
   END;
   RETURN (PDT);
END;

/* 11-18: */
ONENRM:  PROCEDURE (N,A);
   DECLARE A(*,*);
   BIGGEST = 0;
   DO I = 1 TO N;
      SUM = 0;
      DO J = 1 TO N;
         SUM = SUM + ABS(A(I,J));
      END;
      BIGGEST = MAX(BIGGEST, SUM);
   END;
   RETURN (BIGGEST);
END;

/* 11-20: */
GAME:  PROCEDURE OPTIONS (MAIN);
     DECLARE A(100, 100), (MINROWMAX, MAXCOLMIN) FLOAT;
  IN:  GET LIST (M, N, ((A(I,J) DO J = 1 TO N) DO I = 1 TO M));
     CALL MTXOUT (M, N, A);
     MINROWMAX = 1E30;
     DO I = 1 TO M;
        ROWMAX = - 1E30 ;
        DO J = 1 TO N;
           ROWMAX = MAX(ROWMAX, A(I,J));
        END;
        MINROWMAX = MIN(MINROWMAX, ROWMAX);
     END;
     MAXCOLMIN = -1E30;
     DO J = 1 TO N;
        COLMIN = 1E30;
        DO I = 1 TO M;
           COLMIN = MIN(COLMIN, A(I,J));
        END;
        MAXCOLMIN = MAX(MAXCOLMIN, COLMIN);
     END;
     IF MINROWMAX = MAXCOLMIN  THEN PUT LIST ('VALUE = ',MINROWMAX);
        ELSE PUT LIST ('NO SADDLEPOINT');
     GO TO IN;
END;
```

```
/* 11-21: */
CONTCT:  PROCEDURE (R, Q);
   DECLARE R(4,4,4,4), Q(4,4);
   DO J = 1 TO 4;
      DO L = 1 TO 4;
         SUM = 0;
         DO K = 1 TO 4;
            SUM = SUM + R(K,J,K,L);
         END;
         Q(J,L) = SUM;
      END;
   END;
END;

/* 11-26: */
MNYMRG:  PROCEDURE (L, X, Z);
     DECLARE X(*,*)  /* L ROWS, EACH ARRANGED IN ASCENDING ORDER.
                        LAST VALID VALUE IN EACH ROW AND Z IS FOLLOWED
                        BY A DUMMY VALUE OF 1E30.  */,
              Z(*)    /* RESULTANT MERGED LIST.  */,
              K(L) INITIAL ((L)1);
   DO J = 1 BY 1;
      SMALLEST = 1E30;
      DO I = 1 TO L;
         IF X(I,K(I)) < SMALLEST  THEN DO;  ISMALL = I;
            SMALLEST = X(J,K(I));   END;
      END;
      Z(J) = SMALLEST;
      IF SMALLEST = 1E30  THEN RETURN;
      K(ISMALL) = K(ISMALL) + 1;
   END;
END;

/* 11-27: */
MIXEXP:  PROCEDURE (N, A, DT, RETOL, B);
     DECLARE A(*,*), B(*,*), TERMOLD(N,N), TERMNEW(N,N);
   S10:  DO I = 1 TO N;
      DO J = 1 TO N;
         TERMOLD(I,J), B(I,J) = DT*A(I,J);
      END;
      TERMOLD(I,I), B(I,I) = TERMOLD(I,I) + 1;
   END;
   DO K = 2 TO 10;
      ICHANGE = 0;
      DO I = 1 TO N;
         SUM = 0;
         DO L = 1 TO N;
            SUM = SUM + A(I,L)*TERMOLD(L,J)*DT/K;
         END;  /* DOUBLE PRECISION ACCUMULATION MAY BE ADVISABLE */
         TERMNEW(I,J) = SUM;
         B(I,J) = B(I,J) + SUM;
         IF ABS(SUM) > ABS(RELTOL*B(I,J))   THEN ICHANGE = 1;
      END;
      IF ICHANGE = 0  THEN RETURN;
      TERMOLD = TERMNEW;
   END;
   DT = .5*DT;   GO TO S10;
END;
```

```
/* 11-28: */
EQINTP: PROCEDURE (M, N, XIN, XMIN, DX, YIN, XMIN, DY, Z, ZOUT, IER);
   DECLARE Z(*,*);  /* LOWER BOUNDS PRESUMED ZERO   */
   IER = 0;   DIFFX = (XIN - XMIN)/DX;   I = DIFFX;
   DIFFY = (YIN - YMIN)/DY;   J = DIFFY;
   IF (I<0 | J<0 | I>M-1 | J>N-1) THEN DO;   IER = 1;
      RETURN;   END;
   FRACTION = DIFFX - X;
   ZA = Z(I,J) + FRACTION*(A(I+1,J) - Z(I,J));
   ZB = Z(I,J+1) + FRACTION*(Z(I+1, J+1) - Z(I,J+1));
   FRACTION = DIFFY - J;
   ZOUT = ZA + FRACTION*(ZB - ZA);
END;

/* 12-1, CHANGE I LOOP OF THE ELIMINATION AND FOWARD SUBSTITUTION TO:
      DO I = JP1 TO N;
         F = A(I,J)/PIVOT;   IF F = 0 THEN GO TO SKIP;
         DO K = JP1 TO N;
            A(I,K) = A(I,K)-F*A(J,K);
         END;
      SKIP:  END;   */

/* 12-4: */
GAUSS:   PROCEDURE (N, A, BX);   /*  THIS SUBROUTINE SOLVES N LINEAR
      EQUATIONS IN N UNKNOWNS USING GAUSSIAN ELIMINATION.  */
   DECLARE A(*,*),  /*  GIVEN COEFFICIENTS   (ALTERED BY SOLUTION) */
           BX(*);       /* GIVEN CONSTANTS WHICH ARE TRANSFORMED INTO
                                  SOLUTIONS */
   /* ELIMINATION AND FORWARD SUBSTITUTION:   */
   DO J = 1 TO N-1;    /*   SET COLUMN   */
      JP1 = J+1;   PIVOT = A(J,J);
      DO I = JP1 TO N;     /*   SET ROW   */
         F = A(I,J)/PIVOT;   /* CALCULATE MULTIPLE   */
          DO K = JP1 TO N; /* SUBTRACT MULTIPLE OF ROW J FROM ROW I  *
         A(I,K) = A(I,K) - F*A(J,K);
         END;
         BX(I) = BX(J) - F*BX(J);
      END;
   END;
   /* BACK SUBSTITUTION:   */
   DO I = N TO 1 BY -1;   /* SET ROW   */
      SUM = BX(I);
      DO J = I+1 TO N;   /*   SUBTRACT TERM   */
         SUM = SUM - A(I,J)*X(J);
      END;
      BX(I) = SUM/A(I,I);
   END;
END;


/* 12-6: */
GAUSS:   PROCEDURE (N, ABX);
   DECLARE ABX(*,*);
   NP1 = N+1;
   DO J = 1 TO N-1;
      JP1 = J+1;   PIVOT = ABX(J,J);
      DO I = JP1 TO N;
         F = ABX(I,J)/PIVOT;
         DO K = JP1 TO NP1;
            ABX(I,K) = ABX(I,K) - F*ABX(J,K);
         END;
      END;
   END;
```

```
      DO I = N TO 1 BY -1;
         SUM = ABX(I,NP1);
         DO J = I+1 TO N;
            SUM = SUM - ABX(I,J)*ABX(J,NP1);
         END;
         ABX(I,NP1) = SUM/ABX(I,I);
      END;
END;

/* 12-8: */
GAUSS:  PROCEDURE (N, L, A, B, X);
   DECLARE A(*,*), B(*,*), X(*,*);
   DO J = 1 TO N-1;
      JP1 = J+1;  PIVOT = A(J,J);
      DO I = JP1 TO N;
         F = A(I,J)/PIVOT;
         DO K = JP1 TO N;
            A(I,K) = A(I,K) - F*A(J,K);
         END;
         DO K = 1 TO L;
            B(I,K) = B(I,K) - F*B(J,K);
         END;
      END;
   END;
   DO I = N TO 1 BY -1;
      DO K = 1 TO L;
         SUM = B(I,K);
         DO J = I+1 TO N;
            SUM = SUM - A(I,J)*X(J,K);
         END;
         X(I,K) = SUM/A(I,I);
      END;
   END;
END;

/* 12-10: */
SWISS: PROCEDURE (L, M, N, ITMAZ, ABTOL, T, ID);
   DECLARE T(*,*,*), ID(*,*,*);
   DO IT = 1 TO ITMAX;
      BIGGEST = 0;
      DO I = 1 TO L;
         DO J = 1 TO M;
            DO K = 1 TO N;
               IF ID(I,J,K) ¬= 1  THEN GO TO SKIP;
               TNEW = .166667*(T(I,J,K+1) + T(I,J,K-1) + T(I,J+1,K) +
                  T(I,J-1,K) + T(I+1,J,K) + T(I-1,J,K));
               BIGGEST = MAX (BIGGEST, ABS(TNEW-T(I,J,K)));
               T(I,J,K) = TNEW;
         SKIP:  END;
         END;
      END;
      IF BIGGEST > ABTOL   THEN RETURN;
   END;
   ABTOL = BIGGEST;
END;
```

```
/* 12-13: */
INITL:  PROCEDURE (L, M, N, T);
   DECLARE T(*,*,*);
   SUM = 0;
   DO I = 0 TO L+1;
      DO J = 0 TO M+1;
         SUM = SUM + T(I,J,0) + T(I,J,N+1);
      END;
      DO K = 1 TO N;
         SUM = SUM + T(I,0,K) + T(I,M+1,K);
      END;
   END;
   DO J = 1 TO M;
      DO K = 1 TO N;
         SUM = SUM + T(0,J,K) + T(L+1,J,K);
      END;
   END;
   SUM = SUM/(2*(L*M+L*N+M*N) + 4*(L+M+N) + 8);
   DO I = 1 TO L;
      DO J = 1 TO M;
         DO K = 1 TO N;
            T(I,J,K) = SUM;
         END;
      END;
   END;
END;

/* 12-16: */
INPUT:  PROCEDURE (T);
      DECLARE T(*,*,*);
   IN:  GET LIST (I,J,K,TEMP);
      IF I < 0 THEN RETURN;  ELSE T(I,J,K) = TEMP;  /* A NEGATIVE
         I SUBSCRIPT IS A SENTINEL VALUE DESIGNATING THE END OF THE
         SURFACE TEMPERATURES.  USE THE NULL FIELD WHEN A SUBSCRIPT OR
         TEMPERATURE IS THE SAME FOR TWO SUCCESSIVE SURFACE POINTS.  */
      GO TO IN;
END;

/* 12-17: */
BRICK:  PROCEDURE OPTIONS (MAIN);
   GET LIST (L, M, N, ITMAX, RELTOL);
   BEGIN;
      DECLARE T(L,M,N);
      CALL INPUT(T);
      CALL INITL(L, M, N, T);
      CALL TEMP (L, M, N, ITMAX, ABTOL, T);
      PUT LIST (ABTOL);
      OUT:   GET LIST (I, J, K);
          PUT SKIP DATA (T(I,J,K));
          GO TO OUT;
   END;
END;
```

```
/* 12-19: */
GSSSDL:   PROCEDURE (N, A, B, ITMAX, ABTOL, X);
          /* THIS SUBROUTINE SOLVES THE LINEAR EQUATIONS A*X = B BY THE
             GAUSS-SIEDEL METHOD. IF CONVERGENCE TO A MAXIMUM ABSOLUTE ERROR
             OF ABTOL IS NOT ACHEIVED WITHIN ITMAX ITERATION, ABTOL IS RESET
             TO THE ABSOLUTE ERROR THAT IS ACHEIVED.  SET ITMAX TO LESS THAN
             T/(DT*N**2), WHERE DT IS THE TIME TO EXECUTE 1 CYCLE OF THE INNER
             LOOP AND T IS THE TOTAL ALLOWABLE EXECUTION TIME.  X SHOULD BE
             INITIALIZED TO A CRUDE GUESS FOR THE SOLUTION SUCH AS X = 0 OR
             X(I) = B(I)/A(I,I).  */
   DECLARE A(*,*), B(*), X(*);
   DO IT = 1 TO ITMAX;
      BIGGEST = 0;
      DO I = 1 TO N;
         SUM = 0;
         DO J = 1 TO I-1, I+1 TO N;
            SUM = SUM + A(I,J)*X(J);
         END;
         TEMP = (B(I) - SUM)/A(I,I);
         BIGGEST = MAX (BIGGEST, ABS(TEMP-X(I)));
         X(I) = TEMP;
      END;
      IF BIGGEST < ABTOL  THEN RETURN;
   END;
   ABTOL = BIGGEST;
END;

/* 12-23: */
SPSNRM:   PROCEDURE (M, IB, B);
   DECLARE IB(*), B(*);   /* IB AND B HAVE M ELEMENTS  */
      BIGGEST = 0;  SUM = ABS(B(1));
   DO K = 2 TO M;
      IF IB(K) = IB(K-1)  THEN SUM = SUM + ABS(B(K));
         ELSE DO;  BIGGEST = MAX(SUM,BIGGEST);   SUM = ABS(B(K));  END;
   END;
   RETURN (MAX(SUM,BIGGEST));
END;

/* 12-25: */
TRIGSS:   PROCEDURE (N, D, E, F, B, X);
   DECLARE D(*), E(*), F(*), B(*), X(*);
   DO I = 2 TO N;
      Q = D(I)/E(I-1);
      E(I) = E(I) - Q*F(I-1);
      X(I) = B(I) - Q*X(I-1);
   END;
   DO I = N-1 TO 1 BY 1;
      X(I) = (X(I) - F(I)*X(I-1))/E(I);
   END;
END;

/* 12-26: */
BNDGSS:   PROCEDURE (N, M, C, B, X);
   DECLARE C(*,*), B(*), X(*);
   DO J = 1 TO N-1;
      JP1 = J+1;  LIM = MIN(J+M, N);   PIVOT = C(J,0);
      DO I = JP1 TO LIM;
         F = C(I,J-I)/PIVOT;
         DO K = JP1 TO LIM;
            C(I,K-I) = C(I,K-I) - F*C(J,K-J);
         END;
         B(I) = B(I) - F*B(J);
      END;
   END;
```

```
   DO I = N TO 1 BY -1;
      SUM = B(I);
      DO J = I+1 TO N;
         SUM = SUM - A(I,J-I)*X(J);
      END;
      X(I) = SUM/A(I,0);
   END;
END;

/* 13-1: */
NQUAD: PROCEDURE (X, Y);
      DECLARE YLTO BIT(1);
      YLTO = Y < 0;
      IF X < 0  THEN GO TO S10;
      IF YLTO   THEN RETURN (4);
      RETURN (1);
   S10: IF YLTO  THEN RETURN (3);
      RETURN (2);
END;

/* 13-3: */
NQUAD: PROCEDURE (X, Y);
   DECLARE N(0:3) INITIAL(1,2,4,3) STATIC;
   RETURN (N(Y<0 || X<0));
END;

/* 13-6: */
IN:  PROCEDURE (X, Y) BIT(1);
   RETURN  (Y<=.5*X+2 & Y>=-.5*X+3 & Y>=2*X-8);
END;

/* 13-8: */
ROADS:  PROCEDURE  OPTIONS (MAIN);
   DECLARE (R1,R2,R3,R4,R5,R6,R7,R8,R9,Q) BIT(1);
   R1, R2, R3, R4, R5, R6, R7, R8, R9 = '1'B;
   GET DATA;
   Q = R9 & (R5 | R8) & R7;
   PUT LIST (((R1 | (R3 & R2)) & R6) | ((R3 | (R1 & R2)) & R4));
END;

/* 13-10: */
HYD:  PROCEDURE  OPTIONS (MAIN);
      DECLARE FLOW (200,200) BIT(1);
      FLOW = '0'B;
      ON ENDFILE (SYSIN)  GO TO S20;
      GET LIST (N);
      DO K = 1 TO N;
         FLOW(K,K) = '1'B;
      END;
   S10: GET LIST (I,J);
      FLOW(I,J), FLOW(J,I) = '1'B;
      GO TO S10;
   S20: CALL WATER1 (N, FLOW);
      PUT DATA (((FLOW(I,J) DO J = 1 TO I-1) DO I = 1 TO N));
END;
```

```
/* 13-13: */
WTRSYM:   PROCEDURE (N, FROW);
   DECLARE (CHANGE, PATH) BIT(1), (ZEROS, FROW(N)) BIT(N);
      ZEROS = '0'B;
   S10:  CHANGE = '0'B;
      DO I = 1 TO N;
         DO J = 1 TO N;
            PATH = FROW(I) & FROW(J) ¬= ZEROS;
            IF SUBSTR(FROW(I), J, 1) ¬= PATH  THEN DO;
               SUBSTR(FROW(I),J,1), SUBSTR(FROW(J),I,1), CHANGE = '1'B;
               END;
            END;
         END;
      IF CHANGE  THEN GO TO S10;
END;

/* 13-17:
ALTHOUGH MORE COMPACT, THE 2 BIT SCHEME RESULTS IN A GREATER
 "DIFFERENCE" BETWEEN NORMAL AND MODERATELY HIGH THAN BETWEEN NORMAL
 AND SEVERELY HIGH.
*/

/* 13-20:
 BOOL(A, B, '1000'B)
*/

/* 13-21:   LET NOR = #
BOOL (A, B, '0000'B) = '0'B
BOOL (A, B, '0001'B) = (A#A)#(B#B)
BOOL (A, B, '0010'B) = (A#A)#B
BOOL (A, B, '0011'B) = A
BOOL (A, B, '0100'B) = A#(B#B)
BOOL (A, B, '0101'B) = B
BOOL (A, B, '0110'B) = ((A#A)#(B#B))#(A#B)
BOOL (A, B, '0111'B) = (A#B)#(A#B)
BOOL (A, B, '1001'B) = ((A#A)#B)#(A#(B#B))
BOOL (A, B, '1010'B) = B#B
BOOL (A, B, '1011'B) = (A#(B#B))#(A#(B#B))
BOOL (A, B, '1100'B) = A#A
BOOL (A, B, '1101'B) = ((A#A)#B)#((A#A)#B)
BOOL (A, B, '1110'B) = ((A#A)#(B#B))#((A#A)#(B#B))
BOOL (A, B, '1111'B) = '1'B
*/

/* 13-25: */
HAFADD:  PROCEDURE (A, B, CARRY, SUM);
DECLARE (A, B, CARRY, SUM) BIT(1);
   CARRY = A & B;
   SUM = BOOL(A, B, '0110'B);
END;
```

```
/* 13-27: */
PLUS:  PROCEDURE (A, B) BIT(40);
   DECLARE (A, B, ANSW) BIT(40), (CKM1, SUM) BIT(1);
   CALL HAFADD (SUBSTR(A,1,1), SUBSTR(B,1,1), CKM1, SUM);
   SUBSTR(ANSW,1,1) = SUM;
   DO K = 2 TO 40;
      CALL FULADD (CKM1, SUBSTR(A,K,1), SUBSTR(B,K,1), SUM, CKM1);
      SUBSTR(ANSW, K, 1) = SUM;
   END;
   IF CKM1  THEN SIGNAL FIXEDOVERFLOW;
   RETURN (ANSWER);
END;

/* 14-3: */
SPELDT:  PROCEDURE  CHARACTER(18);
   DECLARE MONTH(12) CHARACTER(9) INITIAL ('   JANUARY', ' FEBRUARY',
      '     MARCH', '     APRIL', '       MAY', '      JUNE', '      JULY',
      '    AUGUST', 'SEPTEMBER', '   OCTOBER', '  NOVEMBER', '  DECEMBER'),
      DT CHARACTER(6);
   DT = DATE;   NDAY = SUBSTR(DT,5,2);
   RETURN (MONTH(SUBSTR(DT,3,2)) || NDAY || ', 19' || SUBSTR(DT,1,2));
END;

/* 14-5: */
ANALST: PROCEDURE  OPTIONS (MAIN);
   GET LIST (M, N);
   BEGIN;
      DECLARE (STIMULUS(M), SIGNIFICANT(M,N), RESPONSE) CHARACTER(12);
      GET LIST (STIMULUS, SIGNIFICANT);
      DO I = 1 TO M;
         GET LIST (RESPONSE);
         DO J = 1 TO N;
            IF RESPONSE = SIGNIFICANT(I,J)  THEN DO;  PUT SKIP LIST
               (STIMULUS(I), RESPONSE);  GO TO ENDI;  END;
         END;
      ENDI:  END;
   END;
END;

/* 14-7: */
INDUS:   PROCEDURE  OPTIONS (MAIN);
   DECLARE ALPHABET(256) CHARACTER(1)   /* THERE ARE 256 EBCDIC
      CHARACTERS */, TEXT CHARACTER (32727) VARYING, PAIR CHARACTER(2);
   GET LIST (M, (ALPHABET(I) DO I = 1 TO M), TEXT);
   DO I = 1 TO M;
      DO J = 1 TO M;
         PAIR = ALPHABET(I) || ALPHABET(J);   KOUNT = 0;
         S10:  L = INDEX(TEXT, PAIR);
            IF L ¬= 0  THEN DO;  KOUNT = KOUNT + 1;
               SUBSTR(TEXT, L, 2) = '  ';  GO TO S10;  END;
            IF KOUNT ¬= 0  THEN PUT LIST (PAIR, KOUNT);   /* IT IS
               ASSUMED THAT THE OUTPUT MEDIUM CAN REPRESENT EVERY
               CHARACTER THAT IS USED.  THIS MAY REQUIRE AN UNPRINTED
               MEDIUM SUCH AS A CARD PUNCH OR MACHINE LANGUAGE
               OUTPUT.  */
      END;
   END;
END;
```

```
/* 14-8: */
LINE:   PROCEDURE (M, N, INSTR, PATTERN, ALPHABET) CHARACTER(80);
   /* THIS FUNCTION RETURNS A LINE OF UP TO 80 CHARACTERS */
   DECLARE INSTR BIT(*) /* INPUT SCAN WITH N BITS */,
      PATTERN(*) BIT(49)    /* BIT PATTERNS FOR M CHARACTERS -- PADDED TO
         49 BITS WITH ZEROS FOR SKINNY CHARACTERS */,
      ALPHABET(*) CHARACTER(1)   /* THE M CHARACTERS CORRESPONDING TO
         THE M PATTERNS */,
      OUTSTR CHARACTER(80) /* THE RETURNED STRING -- PADDED WITH BLANKS
         AS NECESSARY */,
      SLICE CHARACTER(7), CTR CHARACTER(49) VARYING;
   L, NB = 0;  OUTSTR = ' ';   CTR = '';
   DO K = 1 TO N BY 7;
      SLICE = SUBSTR(INSTR, K, 7);
      IF SLICE ¬= (7)'0'B  THEN DO;   /* ADD ON SLICE */
         CTR = CTR || SLICE;  GO TO ENDK;  END;
      IF CTR = '' THEN DO;
         NB = NB+1;  IF NB = 4   THEN DO; /* 4 BLANK SLICES = 1 SPACE */
            NB = 0;  L = L+1;  SUBSTR(OUTSTR,L,1) = ' ';  END;
         GO TO ENDK;  END;
      NB = 1;   L = L+1; /* END OF CHARACTER -- DETERMINE IDENTITY: */
      DO J = 1 TO M;
         IF CTR = PATTERN(J)   THEN DO;
            SUBSTR(OUTSTR,L,1) = ALPHABET(J);  GO TO ENDK;  END;
      END;
      SUBSTR(OUTSTR,L,1) = '?';
   ENDK:  END;
   RETURN (OUTSTR);
END;

/* 14-13: */
MORSIN:  PROCEDURE  OPTIONS (MAIN);
   DECLARE INSTR BIT(8056) VARYING,
      OUTSTR CHARACTER(1007) VARYING,
      PATTERN(28) BIT(16)  INITIAL('1'B,'11'B,'1011'B,'11011011'B,
         '1101'B,'101'B,'101101'B,'10101'B,'1010101'B,'110101'B,
         '10110101'B,'101101'B,'110101101'B,'11011'B,'101101101'B,
         '10101101'B,'1101011011'B,'1011011'B,'1101101'B,'11010101'B,
         '10101011'B,'1101101011011'B,'101101101011'B,'1011011011'B
         ,'1101011'B,'1101101011'B,'110101011'B,'110110101'B),
      ALPHABET CHARACTER(29) INITIAL('ETAONIRSHDLUCMPFYWGBV,.JKQXZ '),
      CTR BIT(16) VARYING, B BIT(1);
   NEXT:   GET LIST (INSTR);   NZ = 2;   OUTSTR,CTR = '';
      DO K = 1 TO LENGTH(INSTR);
         B = SUBSTR(INSTR,K,1);
         IF B THEN GO TO S20;
         NZ = NZ +1;
         IF NZ = 1 THEN GO TO S30;
         J = 29;
         IF NZ = 2   THEN DO J = 1 TO 28;
            IF CTR = PATTERN(J)   THEN GO TO S10;   END;
   S10:   OUTSTR = OUTSTR || SUBSTR(ALPHABET,J,1);   CTR = '';
         GO TO ENDK;
   S20:   NZ = 0;
   S30:   CTR = CTR || B;
      ENDK:  END;
   PUT SKIP LIST (OUTSTR);  GO TO NEXT;
END;
```

```
/* 14-18: */
GRAPH:    PROCEDURE (Y, YMIN, YMAX, XMIN, XMAX, XINC, TITLE, YUNITS,
          XUNITS);
   /* FOR INTEGER GRID COORDINATES MAKE (YMAX-YMIN) A MULTIPLE OF 13
      AND XINC A MULTIPLE OF 1/6 */
   DECLARE Y ENTRY, TITLE CHARACTER(*), XUNITS CHARACTER(*), YUNITS
      CHARACTER(*), HGRID CHARACTER(131) INITIAL((14)'+---------'),
      VGRID CHARACTER(131) INITIAL ((14)'|             '),
      LINE CHARACTER(131), COORD FLOAT(3);
   ON ENDPAGE (SYSPRINT);
   DIFFY = YMAX - YMIN;
   LINE = ' ';   SUBSTR(LINE,100) = YUNITS;   PUT LIST (LINE);
   LINE = ' ';   YINC = DIFFY/13;
   DO K = 1 TO 12;
      COORD = YMIN + K*YINC;   SUBSTR(LINE,10*K-4,9) = COORD;
   END;
   PUT SKIP LIST (LINE);
    DY = YINC/20;   DX = XINC/2;
   DO K = 0 TO 13;
      IF ABS(YMIN + K*YINC) <= DY  THEN DO;
         SUBSTR(HGRID,1+10*K,1), SUBSTR(VGRID,1+10*K,1) = '$';
         GO TO S5;   END;
   END;
   S5:   DO LINE# = 0 BY 1;
      X = XMIN + LINE# * XINC;
      IF X > XMAX   THEN GO TO S10;
      IF MOD(LINE#,6) = 0   THEN LINE = HGRID;   ELSE LINE = VGRID;
      IF ABS(X) <= DX   THEN LINE = (131)'$';
      K = MOD(130*(Y(X)-YMIN)/DIFFY, 130) + 1;
      SUBSTR(LINE, K, 1) = '*';
      IF MOD(LINE#,6) = 0   THEN DO;   COORD = X;
         SUBSTR(LINE,1,4) = COORD;   END;
      PUT SKIP LIST (LINE);
   END;
   S10:   PUT SKIP LIST (XUNITS);
      PUT SKIP(2) LIST (TITLE);   PUT PAGE;
END;


/* 14-19: */
GRAPH9:   PROCEDURE (N, YSUB, YMIN, YMAX, XMIN, XMAX, XINC, TITLE,
      YUNITS, XUNITS);
   /* FOR INTEGER GRID COORDINATES MAKE (YMAX-YMIN) A MULTIPLE OF 13
      AND XINC A MULTIPLE OF 1/6 */
   DECLARE YSUB ENTRY, Y(N), (YMIN(*), YMAX(*), COORD) FLOAT(3),
      (TITLE, XUNITS, YUNITS(*))
      CHARACTER(*), HGRID CHARACTER(131) INITIAL((14)'+---------'),
      VGRID CHARACTER(131) INITIAL ((14)'|             '),
      LINE CHARACTER(131);
   ON ENDPAGE (SYSPRINT);
   DO I = 1 TO N;
      PUT SKIP LIST ('CURVE(' || I || ') GOES FROM ' || YMIN(K) ||
         ' TO ' || YMAX(K) || YUNITS(K));
   END;
   DO LINE# = 0 BY 1;
      X = XMIN + LINE# * XINC;
      IF X > XMAX   THEN GO TO S10;
      IF MOD(LINE#,6) = 0   THEN LINE = HGRID;   ELSE LINE = VGRID;
   CALL YSUB(N, X, Y);
   DO I = 1 TO N;
      K = MOD(130*(Y(I)-YMIN(I))/(YMAX(I)-YMIN(I)),130) + 1;
         SUBSTR(LINE,K,1) = 'I';
   END;
```

```
            IF MOD(LINE#,6) = 0   THEN DO;   COORD = X;
                SUBSTR(LINE,1,4) = COORD;   END;
            PUT SKIP LIST (LINE);
        END;
    S10:    PUT SKIP LIST (XUNITS);
            PUT SKIP(2) LIST (TITLE);   PUT PAGE;
END;

/* 14-20: */
POINTS:    PROCEDURE (XMIN,XMAX,NXINCHES,YMIN,YMAX,NYINCHES,N,X,Y);
        /*  THIS SUBROUTINE PLOTS THE N PAIRS OF POINTS (X, Y).   TO
            MINIMIZE THE MEMORY REQUIREMENTS, X AND Y ARE REARRANGED ON THE
            BASIS OF DECREASING Y.   XMIN, XMAX, YMIN, AND YMAX MAY BE EITHER
            CONSERVATIVELY GUESSED OR DETERMINED BY SEARCH PRIOR TO ENTRY.
            THE WIDTH OF THE GRAPH IS GIVEN BY NXINCHES UP TO 13, ASSUMING 10
            SPACES PER INCH.   THE HEIGHT OF THE GRAPH IS GIVEN BY NYINCHES UP
            TO THE LENGTH OF THE PAPER SUPPLY, ASSUMING 6 LINES PER INCH.   */
        DECLARE X(*), Y(*), (LINE, HGRID, VGRID) CHARACTER(10*NXINCH+1),
            COORD FLOAT(3);
        ON ENDPAGE (SYSPRINT);

        /*  SORT X AND Y INTO DESCENDING ORDER OF Y:   */
        DO J = 1 TO N-1;
            KBIG = J;   YBIG = Y(KBIG);
            DO K = J+1 TO N;
                IF Y(K) > YBIG   THEN DO;   YBIG = Y(K);   KBIG = K;   END;
            END;
            Y(KBIG) = Y(J);   Y(J) = YBIG;
            YBIG = X(KBIG);   X(KBIG) = X(J);   X(J) = YBIG;
        END;

        /*  PUT OUT THE GRAPH:   */
        HGRID = (14)'+---------';   VGRID = (14)'|            ';
        NXT10 = NXINCHES*10;   NYT6 = NYINCHES*6;
        XINC = (XMAX - XMIN)/NXT10;   YINC = (YMAX - YMIN)/NYT6;   K = 1;
        DO J = 0 TO NYT6;
            YCOMP = YMAX - (J+.5)*YINC;
            IF MOD(J, 6) = 0   THEN LINE = HGRID;   ELSE LINE = VGRID;
    S10:    IF K > N | Y(K) > YCOMP   THEN GO TO S20;
            L = 1.5 + (X(K) - XMIN)/XINC;
            IF L > 0 & L <= NXT10 + 1   THEN SUBSTR(LINE,L,1) = '*';
            K = K + 1;
            GO TO S10;
    S20:    IF MOD(J,6) = 0   THEN DO;   COORD = ROUND(COMP+.5*YINC,3);
                SUBSTR (LINE,1,9) = COORD;   END;
            PUT SKIP LIST (LINE);
        END;

        /*  PUT OUT THE X COORDINATES:   */
        LINE = ' ';   STEP = 10*XINC;
        DO J = 1 TO NXINCHES-1;
            COORD = ROUND(XMIN+J*STEP,3);   SUBSTR(LINE, 10*J-4, 9) = COORD;
        END;
        PUT SKIP LIST (LINE);

        /*  AT THE EXPENSE OF STORAGE SPACE, IT IS POSSIBLE TO WRITE A
            SIMPLER SUBROUTINE WHICH STORES THE ENTIRE GRAPH .   */
END;
```

```
/* 14-22: */
LINES:  PROCEDURE(XMIN,XMAX,NXINCHES,YMIN,YMAX,NYINCHES,NSEG,NL,NR,X,Y);
    /*  THIS SUBROUTINE PUTS OUT THE DRAWING OF NSEG LINE SEGMENTS.
    THE END COORDINATES OF THE K'TH SEGMENT ARE GIVEN BY X(NL(K)),
    Y(NL(K)), X(NR(K)), AND Y(NR(K)).   */
    DECLARE NL(*), NR(*), X(*), Y(*), LINE CHARACTER(10*NXINCHES+1);
    ON ENDPAGE (SYSPRINT);
    NXT10 = NXINCHES*10;   NYT6 = NYINCHES*6;
    XINC = (XMAX - XMIN)/NXT10;   YINC = (YMAX - YMIN)/NYT6;
    XINCON2 = XINC/2;   YINCON2 = YINC/2;
    DO J = 0 TO NYT6;
        LINE = ' ';   YP = YMAX - J*YINC;
        DO I = 0 TO NXT10;
            XP = XMIN + I*XINC;
            DO K = 1 TO NSEG;
                KL = NL(K);   KR = NR(K);
                Q = XP - X(KL);
                IF Q*(XP-X(KR)) > 0   THEN GO TO ENDK;
                R = YP - Y(KL);
                IF R*(YP-Y(KR)) > 0   THEN GO TO ENDK;
                DX = X(KR) - X(KL);   DY = Y(KR) - Y(KL);
                RESIDUAL = ABS(R*DX - Q*DY);
                IF RESIDUAL > DX*YINCON2 & RESIDUAL > DY*XINCON2
                    THEN GO TO ENDK;
                SUBSTR (LINE, J+1, 1) = '*';
                GO TO ENDJ;
            ENDK:   END;
        ENDJ:   END;
        PUT SKIP LIST (LINE);
    END;
END;

/* 15-1: */
BEARNG:   PROCEDURE   OPTIONS (MAIN);
    DECLARE 1 BEARING, 3 # FIXED(9),
                       3 LOAD, 5 (RADIAL, AXIAL) FIXED(5),
                       3 SPEED FIXED(5),
                       3 DIMENSION, 5 (ID, OD, T) FIXED(5,3),
                       3 COST FIXED(5,2);
    PUT LIST ('BEARING NUMBER', 'RADIAL LOAD', 'AXIAL LOAD',
        'MAXIMUM RPM', 'INSIDE DIAM.', 'OUTSIDE DIAM.', 'THICKNESS',
        'PRICE/UNIT');
    NEXT:   GET LIST (BEARING);
        IF RADIAL>=800 & AXIAL>=500 & COST<=6
            THEN PUT SKIP(2) LIST (BEARING);
        GO TO NEXT;
END;

/* 15-4:   DECLARE
1 ANIMAL, 2 CHORDATA, 3 VERTEBRATA, 4 MAMALIA, 5 CARNIVORA, 6 CANIDAE
    7 CANIS, 8 LUPUS CHARACTER(255), 8 ... 7 ... 6 ... 5 ... 4 ...
    3 ... 2 COLENTERATA, 3 ... ETC.   */

/* 15-6:   DECLARE
    2 VP_SALES, 3 NAME CHARACTER(20),
        3 WESTAERN_MANAGER, 4 NAME CHARACTER(20),
            4 STAFF(15) CHARACTER(20),
        3 EASTERN_MANAGER, NAME CHARACTER(20),
            4 STAFF(15) CHARACTER(20),
    2 VP_DEVELOPMENT, 3 NAME CHARACTER(20),
        3 PACKAGING_SUPERVISOR, 4 NAME CHARACTER(20),
            ETC.      */
```

```
/* 15-8
   A:  BEAMS.CONCRETE.QUANTITY = 1.1*(BEAMS.CONCRETE.QUANTITY +
                                      COLUMNS.CONCRETE.QUANTITY);
       BEAMS.CONCRETE.UNIT_COST = 1.1*(BEAMS.CONCRETE.UNIT_COST +
                                       COLUMNS.CONCRETE.UNIT_COST);
       BEAMS.STEEL.QUANTITY = 1.1*(BEAMS.STEEL.QUANTITY +
                                   COLUMNS.STEEL.QUANTITY);
       BEAMS.STEEL.UNIT_COST = 1.1*(BEAMS.STEEL.UNIT_COST +
                                    COLUMNS.STEEL.UNIT_COST);
   G:  TOTAL.CONCRETE.QUANTITY = BEAMS.CONCRETE.QUANTITY +
           COLUMNS.CONCRETE.QUANTITY + FLOORS.CONCRETE.QUANTITY;
       TOTAL.CONCRETE.UNIT_COST = BEAMS.CONCRETE.UNIT_COST +
          COLUMNS.CONCRETE.UNIT_COST + FLOORS.CONCRETE.UNIT_COST;
       TOTAL.STEEL.QUANTITY = BEAMS.STEEL.QUANTITY +
           COLUMNS.STEEL.QUANTITY + FLOORS.STEEL.QUANTITY;
       TOTAL.STEEL.UNIT_COST = BEAMS.STEEL.UNIT_COST +
           COLUMNS.STEEL.UNIT_COST + FLOORS.STEEL.UNIT_COST;    */

/* 16-1:
     D E C B G I H J F L M K A
*/

/* 16-5: */
VSUM:  PROCEDURE (HEADX, HEADY, HEADZ);
       DECLARE (HEADX, HEADY, HEADZ, TAILZ) POINTER,
          1 X BASED(XPNT), 2 VALUE FLOAT, 2 SUBSCRIPT FIXED BINARY,
             2 LINK POINTER,
          1 Y BASED(YPNT) LIKE X,
          1 Z BASED(ZPNT) LIKE X;
       HEADZ = NULL;
       IF HEADX = NULL & HEADY = NULL  THEN RETURN;
          ALLOCATE Z;   HEADZ, TAILZ = ZPNT;   XPNT = HEADX;   YPNT = HEADY;
   S5:    IF X.SUBSCRIPT = Y.SUBSCRIPT   THEN DO;
             Z.VALUE = X.VALUE + Y.VALUE;   Z.SUBSCRIPT = X.SUBSCRIPT;
             XPNT = X.LINK;   YPNT = Y.LINK;   GO TO S10;   END;
          IF X.SUBSCRIPT > Y.SUBSCRIPT   THEN DO;   Z.VALUE = Y.VALUE;
             Z.SUBSCRIPT = Y.SUBSCRIPT;   YPNT = Y.LINK;   GO TO S10;   END;
          Z.VALUE = X.VALUE;   Z.SUBSCRIPT = X.SUBSCRIPT;   XPNT = X.LINK;
   S10:   IF X.LINK = NULL & Y.LINK = NULL   THEN RETURN;
          ALLOCATE Z;   TAILZ->Z.LINK = ZPNT;   TAILZ = ZPNT;   GO TO S5;
END;


/* 16-7: */
TAILAD:  PROCEDURE (TAIL, XPNT);   /* XPNT IS PRESUMED TO POINT TO
            THE NEW VALUE    */
   DECLARE (TAIL, XPNT) POINTER,
       1 X BASED (XPNT), 2 VALUE FLOAT, 2 LINK POINTER;
   IF TAIL = NULL   THEN LINK = XPNT;   ELSE DO; LINK = TAIL->LINK;
                                                TAIL->LINK = XPNT;   END;
   TAIL = XPNT;
END;

/* 16-10: */
DBLADD:  PROCEDURE (HERE, V);
   DECLARE HERE POINTER,
       1 X BASED (XPNT), 2 VALUE FLOAT, 2 (LEFT, RIGHT) POINTER;
   ALLOCATE X;   VALUE = V;   LEFT = HERE;
   RIGHT = HERE->RIGHT;   HERE = RIGHT;   HERE->LEFT = XPNT;
   HERE = LEFT;   HERE->RIGHT = XPNT;
END;
```

```
/* 16-12: */
TRACE:  PROCEDURE (N, HEAD_ROW);
   DECLARE HEAD_ROW(*) POINTER,
        1 X BASED(XPNT), 2 (VALUE, I, J), 2 (RIGHT, DOWN) POINTER;
   SUM = 0;
   DO K = 1 TO N;
        XPNT = HEAD_ROW(K);
        S10:  IF XPNT = NULL   THEN GO TO ENDK;
              IF I = J   THEN DO;   SUM = SUM + VALUE;   GO TO ENDK;   END;
              IF I < J   THEN GO TO ENDK;
              XPNT = RIGHT;   GO TO S10;
   ENDK:  END;
   RETURN (SUM);
END;

/* 16-16:   HIGABKJFEDC   */
ENDORD:  PROCEDURE (ROOT);
     DECLARE ROOT POINTER,
          1 PREDECESSOR CONTROLLED, 2 NODE POINTER, 2 FLAG BIT(1),
          1 X BASED(XPNT), 2 VALUE CHARACTER(16), 2 COUNT FIXED BINARY,
              2 (LEFT, RIGHT) POINTER;
     XPNT = ROOT;
     S40:  IF XPNT = NULL   THEN GO TO S50;
           ALLOCATE PREDECESSOR;   NODE = XPNT;   FLAG = '1'B;
           XPNT = LEFT;   GO TO S40;
     S50:  IF ¬ ALLOCATION(PREDECESSOR)   THEN RETURN;
           IF FLAG  THEN DO;   FLAG = '0'B;   XPNT = NODE->RIGHT;
               GO TO S40;   END;
           XPNT = NODE;   FREE PREDECESSOR;
           PUT SKIP LIST (VALUE, COUNT);
           GO TO S50;
END;


/* 16-18: */
PLANAR:  PROCEDURE   OPTIONS (MAIN);
     /* NUMBER THE NODES FROM 1 TO N WITH 1 BEING THE STARTING      2--3   9
        NODE AND N BEING THE FINISH NODE.  USING LIST-DIRECTED      |   |
        DATA, PRECEDE EACH NODE FROM 1 THRU N-1 BY A LIST OF        1--4--5
        ITS NEIGHBORS IN CLOCKWISE ORDER, BEGINNING WITH AN         |  |  |
        ARBITRARY NEIGHBOR.  FOR EXAMPLE, THE INPUT FOR THE         6--7  8
        ADJACENT MAZE WOULD BE  6 2 4 1,  1 3 2,  2 3,
        7 1 5 4,  8 4 9 5,  1 7 6,  6 4 7,  5 8    . ALL NODES ARE
        PRESUMED TO HAVE AT LEAST ONE CONNECTION.  THE OUTPUT IS THE NODE
        SEQUENCE OF A PATH THRU THE MAZE -- FOR EXAMPLE 1 6 7 4 5 8 5 9 .
        IF THE SEQUENCE ENDS AT NODE 1, THE MAZE IS NOT PLANAR OR THE
        START AND FINISH CANNOT BE JOINED BY AN IMAGINARY LINE WHICH
        CROSSES NO OTHERS. */
     DECLARE (TAIL(1000), HERE) POINTER,
          1 NEIGHBOR BASED(NPNT), 2 NODE, 2 LINK POINTER,
          (FROM, TO, SAVE) FIXED BINARY;

     /* INPUT STAGE -- STORE NEIGHBORS AS COUNTERCLOCKWISE LISTS: */
     ON ENDFILE (SYSIN)   GO TO S20;
     DO N = 1 BY 1;
          ALLOCATE NEIGHBOR;   GET LIST (NODE);   TAIL(N), LINK = NPNT;
          S10:  GET LIST (K);
                IF K ¬= N   THEN DO;   ALLOCATE NEIGHBOR;   NODE = K;
                HERE = TAIL(N);   LINK = HERE->LINK;   HERE->LINK = NPNT;
                    GO TO S10;   END;
     END;
```

```
        S20:    /* TAKE FIRST STEP: */   FROM = 1;   NPNT = TAIL(FROM);
            SAVE, TO = NODE;   PUT LIST (FROM, TO);
        S25:    NPNT = TAIL(TO);
        S30:    /* LOCATE 'FROM' IN CIRCULAR LIST TO DETERMINE EXTREME RIGHT
            PATH: */   IF NODE = FROM   THEN GO TO S40;
            NPNT = LINK;   GO TO S30;
        S40:    /* TAKE STEP: */   FROM = TO;   HERE = LINK;   TO = HERE->NODE;
            /* PREVENT ENDLESS REPETITION OF SEARCH IN INVALID MAZE: */
            IF FROM = 1 & TO = SAVE   THEN STOP;
            PUT LIST (TO);   IF TO = N   THEN STOP;
            GO TO S25;
END;

/* 16-21: */
FORD:   PROCEDURE OPTIONS (MAIN);
    /* NUMBER THE NODES FROM 1.  USING LIST-DIRECTED DATA,     1-82-2-13-3
       MAKE A LIST OF NEIGHBORS AND THEIR DISTANCES FOR            |  /|
       EACH NODE, BEGINNING WITH 2.  FOLLOW EACH LIST WITH        41 9 38
       THE NUMBER OF THE NODE.  FOR EXAMPLE, THE INPUT FOR         |/   |
       THE ADJACENT NETWORK MAY BE 1 82 2,   2 13 4 9 5 38 3,     4-19-5
       2 41 3 9 5 19 4,   3 38 4 19 5  */
    DECLARE HEAD(500) POINTER, TTOTAL(500),
        1 NEIGHBOR BASED (NPNT), 2 (NODE, TDIFF, LINK POINTER);

    /* INPUT STAGE -- STORE NEIGHBORS AS LINEAR LISTS: */
    ON ENDFILE (SYSIN)   GO TO S20;   SUMT = 0;
    DO N = 2 BY 1;
        HEAD(N) = NULL;
        S10:    GET LIST (K);
            IF K ¬= N   THEN DO;   ALLOCATE NEIGHBOR;   NODE = K;
                GET LIST (TDIFF);   SUMT = SUMT + TDIFF;
                LINK = HEAD(N);   HEAD(N) = NPNT;   GO TO S10;   END;
    END;

    S20:    /* INITIALIZE TTOTAL: */   N = N-1;   TTOTAL(1) = 0;
        DO K = 2 TO N;   TTOTAL(K) = SUMT;   END;

    S30:    /* ITERATE UNTIL THERE IS NO CHANGE IN ANY TTOTAL: */
        CHANGE = '0'B;
        DO K = 2 TO N;
            NPNT = HEAD(K);   SUMT = TTOTAL(K);
            S40:    IF NPNT = NULL   THEN GO TO S50;
                SUMT = MIN (SUMT, TTOTAL(NODE)+TDIFF);
                NPNT = LINK;   GO TO S40;
            S50:    IF SUMT ¬= TTOTAL(K)   THEN DO;
                CHANGE = '1'B;   TTOTAL(K) = SUMT;   END;
        END;
        IF CHANGE   THEN GO TO S30;
        PUT LIST ((K, TTOTAL(K) DO K = 1 TO N));
END;

/* 16-22: */
KRUSKL:   PROCEDURE OPTIONS (MAIN);
    DECLARE PARENT(1000) FIXED BINARY;
    GET LIST (N);
    DO I = 1 TO N;   PARENT(K) = 0;   END;
    DO I = 1 TO N-1;
        S10:    GET LIST (J, K);   JCOPY = J;   KCOPY = K;
            DO WHILE (PARENT(K) ¬= 0);   K = PARENT(K);   END;
            IF J = K   THEN GO TO S10;
            PARENT(J) = K;   PUT SKIP LIST (JSAVE, KSAVE);
    END;
END;
```

```
/* 16-23: */
CASCAD:   PROCEDURE  OPTIONS (MAIN);
   /* NUMBER THE NODES FROM 1 AT THE BEGINNING OF THE             1--12->2
      CASCADE TO N AT THE END.  USING LIST-DIRECTED DATA,         |    ¬|
      PRECEDE EACH NODE FROM 1 THRU N-1 BY ITS NEIGHBORS          |   / |
      TOGETHER WITH THEIR BRANCH GAINS.  FOR EXAMPLE, FOR        13  19 36
      ADJACENT GRAPH WE MAY USE THE INPUT  3 13 2 12 1,           | /   |
      4 36 2,   2 19 4 22 3         BE SURE TO TAKE PROPER        V/    V
      ACCOUNT OF THE ARROWS BECAUSE A CIRCUIT WILL CAUSE         3--22->4
      ENDLESS LOOPING. */
   DECLARE (HEAD(1000), PREDECESSOR CONTROLLED) POINTER,
      1 NEIGHBOR BASED(NPNT), 2 (NODE, BRANCHGAIN, LINK POINTER);

/* INPUT STAGE -- STORE NEIGHBORS AS LINEAR LISTS: */
   ON ENDFILE (SYSIN)  GO TO S20;
   DO N = 1 BY 1;
         HEAD(N) = NULL;
      S10:   GET LIST (K);
         IF K ¬= N  THEN DO;  ALLOCATE NEIGHBOR;   NODE = K;
         GET LIST (BRANCHGAIN);   LINK = HEAD(N);   HEAD(N) = NPNT;
            GO TO S10;   END;
   END;

   S20:   /* START WITH FIRST NEIGHBOR OF FIRST NODE: */
      NODE, PRODUCT = 1;   GAIN = 0;   NPNT = HEAD(NODE);
   S30: /* ADD BRANCH AND START AT TOP OF NEW NEIGHBOR LIST: */
      ALLOCATE PREDECESSOR;   PREDECESSOR = NPNT;
      PRODUCT = PRODUCT * BRANCHGAIN;   NPNT = HEAD(NODE);
   S40:    /* CHECK FOR END OF NEIGHBOR LIST: */
      IF NPNT = NULL   THEN GO TO S50;
      /* CHECK FOR COMPLETION OF A PATH: */
      IF NODE ¬= N THEN GO TO S30;
      GAIN = GAIN + PRODUCT * BRANCHGAIN;   NPNT = LINK;   GO TO S40;
   S50:   /* CHECK FOR COMPLETION OF ALL PATHS: */
      IF ¬ ALLOCATION(PREDECESSOR)    THEN GO TO S60;
      /* DELETE LAST BRANCH AND CHECK NEXT NEIGHBOR: */
      NPNT = PREDECESSOR;   PRODUCT = PRODUCT/BRANCHGAIN;
      FREE PREDECESSOR;   NPNT = LINK;   GO TO S40;
   S60:   PUT LIST (GAIN);
END;

/* 17-4: */
BESSEL:   PROCEDURE (X, BESNEW, IER);   /*  THIS SUBROUTINE CALCULATES
      BESNEW = THE COMPLEX ZERO'TH ORDER BESSEL FUNCTION OF COMPLEX X.
      IER IS SET TO 1 IF TOO LARGE A VALUE OF X IS ATTEMPTED.
      OTHERWISE IER IS SET TO ZERO.   */
   DECLARE (X, BESNEW, BESOLD, TERM, XSQ) COMPLEX;
      IER = 0;   XSQ = X*X;   BESOLD, TERM = 1;
   DO K = 2 TO 20 BY 2;
      TERM = -.25*TERM*XSQ/(K*(K-1))**2;
      BESNEW = BESOLD + TERM;
      IF BESNEW = BESOLD   THEN RETURN;
      BESOLD = BESNEW;
   END;
      IER = 1;
END;
```

```
/* 17-12:
   DECLARE (A(N), B(N), DOT) COMPLEX;
   DOT = SUM(A*B);
*/

/* 17-13: */
DOT:   PROCEDURE (N, A, B) COMPLEX;
   DECLARE (A(*), B(*), SUM) COMPLEX;
   DO J = 1 TO N;
      SUM = SUM + A(J)*B(J);
   END;
   RETURN (SUM);
END;

/* 17-23: */
HRMSUM:   PROCEDURE (N, A) COMPLEX;
   DECLARE (A(*), SUMON) COMPLEX;
   K, SUMON, SUMOFF = 0;
   DO I = 1 TO N;
      DO J = 1 TO I-1;
         K = K + 1;
         SUMOFF = SUMOFF + REAL(A(K));
      END;
      K = K + 1;
      SUMON = SUMON + A(K);
   END;
   RETURN (SUMON + 2*SUMOFF);
END;

/* 17-29: */
SEMI:   PROCEDURE   OPTIONS (MAIN);
      DECLARE (Z, SINZ) COMPLEX;
   IN:   GET LIST (Z);
      SINZ = SIN(Z);
      PUT LIST (IMAG(LOG((SINZ-1)/(SINZ+1)))/3.14159);
      GO TO IN;
END;

/* 17-30: */
ELLIPS:   PROCEDURE OPTIONS (MAIN);
      DECLARE (Z, W) COMPLEX, TAUBAR (500);
      GET LIST (A, N, (TAUBAR(K) DO K = 1 TO N));
      DPHI = 3.14159*2/N;   FACTOR = A - SQRT(A*A-1);
   IN:   GET LIST (Z);
      W = FACTOR*(Z-SQRT(Z*Z-1));
      R = ABS(W);   THETA = ATAN(IMAG(W), REAL(W));
      SUM = 0;
      DO K = 1 TO N;
         PHI = -3.14159 + K*DPHI;
         SUM = SUM + TAUBAR(K)/(1+R*(R-2*COS(THETA-PHI)));
      END;
      PUT SKIP LIST (Z, SUM*DPHI);
      GO TO IN;
END;
```

```
/* 17-31: */
NORMAL:  PROCEDURE  OPTIONS (MAIN);
      DECLARE (Z, W) COMPLEX, G(500);
      GET LIST (A, N, (G(K) DO K = 1 TO N));
      DPHI = 3.14159*2/N;   FACTOR = A - SQRT(A*A-1);
   IN:  GET LIST (Z);
      W = FACTOR*(Z-SQRT(Z*Z-1));
      R = ABS(W);   THETA = ATAN(IMAG(W), REAL(W));
      SUM = 0;
      DO K = 1 TO N;
         PHI = -3.14159 + K*DPHI;
         WTOMINUS2 = EXP(COMPLEX(0,-2*PHI));
   /* NOTE:  DW/DZ = 1/(DZ/DW) = 2/(FACTOR-1/(FACTOR*W*W))     */
         SUM = SUM + 2*G(K)*LOG(1+R*(R-2*COS(PHI-THETA)))/
                           ABS(FACTOR-WTOMINUS2/FACTOR);
      END;
      PUT SKIP LIST (Z, SUM*DPHI);
      GO TO IN;
END;

/* 18-4: */
TRIWLK:  PROCEDURE (N, LORIG, DORIG, L, D);   /* IF LORIG AND DORIG ARE
      NO LONGER NEEDED AFTER DECOMPOSITION, USE THE SAME ARGUMENTS FOR
      L AND D RESPECTIVELY.   */
   DECLARE (LORIG(*), L(*)) FLOAT, DORIG(*), D(*);
   D(1) = DORIG(1);
   DO I = 2 TO N;
      TEMP = LORIG(I);
      L(I) = TEMP/D(I-1);
      D(I) = DORIG(I) - TEMP*L(I);
   END;
END;

TRWKSB:  PROCEDURE (N, L, D, B, X);   /* USE THE SAME ARGUMENT FOR X AND
      B IF THE LATTER IS NO LONGER NEEDD AFTER SOLUTION. */
   DECLARE L(*) FLOAT, D(*), B(*), X(*);

   /* FORWARD SUBSTITUTION AND DIAGONAL DIVISION: */
   X(1) = B(1);
   DO I = 2 TO N;
      IM1 = I-1;
      X(I) = B(I) - L(I)*X(IM1);
      X(IM1) = X(IM1)/D(IM1);
   END;
   X(N) = X(N)/D(N);

   /* BACK SUBSTITUTION: */
   DO I = N-1 TO 1 BY -1;
      IP1 = I+1;
      X(I) = X(I) - L(IP1)*X(IP1);
   END;
END;
```

350   Appendix

```
/* 18-6: */
(NOUNDERFLOW): GJINV:   PROCEDURE (N, A, PERM);
   /* 'A' IS REPLACED BY ITS INVERSE SHIFTED 1 COLUMN TO THE RIGHT AND
      PERMUTED ACCORDING TO THE PERMUTATION VECTOR 'PERM':
      AINVERSE(I,J) = A(PERM(I),PERM(J+1)).  IT IS MOST EFFICIENT TO
      SIMPLY TAKE THIS INTO ACCOUNT DURING OUTPUT, BUT AN ALTERNATIVE
      IS TO REARRANGE THE COLUMNS AND ROWS BEFORE OUTPUT. */
   DECLARE A(*,*), PERM(*) FIXED BINARY, WEIGHT(N);

   /* INITIALIZE PERMUTATION VECTOR, INITIALIZE N+1'TH COLUMN, AND
      DERIVE WEIGHTS: */
   NP1 = N+1;
   DO I = 1 TO N;
      PERM(I) = I;   A(I,NP1) = 0;   BIGGEST = 0;
      DO J = 1 TO N; BIGGEST = MAX (BIGGEST,ABS(A(I,J)));   END;
      WEIGHT(I) = BIGGEST;
   END;

   DO J = N TO 1 BY -1;
      /* SEARCH FOR BIGGEST PIVOT: */
      BIGGEST = 0;
      DO I = J TO 1 BY -1;
         IROW = IPERM(I);   AB = ABS (A(IROW,J)/WEIGHT(IROW));
         IF AB > BIGGEST   THEN DO;  BIGGEST = AB;   JSAVE = I;   END;
      END;
      JROW = PERM(JSAVE);   PERM(JSAVE) = PERM(J);   PERM(J) = JROW;
      A(JROW,J+1) = 1;   PIVOT = A(JROW,J);

      /* DIVIDE PIVOT ROW BY PIVOT: */
      DO K = 1 TO NP1;   A(JROW,K) = A(JROW,K)/PIVOT;   END;

      /* SUBTRACT MULTIPLES OF PIVOT ROW FROM OTHER ROWS: */
      DO I = 1 TO J-1, J+1 TO N;
         IROW = PERM(I);   F = A(IROW,J);
         IF F ¬= 0   THEN DO K = 1 TO N-1;
            A(IROW,K) = A(IROW,K) - F*A(JROW,K);   END;
      END;
      A(JROW,J) = 0;
   END;
END;

/* 18-8: */
BNDWLK:    PROCEDURE (M, N, CORIG, C);
   /* THIS SUBROUTINE DERIVES THE DECOMPOSITION, C, OF A SYMMETRIC
      POSITIVE DEFINATE BAND MATRIX CORIG.  C AND CORIG ARE PRESUMED TO
      HAVE THE BOUNDS (N, -M:0) WHERE N IS THE NUMBER OF ROWS AND M IS
      THE BANDWIDTH FOR THE RELATED SQUARE MATRIX A:   C(I,J-I) = A(I,J)
      USE THE SAME ARGUMENT FOR C AND CORIG IF THE LATTER IS NOT
      NEEDED AFTER DECOMPOSITION. */
   DECLARE CORIG(*,*), C(*,*), T(-M:0), SUM FLOAT(16);

   DO I = 1 TO N;
      DO J = MAX(1,I-M) TO I-1;
         SUM = 0;
         DO K = MAX(1,J-M) TO J-1;
            SUM = SUM - MULTIPLY(T(K-J), C(J,K-J), 16);
         END;
         IF I = J   THEN C(I,0) = CORIG(I,0) - SUM;
            ELSE DO;   T(J-I) = C(I,J-I) - SUM;
               C(I,J-I) = T(J-I)/C(J,0);   END;
      END;
   END;
END;
```

```
BDWKSB:   PROCEDURE (M, N, C, B, X);
   /* USE THE SAME ARGUMENT FOR AND B IF THE LATTER IS NO LONGER
      NEEDED AFTER SUBSTITUTION. */
   DECLARE C(*,*),  B(*),  X(*),   SUM FLOAT(16);

   /* FORWARD SUBSTITUTION: */
   DO I = 1 TO N;
      SUM = 0;
      DO J = MAX(1,I-M) TO I-1;
         SUM = SUM + MULTIPLY (C(I,J-I), X(J), 16);
      END;
      X(I) = B(I) - SUM;
   END;

   /* DIAGONAL DIVISION: */
   DO I = 1 TO N;   X(I) = X(I)/C(I,0);   END;

   /* BACK SUBSTITUTION:   */
   DO I = N-1 TO 1 BY -1;
      SUM = 0;
      DO J = I+1 TO MIN(N,I+M);
         SUM = SUM + MULTIPLY (C(J,I-J), X(J), 16);
      END;
      X(I) = X(I) - SUM;
   END;
END;
```

INDEX

A

ABS built-in function 65
ADD built-in function 105, 109
Adder, half and full 235–236
ADDR built-in function 264, 276, 294
Address 7, 264
ALIGNED attribute 227, 233
ALL built-in function 228, 233
ALLOCATE statement 262–263, 265, 276
ALLOCATION built-in function 263–264, 276
AND operator 72, 221, 233
ANY built-in function 228, 233
Arc-cosine function 45
Arc-sine function 45
Argument 32, 43, 79–80, 89
   agreement with parameter 34–37
   array 155–156, 186
   dummy 125
   function 117–118
   label 137
Array 153–218
   as arguments and parameters . . . 155–156, 162, 186
   banded 218, 311
   bounds and extent . . 154–155, 158–159, 162, 186
      adjustable 170, 178
   cross sections 183–184, 194
   expressions 172–173, 176, 183–184
   I/O . . . 156–157, 163, 171, 176, 184–186, 193–194
   labels 174–175
   manipulation functions 173–174, 176, 183
   natural order 194
   of pointers 273–276
   of structures 258, 259
   symmetric 214–215
   tridiagonal 217, 310
Assignment statement 15
   multiple 82
ATAN built-in function 41
ATAND built-in function 41
ATANH built-in function 44
Attribute factoring 21, 73, 254
AUTOMATIC attribute 160, 262

B

BASE attribute	5, 34
BASED attribute	264–265
BCD code	3
BEGIN statement	171–172, 176, 205, 263
BINARY attribute	5, 34–37
BINARY built-in function	47
Binary constants	231–232, 234
Binary search	164
Binomial coefficient	151
Bit	9
BIT attribute	219–220, 233
BY NAME option	256–257, 259
Byte	9

C

CALL statement	80, 91
CEIL built-in function	66
CHARACTER attribute	237–238, 248
Character-string constant	56–57, 58
Circuit analysis, switching or flow	222–230, 234
Circular list	277–278
Collating sequence	248
Comments	53–54, 58
Comparison operators	61
Compile	8, 13, 49
COMPLEX attribute	282, 289
COMPLEX built-in function and pseudo-variable	283–284, 289
Concatenation operator	222, 239
Condition prefix	136–137, 302
Conformal mapping	285–288, 290–291
CONJ built-in function	283, 289
Continued fraction	45–47, 136
CONTROLLED attribute	262–263
Coordinate transformation	79–80, 91
COPY option	60
COS built-in function	40
COSD built-in function	40
COSH built-in function	44
Crout's method	296–299

D

DATE built-in function ... 249
DECIMAL attribute ... 5, 34
DECIMAL built-in function ... 47
DECLARE statement ... 20–21, 154
Declaration:
   explicit ... 88, 172
   contextual ... 88
   implicit ... 88
Default ... 20–22, 34–38, 159, 262
Determinant ... 306–307
Diagnosis ... 230, 234
Differentiation, numerical ... 165
Dimension attribute ... 182
DIVIDE built-in function ... 105, 109
DO group ... 118–120, 125–126, 128, 172
DO loop ... 139, 152
   multiple specifications ... 143
   nested ... 144
Doolittle's method ... 295–296
Doubly-linked list ... 278–279

E

EBCDIC code ... 3
ELSE clause ... 114–116, 125–126, 128
   null ... 126
END statement ... 17, 88, 120, 139, 151–152, 172
ENDFILE condition ... 51, 97
Endorder ... 279
ENDPAGE condition ... 152, 244
ENTRY attribute ... 35, 37, 118
Equations:
   cubic ... 93
   differential ... 107, 111, 197
   linear ... 92, 199–213, 292–311
      banded ... 218, 311
      tridiagonal ... 217, 310
   quadratic ... 81–83
   quartic ... 93
   transcendental ... 113–125

Index

ERF built-in function 44
ERFC built-in function 44
ERROR condition 137
Errors:
   chopoff 55–56, 100–104, 109, 299–300
   discretization 107, 114–115, 127
Execute 8, 49
EXP built-in function 42
Exponent 19, 41–42
Exponentiation 24
Expressions:
   mixed 22–23, 34–37, 222, 283, 289
   rational 39–40, 45–47
EXTERNAL attribute 90–91

F

Factorial function 112
Factoring of attributes 21, 73, 254
FINISH condition 137
First-letter convention 20
FIXED attribute 20–21, 26, 27
FIXED built-in function 47
FIXEDOVERFLOW condition 23, 25, 27
Fixed-point integer 20, 42
Fixed-point noninteger 72, 73
FLOAT attribute 19, 20–21, 27
FLOAT built-in function 47
FLOOR built-in function 66
Flow charts 63, 141
Flow graph 280
Fraction anomaly 5, 75
FREE statement 262, 276
Function:
   as argument 117–118, 128
   built-in (*see individual names*)
   procedure 31–43
   singularity 67–68

G

Game theory 196
Gamma function 112

Gaussian elimination 200–209, 217, 293–295
Gauss-Jordan elimination 210, 310
Gauss-Siedel method 210–213, 215–217
GET statement 48–50, 193–194, 196–197
GO TO statement ...50–51, 59, 88–90, 137, 174–175
Greatest common denominator 76
Grey code 231–232

H

Horner's rule 45

I

Identifiers 18, 26
IF statement 61–63, 74, 114–116, 118–120
   nested 125–126
Ill-conditioned equations 206–208
IMAG built-in function and pseudo-variable ..283, 289
INDEX built-in function 238, 248
Indirect address 176
Information retrieval 70–74, 76, 237–238
INITIAL attribute 160
Integration, numerical 99, 106, 110–111, 138
Internal procedures 88
Interpolation 158–162, 198
Inverse matrix 293, 307–308
Iteration factor 160, 223
Iteration specification 156–157, 185
Iterative refinement 303–305

J

Job 10

K

Keyword 15, 26
   abbreviation 57

L

Label:
   parameter 137
   procedure 15, 37–38
   subscripted 174–175
   variable 59, 275
Least squares 98
Length attribute 19, 20, 22, 34–37, 73, 104–106, 220, 237
LENGTH built-in function 230, 233
LIKE attribute 257, 259
List processing 266–276
LOG built-in function 41
LOG2 built-in function 41
LOG10 built-in function 41
LU decomposition 293–295

M

Mantissa 19
Markov system 281
Matrix (*see also* array):
   ill-conditioned 206–208
   multiplication 187–193, 194–195
   output subroutine 186
   positive definite 308–310
   singular 206–208
   sparse 217, 268–269, 277, 278
MAX built-in function 65
Maximization 131
Maze 279–280
Mean 95, 98–99, 177
Merge 169–172, 197
MIN built-in function 65
Minimization 132
MOD built-in function 66
Morse code 250
MULTIPLY built-in function 105, 109, 191, 294

N

Norm	178, 196
Normalized equations	81
Normalized floating-point numbers	19
NOT operator	219
NULL built-in function	266–267, 276
Null ELSE clause	126
Null field	52, 70

O

ON statement	97–98, 109, 129
normal return	122
null	116–117
OR operation	70–71, 221
OVERFLOW condition	27, 117, 124

P

PAGE option	42–43
Parameter (*see* argument)	
Path, shortest or quickest	280
Pattern recognition	240
Pecking order	273–276
Pivoting	209, 299–303
Plotting	241–247, 251
POINTER attribute	264
Pointer qualifier	265
POLY built-in function	180
Polynomial:	
arithmetic	179
differentiation	179
integration	180
roots	81–83, 93, 284
Positive definite	308–310
Precision attribute	20, 22, 73
double	104
PRECISION built-in function	105, 109
Preorder	279
Priority of operators	24, 234

PROCEDURE statement 15, 31, 88
PROD built-in function 173
Product, dot, scalar, or inner 177
Pseudo-variable . . . 228–229, 233, 244, 283–284, 289
PUT statement 16, 25–26, 56–58, 60

Q

Quadrants 41
Queue 268

R

REAL built-in function and pseudovariable . . 283, 289
Repetition factor 220, 223
RETURN statement 32, 63, 80, 91
RETURNS attribute 37–38
ROUND built-in function 66

S

Scale attribute 21
Scale factor 20
Scope 88, 172
Series:
    Fourier 178
    infinite 127–128, 134–136
SET option 265
SIGN built-in function 65
SIN built-in function 40
SIND built-in function 40
Singularity functions 67–68
Singular matrix 206–208
SINH built-in function 44
SKIP option 53, 55, 58, 73, 75, 76
Sort 167–169, 175–176, 247–248, 269–273
Sparse matrices or vectors . . . 217, 268–269, 277, 278
SQRT built-in function 42
Stack 263

STATIC attribute 160, 262
STOP statement 122
STRING built-in function 228
Structures 252, 258
  arrays of 258, 259
  elementary names 252
  expressions 255
  I/O 252, 259
  level numbers 253, 254, 259
  major 253
  minor 253
  qualified names 254
Subordinate procedure 31
Subroutines 79–91, 268–269
  scientific package 83–87, 305
Subscripted variable (*see* array)
SUBSTR built-in function or pseudo-variable 228, 233, 239, 248
SUM built-in function 173, 183

T

TAN built-in function 40
TAND built-in function 40
TANH built-in function 44
Tensor 182, 195, 196
THEN clause 62
TIME built-in function 249
Transcendental 30, 113
Transfer function 285
TRANSLATE built-in function 239–240, 249
Trees 269
  spanning 280
Trigonometry 31, 58–59, 92–93
TRUNC built-in function 66

U

UNDERFLOW condition 27, 116, 117

V

VARYING attribute 229–230, 233
Vector (*see* array)

W

Weather reports 256–258
WHILE clause 147–148
Word 9

Z

ZERODIVIDE condition 41, 44, 117, 124

QA
76.73
P25 S74

QA 76.73 P25 S74	Stoutemeyer
	PL/I programming for engineering and science
DATE DUE	BORROWER'S NAME
6 29 8	BRALEY RHONDA

FEB 28 1972